ENGINEERING MECHANICS

A Revision of "Notes on Machine Design" Prepared by Officers of the Department of Marine Engineering and Naval Construction, U. S. Naval Academy

C. N. OFFLEY

Department of Marine Engineering and Naval Construction, U.S.Naval Academy

MAXWELL PRESS

Chennai Trichy New Delhi

MAXWELL PRESS

This edition has been published in india by arrangement with Carson Books, UK

ISBN 978-93-5528-199-9

Maxwell Press
No. 44, Nallathambi Street,
Triplicane, Chennai 600 005

Printed and bound in India

MJP 1301

Publisher : C. Janarthanan

ENGINEERING MECHANICS

A Revision of "Notes on Machine Design" Prepared by Officers of the Department of Marine Engineering and Naval Construction, U. S. Naval Academy

MAXWELL PRESS

Publisher's Note

The legacy of a country is in its varied cultural heritage, historical literature, developments in the field of economy and science. The top nations in the world are competing in the field of science, economy and literature. This vast legacy has to be conserved and documented so that it can be bestowed to the future generation. The knowledge of this legacy is slowly getting perished in the present generation due to lack of documentation.

Keeping this in mind, the concern with retrospective acquiring of rare books has been accented recently by the burgeoning reprint industry. Maxwell Press is gratified to retrieve the rare collections with a view to bring back those books that were landmarks in their time.

In this effort, a series of rare books would be republished under the banner, "Maxwell Press". The books in the reprint series have been carefully selected for their contemporary usefulness as well as their historical importance within the intellectual. We reconstruct the book with slight enhancements made for better presentation, without affecting the contents of the original edition.

Most of the works selected for republishing covers a huge range of subjects, from history to anthropology. We believe this reprint edition will be a service to the numerous researchers and practitioners active in this fascinating field. We allow readers to experience the wonder of peering into a scholarly work of the highest order and seminal significance.

MAXWELL PRESS

PREFACE

The object of the course in engineering mechanics is the study of the theoretical and mathematical principles involved in the application of mechanics to various engineering problems, and then the application of these principles to the solution of actual problems. With this object in view, the problems are selected to bring out as many principles as possible. A number of recitations are first assigned, leading up to each problem, which are then followed by several practical work periods in the drafting room, where a practical solution is worked out. Certain features of the theoretical study of strength of material are reviewed under the present subject. In particular it is the aim to point out the way in which many of the purely theoretical principles are modified in the solution of practical engineering problems.

In the preparation of the following pages many standard works have been consulted and their methods adopted where it has seemed advisable. Among these are: "The Elements of Machine Design," by Dr. W. C. Unwin; "Elements of Machine Design," by Kimball and Barr; "Strength of Materials," by Professor Morley; "Machine Design, Construction and Drawing," by Prof. H. J. Spooner; "Mechanics Applied to Engineering," by Prof. John Goodman.

The practical work problems are taken from "Notes on the Design of Propelling Machinery for Naval Vessels," 1902. These notes are largely the work of Rear-Admiral J. K. Barton, U. S. N., when on duty as an instructor at the Naval Academy from 1890 to 1893. The article on "Riveting" is the work of Captain F. J. Schell, when at the Academy from 1891 to 1895. The article on the "Theory of the Connecting Rod and Problem IV," in its present form, is the work of Prof. Theo. W. Johnson. Others who contributed to these notes are Captains Bartlett, Kinkaid, Gow and Lieut.-Commander Moody. This collection of notes was revised and published in 1908 as "Notes on Machine Design," by Captain F. W. Bartlett, U. S. N., Head of the Department of Marine Engineering and Naval Construction. They have been thoroughly revised and taken bodily into the present work. It is hoped that the features formerly found to be obscure have been simplified and cleared up.

In all cases the endeavor has been made to adopt, as far as possible, the especial methods of the Bureau of Steam Engineering of the Navy Department.

In the preparation of the book Captain F. W. Bartlett, U. S. Navy, and Lieut.-Commander A. W. Hinds, U. S. Navy, have given invaluable assistance in consultations, in checking the various mathematical formulæ and the solutions to the Practical Problems; they have also read and corrected all the proof sheets.

C. N. Offley, U. S. N.

Navy Yard, Puget Sound, October, 1910.

CONTENTS

CHAPTER X.

CHAPTER XI.

CHAPTER XII.

CHAPTER XIII.

CHAPTER XIV.

CHAPTER XV.

CHAPTER XVI.

CHAPTER XVII.

CHAPTER XVIII.

ENGINEERING MECHANICS

CHAPTER I.

UNITS, DEFINITIONS, QUALITY OF MATERIALS.

1. The following units will be used throughout, unless specifically stated to the contrary:

Dimensions follow the usual practice of the Bureau of Steam Engineering, being given in feet and inches, when greater than 24 inches, and in inches up to and including 24″. For obtaining stresses, moments, etc., the dimensions are reduced to inches.

Loads and forces are given in pounds. When the ton is used it is always the standard ton of 2240 pounds.

Stresses are given in pounds per square inch.

Pressures are in pounds per square inch.

Velocities are in feet per minute, or feet per second.

Accelerations are in feet per second per second.

Work is in foot pounds.

Power is in foot pounds per minute.

Speed of rotation is in revolutions per minute, or in angular velocity per second.

Moments (such as bending and twisting or turning moments) are in inch-pounds.

There is frequently a tendency to use the same terms to express the *forces* acting on the parts of a machine or structure and the *deformations* produced by these forces; that is to say, the distinction between *stress* and *strain* is not clearly made. In the same way the same term is often used to express either a quantity or an intensity. This common error may be avoided by constantly having in mind the following relations:

$$\frac{\text{Stress}}{\text{Strain}} = \frac{\text{tension}}{\text{extension}}, \text{ or } \frac{\text{pressure}}{\text{compressure}}, \text{ or } \frac{\text{shearing stress}}{\text{shearing strain}} = \text{elasticity}.$$

$$\text{Extension (e)} = \frac{\text{elongation }(\lambda)}{\text{original length (l)}} \text{ or } e = \frac{\lambda}{l}.$$

$$\text{Compression} = \frac{\text{contraction}}{\text{original length}}.$$

2. *Strain and Stress.*—A *strain* is any definite alteration of form or dimensions experienced by a solid body. It is the change of shape or size of a body, especially of a solid, produced by the action of a *stress;* therefore, it is deformation, and it may be temporary or permanent. When the strain is temporary and disappears on the removal of the force by which it is produced, it is called the *elastic deformation;* when it remains after the removal of the producing force, it is a permanent deformation or *permanent set.*

Stress is the internal reaction between the molecules of a material produced by the application of external loads or forces. Thus stress is of the nature of a *force,* while strain is a *change* of *dimension.*

Elasticity is the relation between *stress* and *strain,* within the elastic limit, and the ratio $\left\{\dfrac{\textit{stress per unit of cross sectional area}}{\textit{strain per unit of length}}\right\}$ is the *modulus of elasticity.* This will be denoted by the symbol E for direct tension or compression. Thus:

$$E=\frac{f}{e}.$$

Elastic and Plastic Conditions.—An *elastic* material is one that, when deformed or strained by the action of a force, recovers its original size and shape when the straining force is removed. A *plastic* material is one which does not completely recover its original size and shape on the removal of the straining force, but takes a permanent set. In general, every strain consists of (a) an elastic deformation, which is proportional to the straining force, and (b) a plastic deformation or *set.* For the materials used in the construction of machinery and other structures there is a range of straining force within which the strain is practically wholly elastic. This strain can, therefore, be allowed for in the design, by providing that the working stress produced by the forces acting on the several parts of the machine does not exceed the elastic limit of the material employed.

Solid materials which possess *plasticity* exhibit the phenomenon *of flow* under unequal stresses in different directions, much in the same way as liquids. This property of *flowing* is utilized in the manufacture of lead pipe, the drawing of wire, the stamping of coins, forging, etc.

Ductility is that property of a material by virtue of which it may be drawn out by tension to a smaller section, as when wire is made by drawing out metal through a hole in a die plate. During ductile extension a material shows a small degree of elasticity together with a much greater degree of plasticity.

Brittleness is the lack of ductility.

Malleability is that property of a material by virtue of which it can be hammered or rolled into the form of a plate. This property is very similar to ductility.

3. *The Relation of Stress and Strain.*—The relation between stress and strain is very clearly shown on the ordinary load-strain or stress-strain diagrams, such as are illustrated by Fig. 2.

Suppose the bar represented by Fig. 1 is subjected to a gradually increasing tensile load and at the same time the increase in length between the fixed points a and b is measured.

A diagram in which the ordinates represent *stress per unit of area* differs from one in which the ordinates represent *total load,* or stress per unit of *original* area of cross section, because, as the load

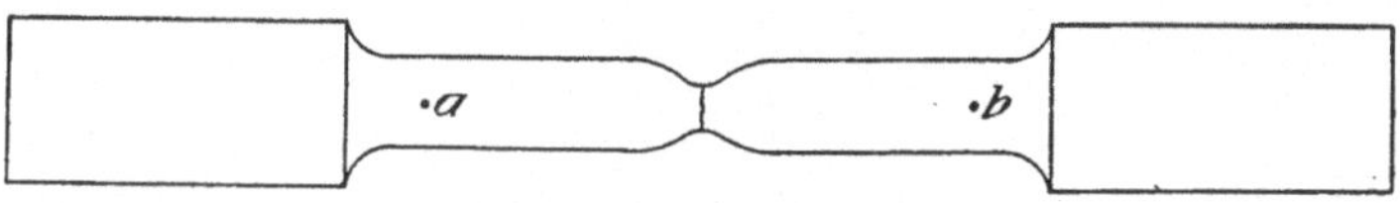

FIG. 1.—Tensile Test Piece.

is increased, the straining action produces a contraction of area. These diagrams are usually made to represent the relation of load, or total force, and strain, and Fig. 2 represents such a diagram.

Let the horizontal axis represent strains and the vertical axis represent loads, or stresses; then the relation between strain and load, or stress, will be represented by the curves. Plot extensions to the right, compressions to the left; tensions upward and pressures downward. The curve for a ductile material, such as mild steel, will then be Oabcde for tension, and Ofg for compression. As the loads in tension are gradually applied the bar stretches, and up to the load represented by the ordinate a_1a the curve will be practically a straight line. That is, the strains are proportional to the loads, and if the load is released the bar will return to its original length. The strains are elastic or temporary. At the point a the proportionality between load and strain ceases, and the strains increase much more rapidly than before, and a part of the strain is now permanent. The point a is, therefore, the *limit* of *proportionality* or the *true elastic limit.* When the material being tested is subjected to rolling, or forging in the process of manufacture, there is generally a second, well-defined point b, at which the test piece takes a great increase of strain from b_1 to c_1, without additional stress. This marks the *yield* point, and is commercially referred to as the elastic limit. As the load is further increased the strains are rapidly increased, as from c to d, until at d the maximum load is reached; further load beyond this point causes a local reduction of area of the test piece, so that it can no longer sustain the maximum load, and it finally breaks with

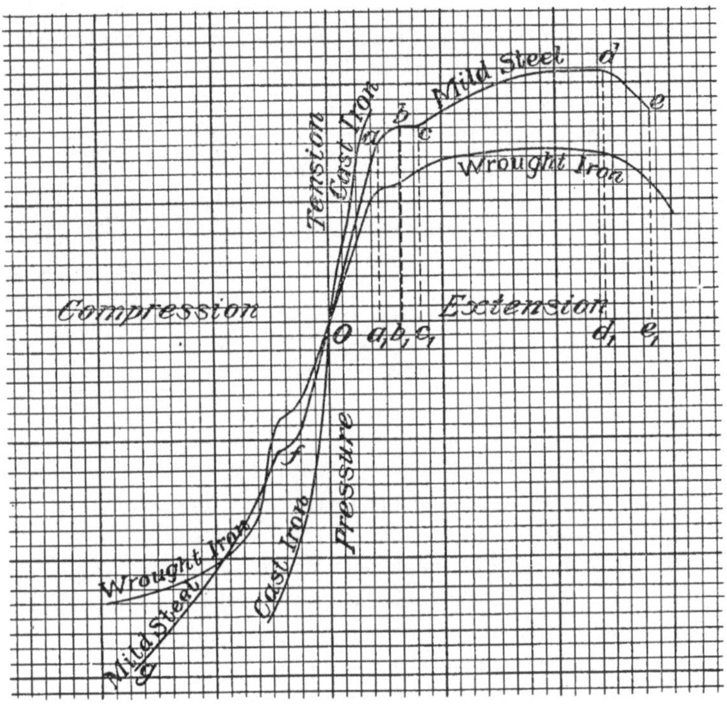

FIG. 2.—Typical Stress-Strain Diagram.

a load ee_1 less than the maximum dd. The stress due to the load dd_1 per unit of area of the original section of the test piece is called the *ultimate strength*, or *breaking stress* of the material.

The intensity of the stress per unit of area of the reduced section of the test piece increases constantly up to the point of rupture, and is much greater than the so-called breaking stress. Under compression the curve is of the nature of Ofg, having a much less defined yield point at f than under tension. Under compression the cross-sectional area of the test piece increases, and, consequently, the straining action increases up to the point of final breaking down of the material.

For such materials as cast iron and the hard varieties of steel the load-strain curve has the general form shown on the diagram, there being no yield point. Such material breaks with much less stress than those having well-defined yield points, and is brittle. Brass and bronze usually do not exhibit an elastic limit, or yield point, but are subject to great plastic deformation before final rupture, and are tough.

4. *The Forces Acting on the Parts of a Machine.*—All materials used in construction are more or less elastic, and, therefore, any part of a machine under the action of a load or force must change its form. This change of form may be very small and temporary; or it may be large and permanent; and if the load is sufficiently great rupture will occur. The character of the straining action and the stresses produced depend on the direction and point of application of the load, and on the form and method of support of the loaded part. Thus the load may produce tension, compression, shearing, bending, torsion, or a combination of these actions. While tension and compression cannot exist at one time between any pair of *molecules,* yet, in the case of bending, tension and compression do occur in different sets of fibers of the same part. This action will be more fully explained later. In tension, compression and bending the molecular stresses act normally to the planes separating the interacting molecules; that is, the stresses increase or decrease the distances between these molecules along the lines joining them. In shearing and torsion the displacement of the molecules is tangentially to the planes between the molecules. In pure shear the motion is rectilinear, while in torsion the adjacent molecules have a relative motion about an axis. Theoretically, only two kinds of strain exist: viz, elongation (compression being considered as negative elongation) and shearing. Similarly there are only two kinds of stresses: viz, direct or normal and tangential or shearing. But for practical convenience it is desirable to treat torsion and bending separately as elementary stresses.

The forces acting on the parts of a machine are summarized by *Unwin as* follows: (1) The useful load due to the effort transmitted from the driving to the working point to accomplish useful work. (2) Prejudicial resistances due to friction in the machine itself or work expended otherwise than at the working point. (3)

The weight of parts of the machine. (4) Reactions of inertia due to changes of velocity of parts of the machine. (5) Centrifugal forces due to changes of direction of motion of parts of the machine. (6) Occasionally there are stresses due to constraint, preventing expansion with changes of temperature.

The total action, on any member of a machine, of these forces may be called the total straining action on that member. The relative importance of the various straining actions is very different in different cases, and generally some of them are small enough to be neglected. The problems of designing are then simplified.

In each part of a machine the straining action varies with the fluctuations of the useful load and with the variations of position and velocity of the different parts of the machine. Each member must be capable of resisting the maximum straining action on that part of the machine. For each part of the machine, therefore, it is necessary to consider under what conditions the straining action is greatest. If in consequence of changes of position or velocity the straining action produces stresses of different kinds at different times, the member must be capable of sustaining the maximum stress of each kind. Lastly, as will be more fully explained presently, the amount of variation of the straining action affects the endurance of the material, and, therefore, requires also to be considered.

Questions and Problems.

Distinguish between stress, strain, tension, extension, pressure, compression. Define elasticity and the modulus of elasticity of extension and compression. Discuss the following properties of materials: elastic and plastic conditions, ductility, brittleness, malleability.

Show, by means of diagrams, the relation between stress and strain for ductile and brittle materials, explaining clearly the following: true elastic limit, yield point, ultimate strength.

Discuss the forces acting on the parts of a machine.

Problems.

1. A steel rod, $3\frac{1}{4}$ feet long, $2\frac{1}{2}$ square inches sectional area, reaches the elastic limit at 125,000 pounds, with an elongation of 0.065 inches. Find the stress and strain at the elastic limit, and the modulus of elasticity and express each in its proper units.

2. A piston rod is 10 feet long and 7 inches diameter. The diameter of the cylinder is 70 inches, and the effective steam pressure 100 pounds per square inch. Find the stress produced and the total alteration in the length of the rod for a complete revolution, E=30,000,000 inch-pound units.

3. A brass pump rod is 5 feet long and 4 inches diameter, and lifts a bucket 28 inches diameter on which is a pressure of 6 pounds per square inch, in addition to the atmosphere. The pressure below the bucket is 2 pounds absolute. What is the stress in the rod and the total extension per stroke? E=9,000,000 inch-pound units.

4. A round bar of mild steel, 18 feet long and $1\frac{1}{2}$ inches diameter, lengthens 1/16 inch under a pull of 7 tons. Find the intensity of the stress, and the value of E.

5. A hollow cylindrical cast-iron column is 10 inches outside and 8 inches inside diameter, and 10 feet long. How much will it shorten under a load of 60 tons? Assume E=18,000,000 inch-pound units.

CHAPTER II.

The General Problems of Design. Straining Actions. Factors of Safety.

5.* *The General Problem of Designing Machinery* may be divided into two parts:

(1) Design of the mechanism to give the required motion.

(2) Proportioning of the parts so that they will carry the necessary loads due to transmitting the energy, without undue distortion or practical departure from the required constrained motion (1) is the application of the principles of mechanism. Very frequently some well-known and thoroughly tried mechanism can be adopted for the problem under solution. In other cases, where some new type of machine is to be built, the problem becomes practically one of invention.

The proportioning of the various parts may conveniently be divided into two parts:

(a) Solution as a whole of the energy and force problem of the mechanism.

When the type and proportions of the mechanism have been fixed the relative velocity of any point in the mechanism may be found. If, then, the energy which the mechanism must transmit is known, it is possible, in general, to find the forces acting, from the law of conservation of energy; or the product of velocity multiplied by force is constant throughout the train.

(b) Assigning the dimensions of the various parts based on the forces acting on them.

When the forces acting can be determined it would seem easy to choose the material and assign proportions for a machine part, based on the laws of mechanics or strength of material, and such is the case when the stresses are simple and the conditions fully known. Thus a machine member, subject to simple tension within known limits, can be intelligently proportioned in this manner. But in many cases the forces acting are very complex, the theoretical design is not always clear, and our knowledge of materials and their

* From Kimball and Barr.

laws is limited in many respects. Recourse must, therefore, often be made to judgment or to empirical data, the result of experience. Even when the conditions are clear, theoretical design must always be tempered with practical modification and by constructive considerations, etc. The logical method of proportioning machine elements where theory is applicable is, therefore, as follows:

(1st) Make as close an analysis as possible of all forces acting and proportion parts according to theoretical principles.

(2d) Modify such design by judgment and a consideration of the practical production of the part.

In the case of details and unimportant parts, judgment and empirical data are commonly the best guides.

Summing up, then, the logical steps in the design of a machine are as follows:

(I) Selection of the mechanism.

(II) Solution of the energy and force problem.

(III) Design of the various machine members so they will not unduly distort or break under the loads carried.

(IV) Specification and drawing.

The last step, specification and drawing, is a necessary and important adjunct to the process of design; it is a powerful aid to the designer's mental process and is the best way of showing the workman what is to be done to construct the machine in question and also of making a record of what has actually been done. It is not machine design of itself, however, as machines may be designed and built without any drawings. It is, nevertheless, an indispensable part of the designer's equipment. Very often written specifications accompanying the drawings are not only useful, but necessary. In fact, the highest skill on the part of the designer is often needed to clearly and fully specify in writing just what is to be done, as the writing of specifications presupposes the most intimate knowledge of theory of design, and selection of materials.

From the above it is seen that the courses already completed, in principles of mechanism, mechanical processes, strength of materials, etc., are a necessary ground work before advancing to machine design.

6. *The Effect of Temperature.*—Certain parts of machines or other structures must work under high temperatures, as, for instance, boilers, steam valves, etc., so the effect of such temperatures must be taken into consideration. Fortunately, iron and steel do not suffer very appreciable change of strength under the temperatures ordinarily reached in practice. The following tables show the relative tenacity of wrought iron and steel at different temperatures, compared to that at ordinary temperature, the tenacity at ordinary temperature being taken as 100 (from Unwin).

TABLE 1.—WROUGHT IRON.

Temp. F.	212°	600°	750°	1000°	1200°
Relative tenacity	104	116	96	75	40

TABLE 2.—MILD STEEL.

Temp. F.	– 4°	212°	400°	570°	750°	900°	1100°
Relative tenacity	106	103	132	123	86	49	28

Professor Morley summarizes the results of experiments on the effect of temperature as follows:

"(1) *The tenacity* (a) at ordinary temperature falls off with increased temperatures until between 200° and 300° F., when it is something of the order of 5 per cent less than at 60° F. (b) It rises from this temperature to a maximum value at some temperature between 400° and 600° F., when it is something of the order of 15 per cent more than 60° F. (c) It falls continuously with further increase of temperature.

"(2) *The elastic limit* falls continuously with increase of temperature.

"(3) *The elongation* (a) falls with increase of temperature above the normal to a minimum value in the neighborhood of 300° F., and then (b) rises again continuously with increase of temperature.

* * * *

"(4) The modulus of direct elasticity (E) decreases steadily with increase of temperature, metals which give a value of about 13,000 tons per square inch at atmospheric temperature falling to about 12,000 tons per square inch at 500° F.

"*Low Temperatures.*—Experiments on a very mild steel at very low temperature show progressive increase of tenacity with decrease of temperature; while the elongation practically vanishes, the material behaving like a very brittle substance. On return to ordinary temperatures no permanent change from the original properties is observed."

The loss of strength with increase of temperature of copper and the copper alloys is much greater than that of iron and steel. Unwin gives the following for such materials:

$$f=a-b(t-60)^2,$$

where f is the tenacity at t° F., in tons per square inch, and a and b are constants having the following values:

TABLE 3.

	a	b
Copper	14.8	.000014
Rolled yellow brass	24.1	.000028
Rolled delta metal	31.3	.000041
Rolled Muntz metal	14.7	.000029
Cast gun metal	12.5	.000021
Cast brass	12.5	.000024
Cast phosphor bronze	16.1	.000026

7. *Stress due to Change of Temperature.*—Nearly all metals expand when heated and contract again when cooled. The expansion of the metal *per unit of length* for a rise of temperature of 1°, is the *coefficient of expansion*, which will be denoted by a.

Then the following are the values of a per degree Farenheit:

TABLE 4.

Hard steel	a=.0000074
Soft steel	a=.0000065
Cast iron	a=.0000062
Wrought iron	a=.0000068
Copper and its alloys	a=.0000010

Suppose a bar of metal to be heated and then to be securely fixed at the ends, so as to entirely prevent its contraction when cooled, there will be a strong pull on the supports, producing a tensional stress in the bar. Let the original length of the bar be l, and the rise of temperature be t° F., and a be the coefficient of expansion; then the length when hot will be $l(1+at)$. After the bar has cooled to the original temperature (but its contraction having been wholly prevented) there is a *strain* per unit of length of at. Let f be the stress produced per unit of area of cross section. Then, remembering the definitions of elasticity and the modulus of elasticity, we have

$$E=\frac{\text{stress}}{\text{strain}}=\frac{f}{at}\therefore f=atE.$$

8. *The Nature of Loads.—Steady or Dead Load.*—A steady load is one which is invariable both in direction and amount during the life time of the machine, and which, therefore, produces a permanent straining action. The weight of the part of a machine is an example of a steady load.

Variable or Live Load.—A variable load is one which is alternately put on and taken off, and which produces a straining action constantly varying in amount. The action of steam on the piston of an engine, and the passage of a train over a bridge are examples of such a load.

The effect of a steady load is generally easily calculated and the dimensions of the part to resist its action can be very exactly determined, and, consequently, a low factor of safety may be safely employed. The live load is more difficult to allow for, and, moreover, its effect is more destructive on the parts resisting it, than the steady load. A higher factor of safety must, therefore, be used in determining the dimensions of the parts subject to such load.

Sudden and Impulsive Loads.—A *sudden* load is one applied without velocity, but at one instant, and continues to act during the deformation produced. Just at first the load is not balanced by the stress, so that it acquires a certain amount of kinetic energy, while the resisting part is being deformed, and this energy is effective in increasing the deformation beyond that due to the same steady load. If the greatest stress thus produced is within the elastic limit of the material, the amount of this stress, momentarily, is double that due to the same load applied gradually, or resting on the part.

If a heavy body impinges with velocity and thus possesses kinetic energy the stress produced is much greater than that due to the same body resting on the part.

Such a load is said to be an *impulsive load,* and the stresses are said to be due to shock. In order to find the value of the stress in such a case, the work of deformation is equated to the kinetic energy of the impinging load.

This subject will be further treated later (see Art. 13).

9. *Fatigue.*—It is a well-known fact that parts of machines have broken, after greater or less length of time, under loads which have previously produced no bad effects. This is especially the case when the load has been such as to produce alternations of tensile and compressive stresses. Thus piston rods and crank shafts have suddenly broken under usual working conditions. The only reason that can be assigned for such failures is that the material under such alternations of stress suffers a deterioration, which renders it incapable of resisting a stress previously safe. This deterioration is called *fatigue.* Microscopic examinations of specimens under test have led to a theory to account for the effect of fatigue. It was found that the crystalline particles of the material are changed in shape by *slips* at the plane of cleavage of the crystals. These slips gradually increase in number, and becoming massed together produce *cracks.* When the specimen is under reversed stresses these cracks are much more rapidly developed by the loss of cohesion due to the grinding action on the cleavage planes, along which slipping *to and fro* takes place.

Another fact having a bearing on this subject is the well-known variation of elastic limit of ductile materials. Thus if a specimen is tested under tension, and the test is discontinued before rupture, and the test then repeated, it is found that the elastic limit has a higher value at the second test—but the ductility is reduced, and the material is more brittle than before. In the same way, if a specimen is subject to a compressive test, and then a second test is applied to the specimen, but this time, under a tensile load, the elastic limit is found to be much lowered.

10. *Experiments on the Action of Live Loads.*—Numerous experiments have been made to determine the effect of live loads, and of loads causing a reversal of stress. Of these Wöhler's researches are generally accepted as giving the most satisfactory working hypothesis. His experiments consisted of a series of a large number of tests in which the stresses were alternately produced and wholly or partially removed; and others in which the stresses were alternately tension and compression. Also series of tests under torsional stresses. The stresses applied at first were very large, in fact, nearly up to the limit of the statical breaking strength, and such stresses produced fracture after a small number of repetitions of load, but the number of repetitions before fracture occurred rapidly increased as the limit of the stresses was reduced, and, finally, for a very low limit of stresses, the number of repetitions necessary to produce fracture approached an infinite number.

These experiments show that the stress, at which a material fractures, when subject to varying strains, depends on the *range of variation* of stress and on the *number of repetitions* of change of load. That is to say, the safe working stress is lower the greater the range of stress produced and the greater the number of repetitions of loading. For example, the crank shaft and the piston rod of an engine suffer a complete cycle of reversal of stress for each revolution of the engine. Therefore, such parts, in order to run safely for indefinitely long periods, must be designed for much smaller working stresses than would be allowable for the same material under a steady load.

The following table shows very clearly the general nature of the results obtained by Wöhler:

TABLE 5.—RUPTURE OF WROUGHT IRON BARS BY TENSION.

Rupture by	1	application of	55,000	lbs. per sq. inch.
" "	800	applications of	51,500	" " " "
" "	107,000	" "	47,000	" " " "
" "	341,000	" "	42,000	" " " "
" "	481,000	" "	38,000	" " " "

A SAMPLE OF SPRING STEEL, SUBJECTED TO BENDING, BROKE:

When subjected to	81,000	applications of	95,000	lbs. per sq. in.
" " "	154,000	" "	85,000	" " " "
" " "	210,000	" "	75,000	" " " "
" "	472,000	" "	65,000	" " " "
" " "	539,000	" "	58,000	" " " "
" "	1,165,000	" "	53,000	" " " "

11. *Algebraic Equation for Wöhler's Law.*—Let C be the statical * breaking strength of a material, and suppose that the stress in a bar, or a part of a machine, varies from a value represented by f_{max} to a value of f_{min}, so that the range of stress is $R = f_{max} - f_{min}$. To

* Some writers, particularly those of Germany, give the breaking strength different names, according to the conditions under which the piece is placed. Thus the stress at which rupture occurs under a steady or very gradually applied load is called the *statical breaking strength*. When the bar returns to its original condition after each application of a repeated load, and if the stresses are all of the same sign (that

use this expression call tension + and pressures −. Then if the two stresses are of different sign, the range, $R=[f_{max}-(-f_{min})]=f_{max}+f_{min}$. If the number of changes of load is indefinitely great, fracture will occur for a value of f_{max} less than C, and so much smaller the greater the value of R. In other words, the breaking strength is less than C. Therefore, in designing for a varying load the working stress must be taken at a value of f less than C, and this value depends upon R. f_{max} is the breaking strength of a material under a variable load ranging between the limits of f_{max} and $\pm f_{min}$ repeated an indefinitely great number of times. f_{min} is + if the stress is the same kind as f_{max} and − if of an opposite kind, and, further, f_{min} is supposed to be less than f_{max}. Then the range $R=f_{max}\mp f_{min}$, the upper sign being taken if the stresses are of the same kind and the lower if they are opposite. Thus R will always be +.

The results of Wöhler's experiments can then be expressed by the equation

$$f_{max}=\frac{R}{2}+\sqrt{C^2-kRC}. \qquad (1)$$

If $R=0$, then $f_{max}=C$, the load being steady. k is a constant. such that when given suitable values, the decrease of f_{max}, as the value of R increases, is made to conform with the values observed in Wöhler's experiments. For the more ductile varieties of iron and steel the average value of k is 1.5, and for the harder varieties $k=2.0$.

There are three principal cases to consider:

(a) The load is steady; then $R=0$.

(b) The load is entirely removed and replaced; then $f_{min}=0$ and $R=f_{max}$.

(c) The load produces alternately a tensile and a compressive stress of equal magnitudes; then f_{max} and f_{min} are equal and of opposite sign, and $R=2f_{max}$.

---

is, all tensions, compressions, or shearing in one direction), the greatest stress that can be sustained for a given number of repetitions is called the *primitive strength*. If the stresses are alternately of opposite signs, that is alternately tension and compression, or shearing in opposite directions, the maximum stress which can be sustained is the *vibration strength*.

Formula (1) then takes the following forms for these cases:

	Greatest stress.	Least stress.	Range of stress.	Value of f_{max}.
(a)	f_{max}	f_{max}	$R=0$	$f_{max}=\frac{0}{2}+\sqrt{C^2-k\times 0\times C}=C;$
(b)	f_{max}	0	$R=f_{max}$	$f_{max}=\frac{f_{max}}{2}+\sqrt{C^2-kf_{max}C};$

$$\left(\frac{f_{max}}{2}\right)^2=C^2-kf_{max}C,$$

which reduces to

$$f_{max}=2C(\sqrt{k^2+1}-k),$$

	Greatest stress.	Least stress.	Range of stress.	Value of f_{max}.
(c)	f_{max}	$-f_{max}$	$R=2f_{max}$	$f_{max}=\frac{2f_{max}}{2}+\sqrt{C^2-2kCf_{max}},$

$$f_{max}=\frac{C}{2k}.$$

By giving k a value of 1.5, which is its average value for the kinds of iron and steel most commonly used for construction, the above equations reduce to the following:

$$\text{(a)}\ f_{max}=C;$$
$$\text{(b)}\ f_{max}=0.6054C;$$
$$\text{(c)}\ f_{max}=\tfrac{1}{3}C.$$

That is to say, in round numbers, the stresses which a bar of given material can sustain are in the ratio of 3 : 2 : 1 for the three kinds of loading.

12. Factor of Safety.—It is now evident that the parts of a machine must be so designed that the stresses produced in each are within the elastic limits of the materials. The *working stress* is the stress at which any machine part is designed to operate, and the ratio of the *breaking stress* to the *working stress* is the *factor of safety*. It is not sufficient generally that the working stress be kept only below the *elastic limit*, but the size of the part must be such that the *strain* produced is so small as not to throw it out of alignment, which might produce unnecessary friction, or even interference with other parts. A particular part might easily be made amply *strong* to resist the forces acting on it, but yet it might distort to such an extent as to render it unfit for its purpose. Thus it is seen that considerations both of *strength* and of *stiffness* must be employed in determining the sizes of the various parts of machinery.

A distinction must be made between the *apparent factor of safety*, as defined above, and the *real factor of safety*. The latter is the ratio of the *carrying strength*, or the maximum allowable stress as determined by Wöhler's law, to the working stress. Thus, suppose we use a factor of safety of 3, to insure that the stress will be within the elastic limit, and to cover defects of material, and uncalculated effects, then the apparent factors of safety will be approximately:

For a steady or dead load, factor of safety $=3$.

For a load alternately removed and replaced, factor of safety $=3\times2=6$.

For a load producing stresses alternately in opposite directions, factor of safety $=3\times3=9$.

The actual selection of the proper factor of safety requires the exercise of the highest skill and trained judgment on the part of the engineer. The following from Kent, on "Factors of Safety," shows the general nature of the considerations necessary:

"The selection of the proper factor of safety or the proper maximum unit stress for any given case is a matter to be largely determined by the judgment of the engineer and by experience. No definite rules can be given. In general, the following circumstances are to be taken into account in the selection of a factor:

"1. When the ultimate strength of the material is known within narrow limits, as in the case of structural steel when tests of samples have been made, when the load is entirely a steady one of a known amount, and there is no reason to fear the deterioration of the metal by corrosion, the lowest factor that should be adopted is 3.

"2. When the circumstances are modified by a portion of the load being variable, as in floors of warehouses, the factor should not be less than 4.

"3. When the whole load, or nearly the whole, is apt to be alternately put on and taken off, as in suspension rods of floors of bridges, the factor should be 5 or 6.

"4. When the stresses are reversed in direction from tension to compression, as in some bridge diagonals and parts of machines, the factor should not be less than 6.

"5. When the piece is subjected to repeated shocks, the factor should be not less than 10.

"6. When the piece is subject to deterioration from corrosion, the section should be sufficiently increased to allow for a definite amount of corrosion before the piece be so far weakened by it as to require removal.

"7. When the strength of the material, or the amount of the load, or both, are uncertain, the factor should be increased by an allowance sufficient to cover the amount of the uncertainty.

"8. When the strains are of a complex character and of uncertain amount, such as those in the crank-shaft of a reversing engine, a very high factor is necessary, possibly, even as high as 40, the figure given by Rankine for shafts in mill work."

In the absence of experience the following table of factors of safety may be used:

TABLE 6.—FACTORS OF SAFETY.

Material.	Dead load.	Repeated stresses in one direction.		Reversed stresses.	
		Gradually applied.	Suddenly applied.	Gradually applied.	Suddenly applied.
Wrought iron, steel and ductile metals..........	3	5	10	6	12
Cast iron and brittle metals..................	4	6	12	10	20

Questions and Problems.

Discuss the general problem of designing a machine.

Explain the effect of temperature on materials and the stresses due to change of temperature.

Define steady load, live load, sudden load, impulsive load, fatigue.

Explain Wöhler's experiments on the action of live loads.

Given Wöhler's formula, $f_{max}=\frac{R}{2}+\sqrt{C^2-kRC}$, explain the meaning of the symbols, and show how the formula becomes modified (1) for a steady load, (2) for a repeated load, (3) for a *vibrating* load. Assuming $k=1.5$, determine the value of f_{max} in terms of C for each of the above cases.

Define factor of safety; distinguish between real and apparent factors of safety. Discuss the general considerations leading to the selection of a factor of safety.

Problems.

1. A driven rivet cools through 650° F. Coefficient of expansion .00001, $E=25{,}000{,}000$. Find the maximum possible tensile stress

per square inch. In practice would the actual stress be as great as this? Why?

2. A bar of steel 1 inch diameter and 10 feet long, is heated 100° F, and the ends are then firmly gripped. Find the tensional stress per square inch, and the total pull of the bar, when it returns to its original temperature, if the end supports of the grips are pulled 1/40 inch nearer together. Coefficient of expansion for steel .0000062; E=26,000,000 lbs per square inch.

3. Two walls, 25 feet apart, are stayed together by a steel bar 1 inch diameter. The bar passes through plates, and is set up by nuts at each end. If the nuts are screwed up to the plates while the bar is at a temperature of 300° F., find the total pull exerted when the bar cools to 60 F. (1) if the ends do not yield, (2) if the total yielding at the two ends is $\frac{1}{4}$ inch. Coefficient of expansion for steel=.0000062, E=27,000,000 pounds per square inch.

4. In the test of a certain sample of steel it is found that the stress at the elastic limit is 71,000 pounds per square inch, and that the ultimate strength is 118,000 pounds per square inch. What factor of safety must be used in order to bring the working stress within the elastic limit?

5. Plot a curve showing the effect of temperature on the tenacity of copper. (Suggestion, let the vertical ordinates represent f, the tenacity and the abcissæ represent temperatures.) Art. 6.

6. Plot a curve showing the effect of temperature on phosphor bronze (Art. 6).

CHAPTER III.

RESISTANCE OF MATERIALS TO STRAINING ACTIONS.

RESILIENCE, SUDDEN AND IMPULSIVE LOADS, TABLES OF STRENGTH, ETC. TENSION, COMPRESSION AND SHEARING.

13. *Resilience.*—This term is commonly used to express the action of a strained body in springing back when the straining force is removed. As used technically, in mechanics, it means the work done in straining a material up to the elastic limit. Fig. 3 represents the portion of a stress-strain diagram, within the elastic limit. If a bar of metal is subjected to a *gradually* increasing stress, the strain increases in proportion to the increase of stress, within the elastic limit. The work done in producing the strain is represented by the shaded area OAB of Fig. 3.

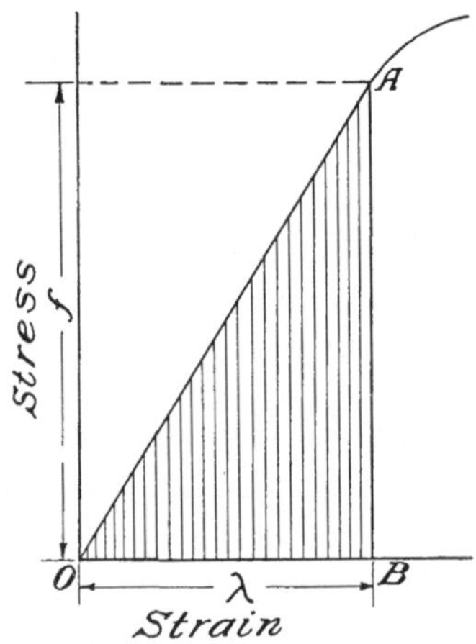

FIG. 3.

Let λ = the stress.

f = the stress per unit of cross-sectional area.

l = the original length of the bar.

A = Area of section of the bar.

E = Modulus of elasticity.

Then $\frac{\lambda}{l} = e =$ strain per unit of length, and

$$E = \frac{\text{stress}}{\text{strain}} = \frac{f}{e} = \frac{lf}{\lambda}.$$

$$\lambda = \frac{lf}{E}.$$

The work done in straining the bar of sectional area A is, then, from Fig. 3.

Work per unit of area=area of triangle $OAB=\frac{1}{2}f\lambda$.

$$\text{Total work}=\frac{Af\lambda}{2}=\frac{f^2Al}{2E}=\frac{Al}{2}\times\frac{f^2}{E}.$$

Now $\frac{Al}{2}$ is $\frac{1}{2}$ the volume of the bar, and $\frac{f^2}{E}$ is called the *modulus of elastic resilience:* therefore, the work done$=\frac{1}{2}$ volume of bar $\times$modulus of elastic resilience.

14. *The Effect of Sudden and Impulsive Loads.*—Suppose that a load W is *gradually* applied to a bar of sectional area A; and that the bar stretches an amount λ (from O to B, Fig. 4). The stress produced will be zero at O, as it is applied gradually, increasing to f at B, and $f=\frac{W}{A}$; and the *mean* stress during the operation is $f_{mean}=\frac{W}{2A}$.

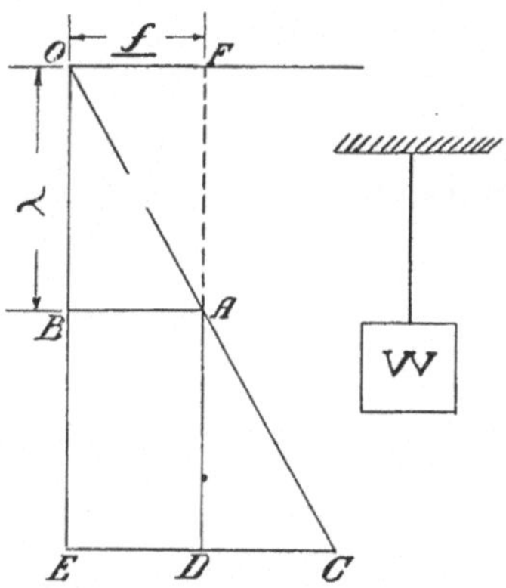

FIG. 4.

Now suppose the load W is *suddenly* applied and that the bar stretches from O to E (Fig. 4).

The work strain caused by weight W falling through distance $\lambda=W\lambda=fA\lambda$.

When the weight reaches B only a part of the energy stored in the weight has been expended in stretching the bar, and the remainder is still stored in the weight (as kinetic energy) and this energy is absorbed in further extending the bar from B to E. The work absorbed by the bar in stretching an amount λ is $\frac{fA\lambda}{2}$. The energy remaining in the weight when it reaches B is, then, the difference between the total energy and the energy absorbed by the bar. This difference is $fA\lambda-\frac{fA\lambda}{2}=\frac{fA\lambda}{2}$.

The bar will continue to stretch until the energy absorbed by it is equal to the total energy in the weight.

This will occur when the OEC=the area OFDE, and then the area ADC=the area OBA.

When the weight reaches E the strain and, therefore, also the stress is double that at B. Thus a suddenly applied load produces *twice* the stress, and twice the strain produced by the same load gradually applied.

In Fig. 5 suppose a weight W falls through a height h and brings up on a stop in such a manner as to produce an axial tension in a bar. As before

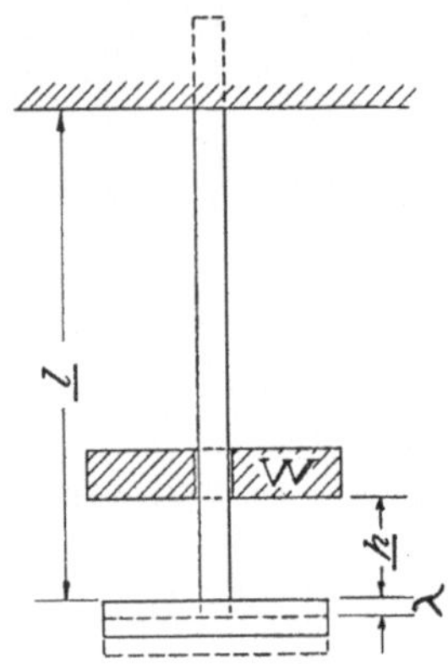

FIG. 5.

Let l=original length of bar.

λ=the stretch.

f=the stress produced.

Then the total work done by the weight $=W(h+\lambda)$: The work absorbed by the bar is $\frac{1}{2}f\lambda A=\frac{Al}{2}\times\frac{f^2}{E}$. Now, since all the *connections* are supposed to be absolutely rigid, the two expressions for work must be equal, or

$$W(h+\lambda)=\frac{Al}{2}\times\frac{f^2}{E}=\frac{f^2}{2E}\times\text{volume of the bar}$$

$$\text{or } f^2=\frac{2EW(h+\lambda)}{\text{volume of bar}},$$

from which f can be easily calculated when E for the given material is known.*

If, in the above equation, we put $h=0$, we have the case of the *sudden* load. Then

$$f^2=\frac{2Ew\lambda}{\text{volume}}=\frac{2EW\lambda}{Al}=2\frac{W}{A}E\frac{\lambda}{l}$$

$$=2\frac{W}{A}Ee=2\frac{W}{A}f$$

$$\text{then } f=2\frac{W}{A},$$

or the stress for a sudden load is double that produced by the same load applied gradually.

15. *Table of Strength and Coefficients of Elasticity.*—The following table, Table 7, taken from Kimball and Barr, with a few altera-

* When h is large compared to λ, which is usually the case, λ may be dropped, and the equation becomes

$$f^2=\frac{2EWh}{\text{volume}}.$$

TABLE 7.—ULTIMATE AND ELASTIC STRENGTH.

POUNDS PER SQUARE INCH.

Material.	Ultimate strength.			Elastic strength.			Direct coefficient of elasticity. E.	Transverse coefficient of elasticity. G.
	Tension.	Compression.	Shear.	Tension.	Compression.	Shear.		
Cast iron	20,000	95,000	20,000	10,000*	25,000	8,000	15,000,000	6,000,000
Malleable iron	35,000	42,000	20,000	......	.	......	......	......
Wrought iron	55,000	......	40,000	30,000	28,000	22,000	28,000,000	10,000,000
Steel, 0.15 c.	63,000	......	48,000	42,000	40 000	......	30,000,000	10,000,000
Steel, 0.50 c.	80,000	......	57,000	48,000	46,000	......	30,000,000	10,000,000
Steel, 0.70 c.	89,000	......	60,000	53,000	53,000	......	30,000,000	10,000,000
Steel, 0.96 c.	118,000	... .	83,000	69,000	71,000	.....	30,000,000	10,000,000
Steel, boiler plate	60,000	.....	48,000	30,000	......	.... .	30,000,000	.....
Crucible steel	116,000	......	......	80,000	80,000	......	31,000,000	12,400,000
Steel castings	50,000	......	40,000	30,000	30,000	.....	25,000,000	......
Nickel steel	90,000 / 100,000	......	......	60,000	......	......	31,000,000	......
Copper castings	22,000	60,000	......	[illegible]	......	......	12,000,000	.....
Rolled copper	31,000	......	......	[illegible]	......	......	15,000,000	......
Brass castings	20,000	12,000	......	......	......	......	10,000,000	......
Bronze, gun metal	35,000	......	......	......	......	.....	12,000,000	......
Bronze, phosphor	50,000	......	......	20,000	......	......	14,000,000	......
Muntz metal	80,000	......	......	55,000	......	......	......	......
Aluminum castings.	[illegible]	12,000	12,000	6,500	3,500	......	11,000,000	......

*Cast iron has, properly speaking, no elastic limit.

tions, shows the ultimate and elastic strength of different materials commonly used in engineering, when the stress is simple tension, compression or shearing. The direct coefficient of elasticity has been defined previously. The coefficient of transverse elasticity, frequently called the *coefficient of rigidity,* is the ratio of the shearing stress f_s per unit of area, to the distortion n, the distortion being measured by the tangent of the difference of the angles of an originally square particle before and after the stress is applied.

In using such a table as the following, it must be understood that the values given are general averages and an amount of judgment must be exercised in deciding how far such averages may be adopted for any particular case.

NOTE.—In the solution of problems the values given in Tables 7 and 8 may be used, when the working stress is not given in the problem.

TABLE 8.—WORKING STRESS.

A.—STEADY OR PERMANENT LOAD.

Material.	Kind of stress.				
	Tension. f_t	Compression. f_c	Bending. f_b	Shear. f_s	Torsion.
Cast iron	4,200	12,000	6,000 to 8,000	4.000	4,000 to 6.000
Bar iron	15,000	15,000	15,000	12,000	7,500
Plate iron, with grain	13,500	....	....	....	.. .
Plate iron, across grain	12,000	....	....	10,000	....
Steel, mild	13,000 to 17,000	13,000 to 17,000	13,000 to 17.000	10,000 to 13,000	8,000 to 12 000
Steel castings	8,000 to 12,000	12,000 to 16,000	10.000 to 14,000	7,000 to 12,000	7 000 to 12,000
Bronze, phosphor	10,000	....	....	7,000	4 200
Gun metal	4,200	....	....	....	....
Rolled copper	6,000	....	.. .	2,400	....
Brass	3,000	...	....	....	....

TABLE 8 (Continued).

B.—Load Varying Frequently from 0 to a Greatest Value.

Material.	Kind of stress.				
	Tension. f_t	Compression. f_c	Bending. f_b	Shear. f_s	Torsion.
Cast iron	2,800	8,500	4,000 to 5,500	2,800	2,500 to 4,000
Bar iron	10,000	10,000	10,000	8,000	5.000
Plate iron, with grain	9,000	...	....	....	....
Plate iron, across grain	8,000	....	....	6,500	....
Steel, mild	8,000 to 12,000	8,500 to 12,000	8,500 to 12,000	6,500 to 8,500	5,500 to 8.000
Steel castings	5,000 to 8,000	8,000 to 10,500	6,500 to 9,500	4.500 to 8,000	4,500 to 8,000
Bronze, phosphor	6,500	....	....	4,500	2,800
Gun metal	2,800	....	....	....	....
Rolled copper	3,000	....	....	1,600	....
Brass	2,000	....	...	....	....

TABLE 8 (Continued).

C.—Load Producing Alternate Stresses of Opposite Sign.

Material.	Tension and compression.	Bending.	Shear.	Torsion.
Cast iron	1,400	2.000 to 2,500	1,400	1,200 to 2,000
Bar iron	5,000	5.000	4,000	2,500
Steel, mild	4.500 to 6,000	4,500 to 6.000	3,500 to 4,500	2,500 to 4,000
Steel castings	2,500 to 4,000	3,500 to 5 000	2 500 to 4.000	2,500 to 4,000
Gun metal	1,400	....	....	

16. *Resistance to Simple Tension.*—A bar is in simple tension when the load tends to elongate the bar, and the stress acts along lines parallel to the axis.

Fig. 6 represents a portion of a bar on which a longitudinal pull is acting. The resultant of the load is supposed to act along the axis of the bar, and the load will then produce a normal tensional stress, uniformly distributed over a cross section AB. The stress on any oblique section CD will also be uniformly distributed and will consist of a direct stress of tension normal to the inclined section and a shearing stress tangential to the section. Let F be the load in pounds, and A be the area of the normal section in square inches. Then the stress on AB will be a tension of intensity

$$f_t = \frac{F}{A} \text{ lbs. per sq. in.} \qquad (1)$$

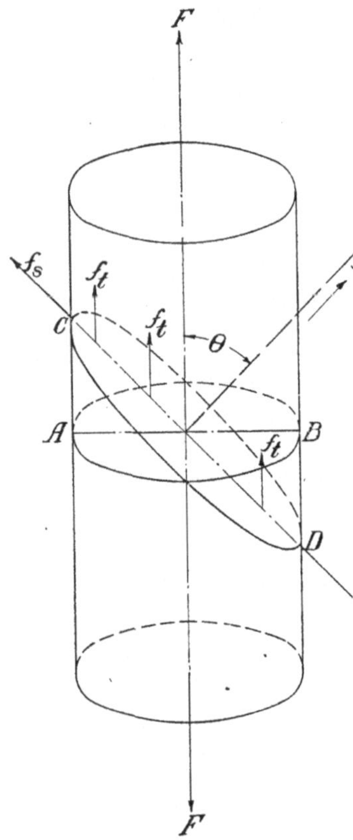

FIG. 6.

Let the section CD be inclined at the angle θ to the normal section. Then the stresses on CD can be found as follows:

The area of the inclined section is A see θ—and the intensity of the stress, in its original direction (along the axis of the bar), over the inclined section is

$$\frac{F}{A \sec \theta} = \frac{F}{A} \cos \theta = f_t \cos \theta. \qquad (2)$$

Then the normal stress

$$f_n = f_t \cos \theta \times \cos \theta = f_t \cos^2 \theta \qquad (3)$$

and the tangential stress

$$f_s = f_t \cos \theta \times \sin \theta \qquad (4)$$

Thus it is evident that the stress on any oblique section is less than on a normal section, also the shearing stress is greatest on a section inclined at an angle of 45° to the normal, and is, then, equal to half the tension on a normal section.* Therefore, the normal stresses only need be taken into consideration unless the resistance of the material to shearing is less than one-half its resistance to tension.

In practice, the problem is generally to find the cross section necessary to carry the given load F. A value for the working stress f_t in tension is selected, from Table 8, depending on the kind of material to be used and the nature of the load. Then the normal cross section of the part is

$$A = \frac{F}{f_t}. \tag{5}$$

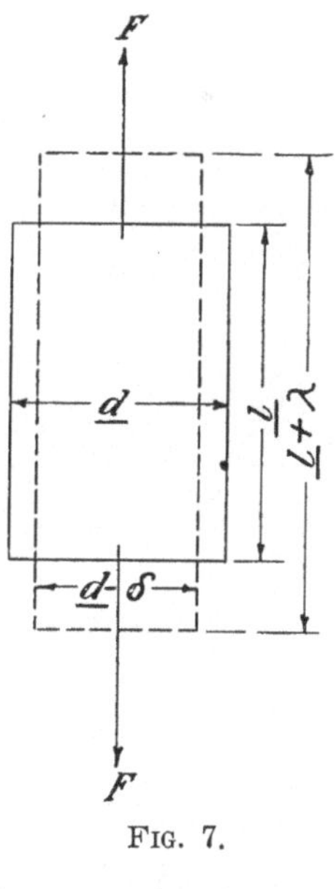

FIG. 7.

The effect of a tensional load on a bar of length l and diameter d, is to produce an elongation λ and a contraction δ. Within the elastic limit the extension per unit of length is $e = \frac{\lambda}{l}$; and since $E = \frac{\text{stress}}{\text{strain}} = \frac{f_t}{e}$, we have

$$e = \frac{\lambda}{l} = \frac{f_t}{E}. \tag{6}$$

In the same way the contraction per unit of width is

$$e_1 = \frac{\delta}{d}. \tag{7}$$

Fig. 7 represents the bar, the full lines representing its unloaded condition, and the broken lines, after the load is applied.

17. *Poisson's Ratio.*—It is found by experiment that the lateral contraction (or expansion, when the load is one of compression) is proportional to the change in length of the bar; that is to say, the ratio between the two is a constant, for a given material. This constant is usually denoted by $\frac{1}{m}$ and is called Poisson's ratio.

* This is obtained by differentiating equation (4) with respect to θ, placing the first derivative $= 0$, and solving for θ.

Thus

$$\frac{1}{m} = \frac{\delta/d}{\lambda/l} = \frac{e_1}{e} \tag{8}$$

Poisson's ratio varies for different materials from $\frac{1}{4}$ to $\frac{1}{3}$ and its average value for metals is about $\frac{3}{10}$.

18. *Simple Compression.*—If the straining action on a bar is an axial thrust it produces a longitudinal compression and a lateral expansion. The intensity of the pressure on a normal cross section of area A with a load P will be $f_c = \frac{P}{A}$. If we call c the longitudinal compression per unit of length we have similarly to equation (6)

$$c = \frac{f_c}{E}. \tag{9}$$

In general, the value of E for compression is the same as for tension.

When the bar is more than about five diameters in length, the tendency to bend, or buckle, must be taken into consideration, and the strength then determined by the laws for long columns.

19. *Thin Cylinders under Internal Pressure.*—Let the figure represent a thin cylindrical shell of length l and diameter d and let t

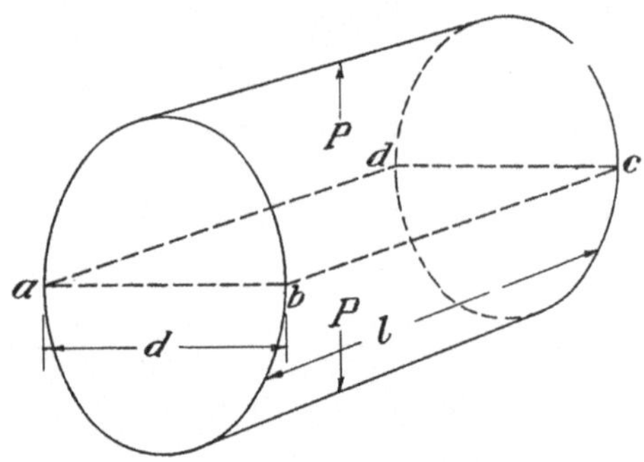

FIG. 8.

be the thickness of the metal of the plates, all in inches. Let the internal pressure be p pounds per square inch. Now suppose the cylinder to be cut by a diametrical plane abcd; then the total pressure or force acting on either side of the plane is p × area abcd. Call this force P; then $P = p \times l \times d$. This force tends to tear the cylinder apart along the lines bc and ad, and produces a tensile stress in the metal the intensity of which we will call f. The area over which this stress is distributed is bc × thickness + ad × thickness, or is $2l \times t$, and, consequently, the total resisting force is 2ltf. Putting the total load and the total resistance equal to each other we have $pld = 2ltf$; so that the circumferential stress is

$$f = \frac{pd}{2t}. \qquad (10)$$

Now consider the total bursting force as acting axially, then the shell tends to tear apart around a circumference as at efgh.

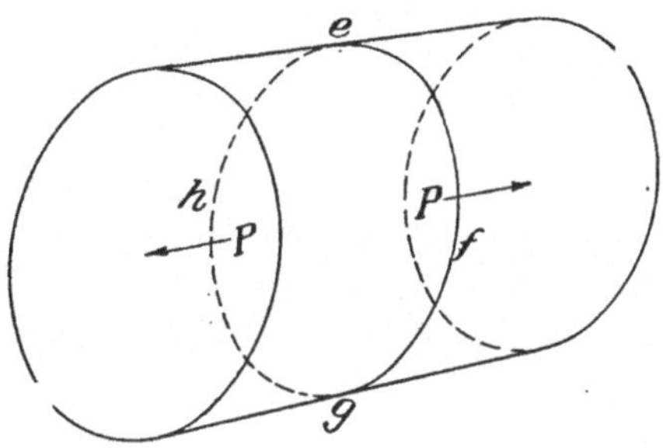

FIG. 9.

The total load P is now $\frac{\pi d^2}{4} \times p$, and the area of metal section resisting rupture is $\pi d \times t$. Let f, as before, represent the intensity of stress produced. Then equating load and resistance we have $\frac{\pi d^2}{4} p = \pi dtf$, and the longitudinal stress is

$$f = \frac{pd}{4t}. \qquad (11)$$

Whence it is seen that the longitudinal stress is half the circumferential stress.

Formula (11) is also applicable to a spherical vessel under an internal pressure.

The above formulæ are used in calculating the thickness of boiler shells. Such shells are made up of several plates riveted together, having both longitudinal and circumferential seams. If the whole shell were made of a single homogeneous plate without seams we would have, from formula (10) $t = \frac{pd}{2f}$, but in the actual case the efficiency of the riveting must be considered. Call this efficiency η. Then

$$t = \frac{pd}{2f\eta}. \qquad (12)$$

Formulæ (10), (11) and (12) are used for *thin* cylinders, where the thickness t is very small compared to the diameter d.

20. *Thick Cylinders.*—When the thickness of the cylinder is an appreciable proportion of the internal diameter, as in the case of hydraulic cylinders, the *mean* stress is unaltered, but the inner layers of the metal are more severely stressed than the outer. In such cases one of the following formulæ are used:

$$t=\frac{d}{2}\left(\sqrt{\frac{f+p}{f-p}}-1\right) \qquad \text{Lamé} \quad (13)$$

$$t=\frac{d}{2}\left(\sqrt{\frac{3f+2p}{3f-4p}}-1\right) \qquad \text{Grashof} \quad (14)$$

In these formulæ, t=thickness in inches, d=internal diameter in inches, f=the working stress allowed and p=the internal pressure in pounds per square inch. In designing such a cylinder it is usual to allow for a test pressure of double the ordinary working pressure, and to use a value of f=9000 for cast iron, or 25,000 for steel casting. That is, these values of stress must not be exceeded under the test pressure.

(For the deduction of these formulæ the student is referred to any complete work on mechanics.)

The fact that the inner layers of thick cylinders are under greater stress than the outer, has led to the practical device of making the cylinder of several concentric tubes, shrunk over each other in such a manner that the inner tube is under an initial compression when the pressure is not acting. Familiar examples of this method are the built-up gun, used in the service, and hydraulic cylinders. The outer tube is bored to a slightly less diameter than the outside of the tube over which it is to be placed; it is then heated until the expansion is sufficient to allow it to be slipped over the inner tube. On cooling, it produces a compression of the inner tube, and a tension in itself. If, now, a pressure is applied inside the cylinder, the first effect is to overcome the initial compression of the inner tube and to increase the tension of the outer; so that the resulting stress in the inner tube is the difference of the initial compression and the working tension, and in the outer tube the tension produced is the sum of the initial and that due to the working pressure. This distributes the stress more equally throughout the walls of the cylinder.

21. *Resistance to Shearing.*—A shearing action is one which causes sliding parallel to the section considered. Thus, in Fig. 10, the action of the blades of a shearing machine are shown. The pressure causes a shearing stress in the plane ab. The intensity of the shearing stress is the pressure P divided by the cross-sectional area A, and is denoted by f_s. Thus

$$f_s = \frac{P}{A}. \qquad (15)$$

Fig. 11 illustrates the action when the force and reaction do not act in the plane of the section. There is now a distinct bending

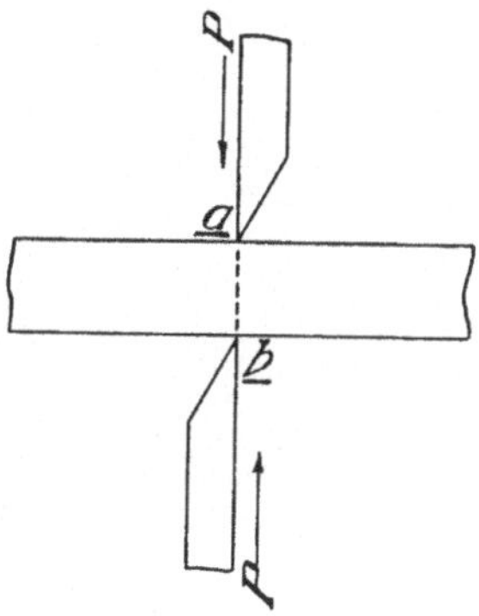

FIG. 10.

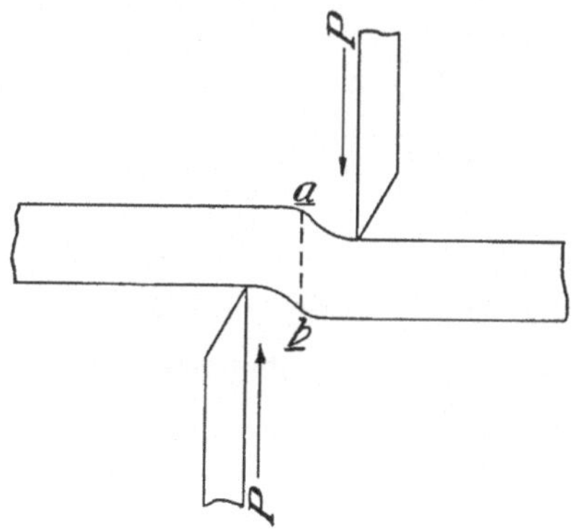

FIG. 11.

action produced, and the effect is to alter the distribution of the stress over the section. At a plane half-way between the upper and lower surfaces the shearing is greater than the average, while at the upper and lower edges of the section there is no shearing action.

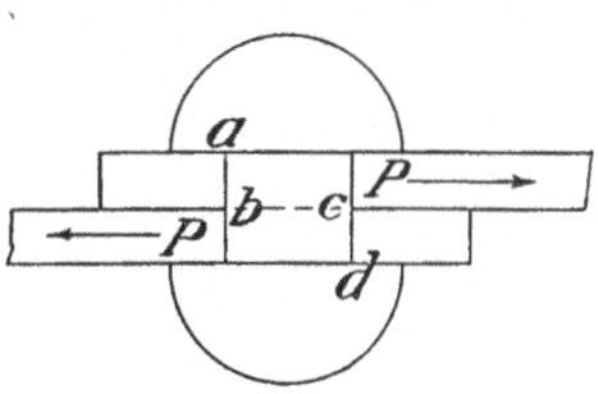

FIG. 12.

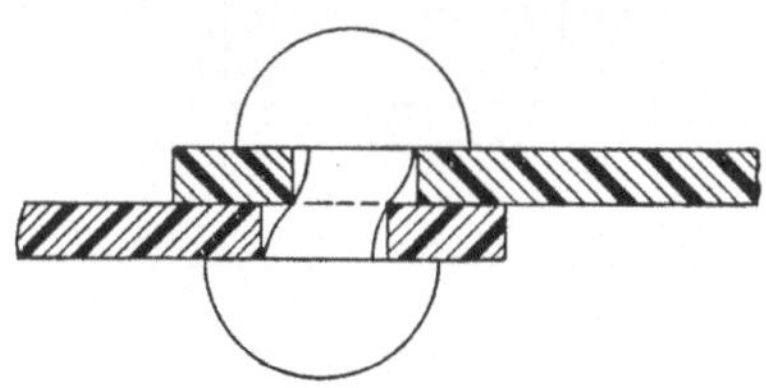

FIG. 13.

Well-driven rivets are in shear, and well-fitted bolts are usually considered to be in shear. On the other hand, pins and cotters are rarely in pure shear, in practice.

Figs. 12 and 13 represent the action in the cases of well-fitting and loose rivets. In Fig. 12 it will be seen that the forces really act along the centers of thickness of the plates, but, on account of the friction and rigidity of the edges ab and cd, the action of the forces is practically across the section bc. In practice, it is assumed that rivets are subjected to pure shearing and their size is calculated from equation (15). In the case of pins and cotters it is usual to calculate the strength on the assumption that they are subject to bending.

Unwin gives the following values of the maximum stress, based on experiment, for such cases:

$f_s = 1.5 \frac{P}{A}$ if the section is rectangular and P perpendicular to one side.

$= 1.33 \frac{P}{A}$ if the section is circular or elliptical.

$= 1.59 \frac{P}{A}$ at half-way between the angle and center, if the section is square, and P acts parallel to a diagonal.

$= 2 \frac{P}{A}$ if the section is a ring of small thickness compared to the diameter.

22. *The Nature of Shearing Stress.**—Consider an originally square particle of thickness unity, to be acted upon by forces P

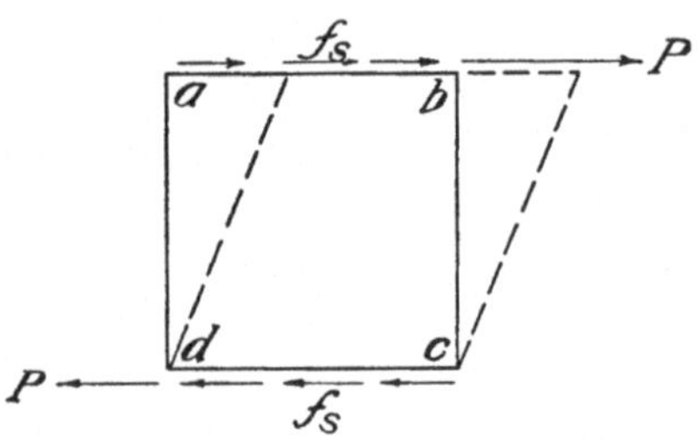

FIG. 14.

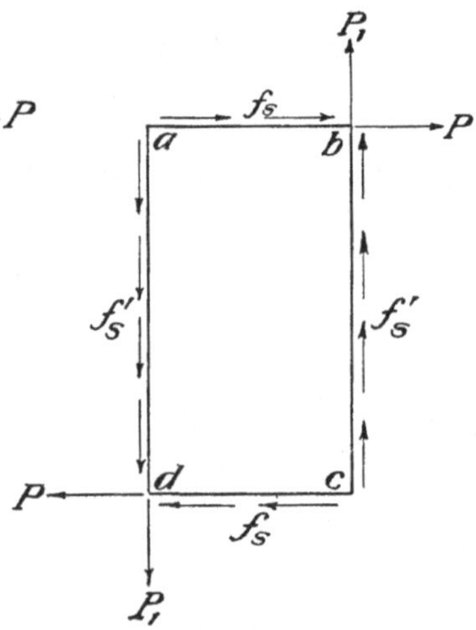

FIG. 15.

parallel to two of its opposite faces. The square will be distorted into a rhombus, Fig. 14, and the shearing stress will be $f_s = \frac{P}{ab}$ (the thickness being unity). The particle would now spin around due to the couple $P \times ad$, or $P \times bc$, unless an equal and opposite couple

* From Goodman.

is applied to it. In order to make the following remarks perfectly general, we will take a rectangular plate, as shown in Fig. 15.

The plate is acted upon by a clockwise couple, $P \times ad$, or $f_s \times ab \times ad$, and a counter clockwise couple, $P_1 \times ab$, or $f_s' \times ad \times ab$, but, since the plate remains in equilibrium, these must be equal; then $f_s \times ab \times ad = f_s' \times ad \times ab$, or $f_s = f_s'$; *i. e.,* the intensity of the stress on the two sides of the plate is the same. In other words, when a force acts so as to produce a shearing stress on opposite sides of a particle, there is a stress of equal intensity induced on the other two faces, at right angles to the first.

Now, for convenience, we will return to our square particle. The forces P and P_1 acting on the two sides may be resolved into forces R and R_1, acting along the diagonals as shown in Fig. 16. The

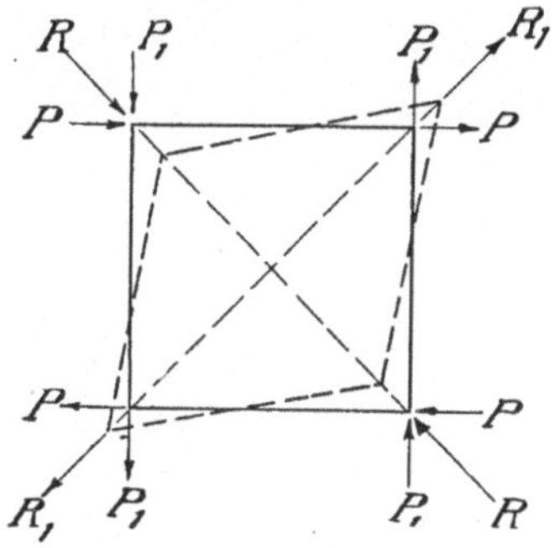

FIG. 16.

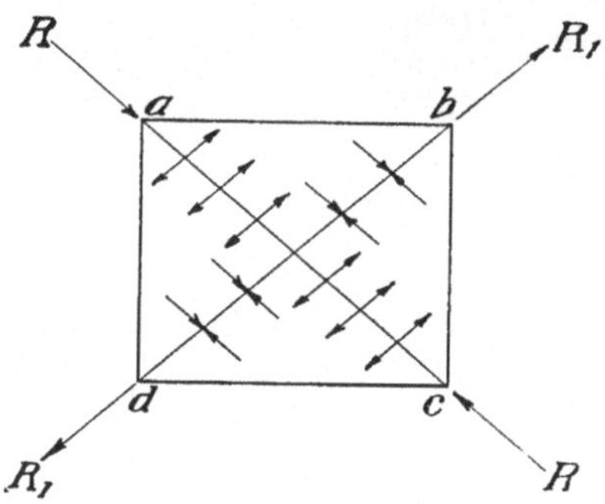

FIG. 17.

effect of these forces will be to distort the square into a rhombus exactly as before.

NOTE.—The rhombus in Fig. 14 is drawn in a wrong position for simplicity.

These two forces act at right angles to one another; hence we see that a shear stress consists of two equal and opposite stresses, a tension and a compression, acting at right angles to one another.

In Fig. 17 it will be seen that there is a tensile stress acting normal to one diagonal, and a compressive stress normal to the other. The one set of resultants, R, tend to pull the two triangles abc, acd apart, and the other resultants to push the two triangles abd, bdc togther.

Let f_o = the stress normal to the diagonal.

Then $f_o \times ac = f_o \times \sqrt{2}ab = f_o \times \sqrt{2}bc = R$.

But $\sqrt{2} \times P = \sqrt{2}f_s \times ab = \sqrt{2}f_s \times bc = R$.

Hence $f_o = f_s = f_s'$.

Thus the intensity of stress is equal on all the four edges and on the two diagonals of a rectangular particle subjected to shear.

23. *Modulus of Transverse Elasticity, or Modulus of Rigidity.*—Fig. 18 represents an originally square particle abcd distorted into the rhombus a′b′cd under the action of a shearing force. The amount of distortion aa′ or bb′ is called the *slide;* let this equal x. Then, calling l the original length of a face, the ratio x/l is the distortion per unit of length, and is called the *sliding.* Let $n = x/l$, then G, the *modulus of rigidity* $= \frac{\text{stress}}{\text{strain}} = f_s/n$.

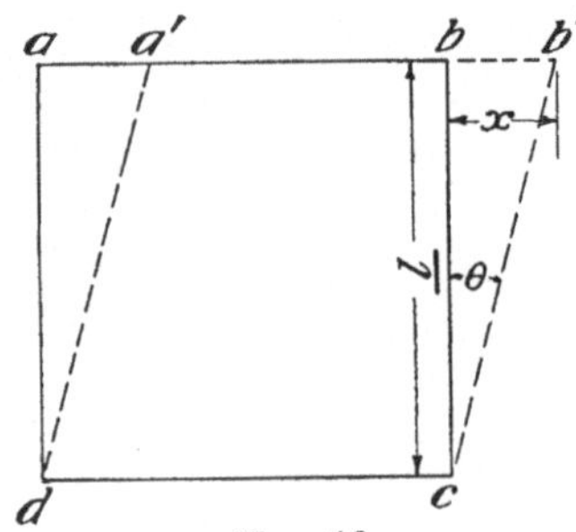

FIG. 18.

From the figure it is seen that n is the tangent of the angle of distortion θ. When the limit of elasticity is not passed, the deformation is small and we then have

$$\theta = \tan\theta = n$$

and

$$G = f_s/n = f_s/\theta,$$

or

$$f_s = G \times \theta$$

(θ being expressed in radians).

The relation between the direct elasticity E and the transverse elasticity G depends on Poisson's ratio, 1/m (see Art. 17).

$$^*G = \frac{mE}{2(m+1)};$$

$$E = \frac{2(m+1)}{m} G.$$

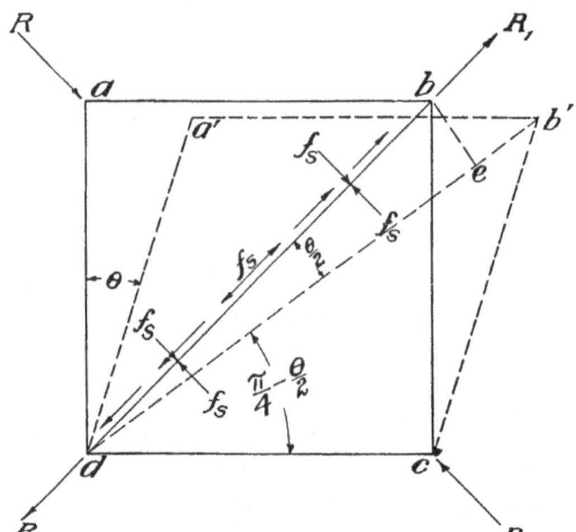

FIG. 19.

The average value of m being $\frac{10}{3}$, we have

$$G = \frac{5}{13}E.$$

* This relation is obtained as follows: It has been shown (Art. 23, Fig. 17) that a shearing action may be resolved into a tension R_1, acting along a diagonal, and a thrust R, perpendicular to it, producing

Questions and Problems.

Define *resilience* and the *modulus of resilience.*

Discuss the effect of sudden and impulsive loads, showing how to find the stresses produced by a load applied suddenly or impulsively compared to that produced by the same steady load.

Discuss simple tension (or compression) and show why, ordinarily, the stress on a normal cross section only need be considered. Show how to find the value of the normal and tangential stresses on any oblique section of a material under simple tension (or compression). What is Poisson's ratio?

Deduce equations for finding the thickness of boiler shells to withstand a given internal pressure p. Explain the reasons for building up thick cylinders on the principle of *initial compression.*

What is shearing? Explain the internal stresses produced when a material is subject to shearing forces. Define *modulus* of *rigidity.*

---

equal stresses, to the stress along the faces of the particle. If the face dc is supposed to be rigidly fixed, the square particle abcd takes the form of the rhombus a′b′cd under the shearing action.

The elongation of db due to $R_1 = f_s \times db/E$.

The elongation of db due to $R = f_s \times db/E \times m$ (see Art. 17). The total elongation of db is then

$$\frac{f_s \times db}{E}\left(1+\frac{1}{m}\right) \text{ but } db = ab \times \sqrt{2}.$$

Now, for simplicity, consider the original length of each face of the particle = unity, then $bd = \sqrt{2}$ and we have

$$\text{elongation} = x = \frac{f_s}{E} \times \sqrt{2}\left(1+\frac{1}{m}\right). \qquad \text{(a)}$$

Draw be perpendicular to db′, and remembering that θ is a very small angle, we have

$$eb' = \text{elongation of the diagonal} = x.$$

$$x = db' - de = db' - dc\sqrt{2} = db' - \sqrt{2}.$$

From trigonometry $c = a \sin c \operatorname{cosec} A$, $C = 180° - (A+B)$

$$\sin C = \sin(A+B) = \sin 2\left(\frac{\pi}{4}-\frac{\theta}{2}\right) = 2\sin\left(\frac{\pi}{4}-\frac{\theta}{2}\right)\cos\left(\frac{\pi}{4}-\frac{\theta}{2}\right)$$

$$\therefore c = db' = 2dc\,\frac{\sin\left(\frac{\pi}{4}-\frac{\theta}{2}\right)\cos\left(\frac{\pi}{4}-\frac{\theta}{2}\right)}{\sin\left(\frac{\pi}{4}-\frac{\theta}{2}\right)} = 2dc\cos\left(\frac{\pi}{4}-\frac{\theta}{2}\right)$$

$$= 2\cos\left(\frac{\pi}{4}-\frac{\theta}{2}\right) \text{ since } dc = 1.$$

$$\therefore x = 2\cos\left(\frac{\pi}{4}-\frac{\theta}{2}\right) - \sqrt{2} = 2\left(\cos\frac{\pi}{4}\cos\frac{\theta}{2} + \sin\frac{\pi}{4}\sin\frac{\theta}{2}\right) - \sqrt{2}$$

PROBLEMS.

1. Find the stress and the extension produced in a bar 10 feet long and $1\frac{1}{2}$ square inches section, by the sudden application of a tensile load of 6 tons. What suddenly applied load would produce an extension of $\frac{1}{20}$ inch? $E=26,000,000$.

2. A load of 500 pounds falls through $\frac{1}{2}$ inch onto a stop at the end of a vertical bar 10 feet long and 1 square inch section. Find the stress produced in the bar. $E=26,000,000$.

3. Find the greatest height from which the load of the above problem could fall, before bringing up on the stop, so as to produce a stress not greater than 14 tons per square inch.

4. A cylindrical vessel with hemispherical ends, diameter 6 feet, carries a pressure of 200 pounds per square inch above the atmosphere. The cylindrical part is constructed of steel rings, riveted together. If $f=7$ tons per square inch, how thick must the metal be, and what is the longitudinal stress in the metal of the ring joint whose section is $\frac{7}{10}$ that of the solid plate.

5. A round iron 1 inch diameter sustains a tension load of 20,000 pounds; find the unit normal and tangential stresses on a section at an angle of 30° with the normal section. Solve also for a section at 45°.

6. From the values of E and G given in the table, Art. 15, find the value of m for steel.

7. An eye bar carries a steady load of 15 tons. What must be the diameter of the wrought-iron pin to resist shearing.

NOTE.—Consider the pin in double shear, and assume this shear uniformly distributed over the area of the pin. Use Table 8.

But $\frac{\theta}{2}$ being very small $\sin\frac{\theta}{2}=\frac{\theta}{2}$ and $\cos\frac{\theta}{2}=1$.

$$x=2\left(\frac{1}{\sqrt{2}}\times 1+\frac{1}{\sqrt{2}}\times\frac{\theta}{2}\right)-\sqrt{2}=\frac{\theta}{\sqrt{2}}.$$

By definition $G=\frac{f_s}{\theta}$, $\theta=\frac{f_s}{G}$.

$$\therefore\ x=\frac{f_s}{G\sqrt{2}}=\frac{f_s}{E}\sqrt{2}\left(1+\frac{1}{m}\right)\text{ from (a).}$$

$$\frac{1}{G}=\frac{2}{E}\left(\frac{m+1}{m}\right).$$

$$E=2\left(\frac{m+1}{m}\right)G.$$

$$G=\frac{Em}{2(m+1)}.$$

CHAPTER IV.

Resistance of Materials to Straining Actions (Continued).

BENDING, LIMITATION OF THE THEORY OF SIMPLE BENDING, MOMENT OF INERTIA, MODULUS OF SECTION. TABLE OF I, Z, ETC. CURVES OF SHEARING FORCE AND B. M. TABLE OF S. F., B. M. AND DEFLECTION DIAGRAMS OF S. F. AND B. M.

24. *Bending.*—For the general theory of beams, see Smith, "Strength of Material."

In order to have simple bending the following conditions must be fulfilled: (1) The axis of the bar is a straight line, joining the

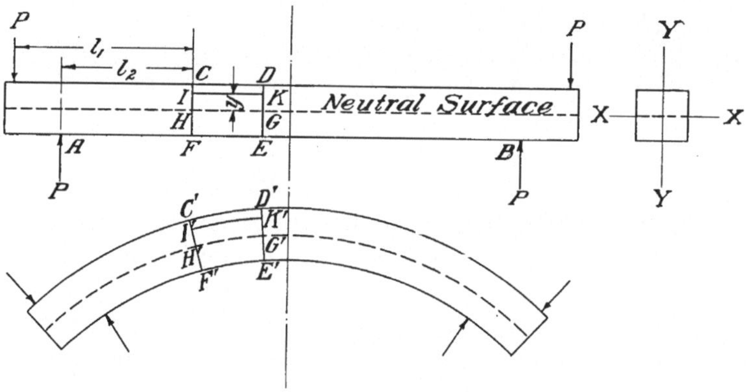

Fig. 20.

centers of figure of parallel transverse sections; (2) the bar is symmetrical about a plane passing through the axis; (3) all the external forces act in such a plane of symmetry, called the plane of bending, normally to the axis (Unwin).

Fig. 20 represents a beam under simple bending, due to the action of the forces, as shown in the figure, the flexure being greatly exaggerated in the lower figure.

In order to obtain a starting point for the consideration of bending actions, certain assumptions must be made, as follows:

(1) Plane transverse sections remain plane and normal to the longitudinal axis after bending.

(2) The material is homogeneous, and obeys Hooke's law, and the strains produced do not exceed the elastic limit.

(3) Every layer of the material is free to extend or contract under stress, as if separate from other layers.

(4) The modulus of elasticity is the same for tension and compression.

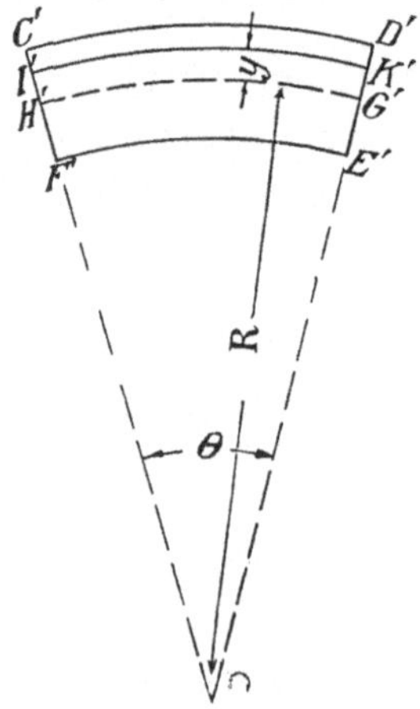

FIG. 21.

Returning now to Fig. 20, since the straining action is the same on every section between A and B, the curvature of the beam will be circular. Consider any two transverse sections, CF and DE, close together; after bending they are no longer parallel, but the layer of material CD is stretched to C′D′, and the layer EF is compressed to E′F′, while the layer HG is neither stretched nor compressed. The surface, then, through HG is not under any longitudinal stress, and is called the *neutral surface,* and its line of intersection XX with a transverse section is called the *neutral axis* of the section.

Suppose the sections C′F′ and D′E′ produced to meet in a line, which being seen end on, is represented by the point O, Fig. 21, the angle of intersection being θ. Call the radius of curvature of the neutral surface R, and let y be the distance from H′G′ of a layer I′K′, originally parallel to the neutral surface. Then

$$\frac{I'K'}{H'G'} = \frac{(R+y)\theta}{R\theta} = \frac{R+y}{R}.$$

The strain of the layer I′G′ is then

$$e = \frac{I'K'-IK}{IK} = \frac{I'K'-H'G'}{H'G'} = \frac{(R+y)\theta - R\theta}{R\theta} = \frac{y}{R}.$$

The intensity of the stress thus produced, f, being by assumption within the elastic limit, is, then,

$$f = E \times e = E\,\frac{y}{R} \text{ (equation 6)}. \qquad (15)$$

The intensity of the compressive stress will have the same value, since by assumption E has the same value for compression as for tension.

The intensity of the longitudinal stress at every point in the cross section is, then, proportional to its distance from the neutral axis.

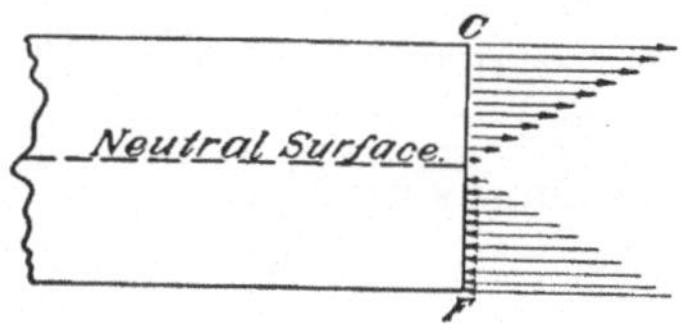

FIG. 22.

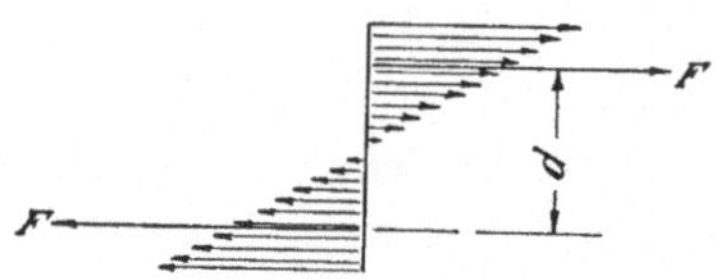

FIG. 23.

Thus, at the neutral axis $(y=0)$ the stress is zero, at unit distance from the axis $(y=1)$ its value is $\frac{E}{R}$, and it is a maximum at the boundary furthest from the neutral axis.

The variation of the longitudinal stress acting on any section CF is shown by Fig. 22. These stresses have resultants FF, which form a couple $F\times d$, as shown by Fig. 23. This may be called the couple of the internal stresses.

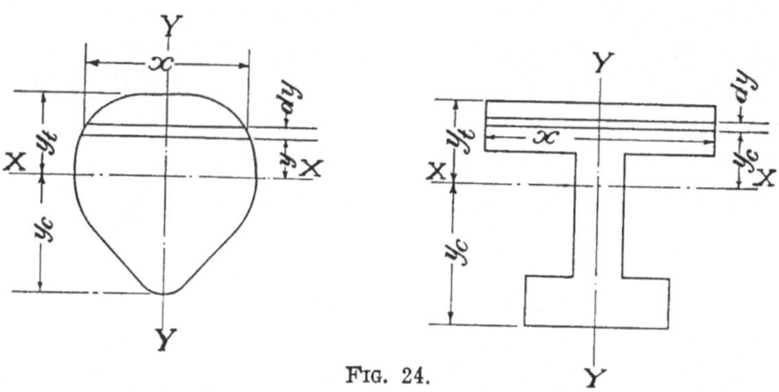

FIG. 24.

Again, returning to Fig. 20, and considering the external forces acting to the left of the section CF, the bending moment is seen to be Pl_1-Pl_2. Since equilibrium is sustained, this bending moment is balanced by the couple of the internal stresses, or

$$M=P(l_1-l_2)=Fd. \qquad (16)$$

In Fig. 20, the section of the beam is shown symmetrical about a horizontal axis, but this is not necessary; the beam may have any section symmetrical about the plane of bending YY as shown in Fig. 24.

The algebraic sum of all the external moments acting must be zero, since the beam is in equilibrium. Also the total of the tensile forces must balance the total of the compressive forces, since they form a couple balancing the total of the external forces. Therefore, the algebraic sum of the internal horizontal forces is equal to zero.

In Fig. 22, let δA, or xdy, be an elementary strip of area, parallel to the neutral axis, x being the variable width of the strip. The stress on this element is $f \times \delta A = \frac{Ey\delta A}{R}$ (from equation 15) and the total stress on the section is

$$\Sigma \frac{Ey\delta A}{R},$$

but since the total stress is zero

$$\Sigma \frac{Ey\delta A}{R} = 0,$$

and, since $\frac{E}{R}$ is a constant,

$$\Sigma y\delta A = 0.$$

But this relation can hold only if the distances y are measured from an axis passing through the center of gravity of the cross section. *Therefore, the neutral axis passes through the center of gravity of the section.*

The moment of the stress acting on the strip of area is $f\delta Ay = \frac{Ey^2\delta A}{R}$. The moment of the total stress acting on the section is then

$$\Sigma\left(\frac{Ey^2\delta A}{R}\right) = \frac{E}{R}\Sigma y^2\delta A = \frac{EI}{R},$$

I being the *moment of inertia* of the cross section. This total moment of the internal stresses is called the *moment of resistance* of the section.

If, now, we equate the external bending moment M to the moment of resistance of the section, we have

$$M = \frac{EI}{R}, \qquad (16)$$

expressing the relation between the bending moment and the curvature of the beam.

If now f is the stress at a distance y from the neutral axis we have from equation 15

$$f=\frac{Ey}{R} \text{ or } \frac{E}{R}=\frac{f}{y}.$$

Equation (16) then takes the form

$$M=\frac{EI}{R}=\frac{fI}{y},$$

whence

$$f=M\frac{y}{I}. \tag{17}$$

Equation (17) enables us to obtain the stress at any point of the section of a beam distant y from the neutral axis, knowing the value of the external bending moment, and the moment of inertia of the section. Since it has been shown above that the greatest stress occurs in the fibers most distant from the neutral axis, the problem is, generally, to find the greatest tensile or compressive stresses produced, and to so fix the dimensions of the section that these greatest stresses do not exceed the safe working stress allowable under given conditions. Let y_t be the distance of the most distant fiber in tension, from the neutral axis, and y_c the distance of the most distant fiber in compression. Then the greatest tensile and compressive stresses are

$$f_t=M\frac{y_t}{I} \text{ and } f_c=M\frac{y_c}{I}.$$

Let

$$\frac{I}{y_t}=Z_t \text{ and } \frac{I}{y_c}=Z_c,$$

then

$$M=f_tZ_t=f_cZ_c. \tag{18}$$

Z_t and Z_c are called the *moduli* of the section with respect to tension and compression. If the section is symmetrical about the neutral axis, $y_t=y_c$ and $Z_t=Z_c$, and it is then necessary to consider only the stress, whether tension or compression, that the material is weakest to resist.

25. *Limitation of the Theory of Simple Bending.*—The theory of *simple bending* explained in the preceding article refers only to cases in which there is no shearing force, but in many cases shearing force is present as well as bending. In such cases the internal stresses acting at any section have not only to balance the bending moment, but also the shearing force, and, consequently, there will be tangential as well as direct longitudinal stresses acting on any section. The case, then, becomes one of *compound stresses,* which will be investigated later.

However, in most practical cases, the theory of simple bending is sufficient to enable the calculation of stresses and strains to be made with a workable degree of approximation. In many such cases the *greatest* bending moment occurs at sections across which the shearing force is zero, or negligible, and very frequently, where the shearing force is greatest, the bending moment is small enough to be neglected. The usual practice of the engineer is to follow the simple bending theory, allowing for doubtful cases by a proper selection of the allowable safe working stress—in other words—by a judicious selection of the factor of safety, being guided by previous successful solutions of similar problems.

26. *Moments of Inertia.*—Properly speaking, the moment of inertia of a body about a given point, or line, is the sum of the products of the elementary masses of which it is composed by the squares of their distances from the point or line. This definition is extended in meaning when applied to plane surfaces as the sum of the products of the elements of area by the squares of their distances from the point or line. Thus, let dA be an element of area at a distance y from the axis; then the moment of inertia of the surface is

$$I = \Sigma dAy^2.$$

If A is the total area of a surface and I its moment of inertia about an axis through the center of gravity, then

$$I/A = k^2,$$

and k is called the *radius of gyration* of the section.

In order to facilitate the use of formula (18), $M = fZ$, all text books, hand books and reference books for the use of engineers contain tables showing the values of moments of inertia, radii of gyration, moduli of sections, etc. Such a table, for a few of the simpler sections, is given in Table 9, Art. 27.

The values given in such tables are necessarily based on a given axis, usually the neutral axis of the section, but it is frequently desirable to have the value of the moment of inertia about some other axis.

27. *Moment of Inertia about Parallel Axes.*—In Fig. 25 suppose it is wished to obtain the moment of inertia of the area about the axis ZZ. Let I be the moment of inertia of the surface about an axis XX through its center of figure obtained from the table. Let I_1 be the moment of inertia about the new axis ZZ, parallel to XX,

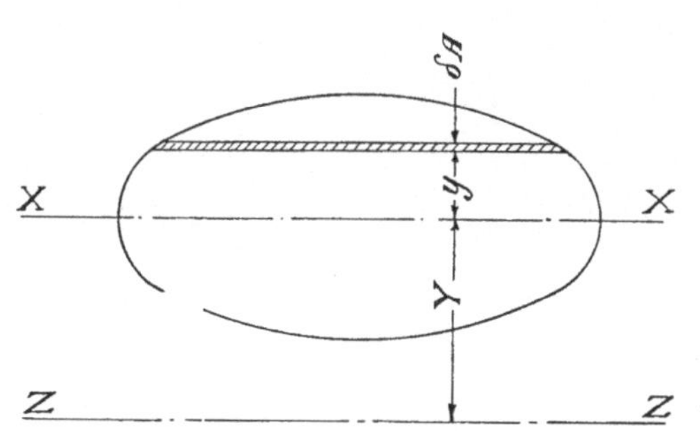

FIG. 25.

and distant Y from it. Let δA be an element of area, distant y from the axis XX, and let A be the total area of the surface. Then

$$I = \Sigma\delta Ay^2,$$

$$I_1 = \Sigma\delta A(Y+y)^2 = \Sigma\delta Ay^2 + 2Y\Sigma\delta Ay + Y^2\Sigma\delta A,$$

but

$$\Sigma\delta Ay = 0,$$

since y is measured from an axis passing through the center of figure, and

$$\Sigma\delta A = A,$$

$$\therefore I_1 = I + AY^2. \qquad (19)$$

Equation (19) is used for finding the moment of inertia of a complex section which can be divided up into a number of rectangles or other simple sections. Thus, suppose we have a beam of I section and wish to find the position of the neutral axis and moment of inertia of the complete section. Divide the section into rectangles, as in Fig. 26. To find the distance of the neutral axis from some fixed line, as, for instance, the base line of the figure, XX, take moments about this line, thus:

Moment of upper flange $= A \times a$.
Moment of web flange $= B \times b$.
Moment of lower flange $= C \times c$.

$$\text{Total moment} = A \times a + B \times b + C \times c. \quad (a)$$

$$\text{Total area} = A + B + C. \quad (b)$$

$$\text{Then, the arm, } y_x = \frac{\text{equation (a)}}{\text{equation (b)}}, \quad (c)$$

which locates the neutral axis xx of the combined section. Now, calling the distances of the neutral axes of each part considered

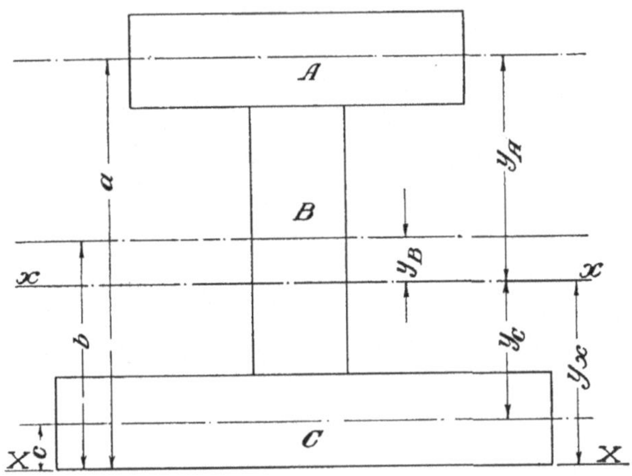

FIG. 26.

separately from the combined neutral axis, y_A, y_B, y_C, and the moment of inertia of each part, I_A, I_B, I_C, we have, from equation (19)

$$I_{total} = I_A + I_B + I_C + Ay^2_A + By^2_B + Cy^2_C. \quad (20)$$

The value of I_{total} having been found thus, it is divided by the distance of the most distant fiber in tension, and in compression, thus giving the value of Z_t or Z_c and these values of Z are used in equation (18).

The axis of moments is the horizontal broken line through the center of figure of the section. I is the moment of inertia about the axis through the center of figure and at right angles to the plane of

TABLE 9.—PROPERTIES OF SOME SIMPLE SECTIONS.

Shape of Section.	Area of Section.	Moment of Inertia. I	Square of Radius of Gyration. $k^2 = \frac{I}{A}$	Modulus of Section. I/y
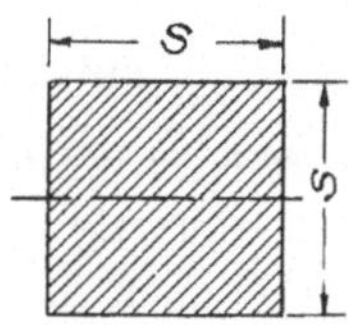	S^2	$\frac{1}{12} S^4$	$\frac{1}{12} S^2$	$\frac{1}{6} S^3$
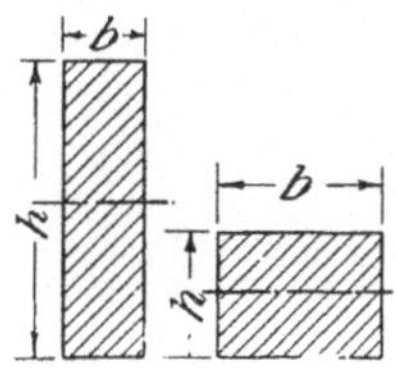	bh	$\frac{1}{12} bh^3$	$\frac{1}{12} h^2$	$\frac{1}{6} bh^2$
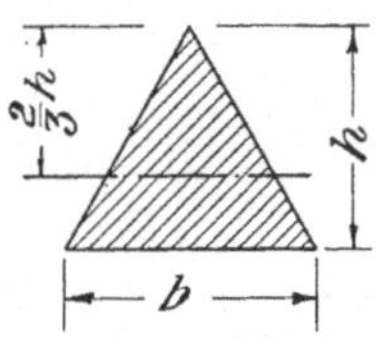	$\frac{1}{2} bh$	$\frac{1}{36} bh^3$	$\frac{1}{18} h^2$	$Z_1 = \frac{1}{24} bh^2$ $Z_2 = \frac{1}{12} bh^2$
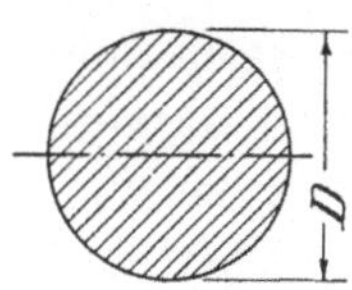	$\frac{\pi D^2}{4}$	$\frac{\pi}{64} D^4 = .0491 D^4$	$\frac{1}{16} D^2$	$\frac{\pi}{32} D^3 = .0982 D^3$
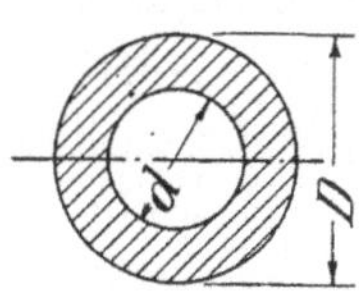	$\frac{\pi}{4} (D^2 - d^2)$	$\frac{\pi}{64} (D^4 - d^4)$	$\frac{1}{16} (D^2 + d^2)$ If $D - d$ is small $= \frac{1}{8} D^2$ nearly	$\frac{\pi}{32} \left(\frac{D^4 - d^4}{D}\right)$
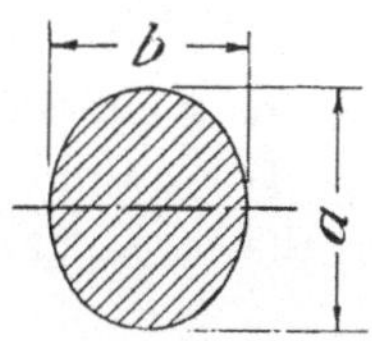	$\frac{\pi}{4} ab$	$\frac{\pi}{64} ba^3$	$\frac{1}{16} a^2$	$\frac{\pi}{32} ba^2$

bending. Where two values of Z are given, Z_1 is the modulus for fibers at top edge of figure, and Z_2 for fibers at bottom edge.

28. *Beam Fixed at One End.*—This case is a little more complex than the case of simple bending considered in Art. 24. Suppose a force W acts on the end of a beam which is solidly fixed at its other end, as in Fig. 27, and that it is required to find the action at the section ab. If we introduce two equal and opposite forces W′ and W″, each equal to W, at ab, the equilibrium of the beam is not distnrbed. Then the action of W on the section ab is equivalent to the couple WW″, and the unbalanced force W′. The moment of the couple is W×l and is balanced by the moment of the internal stresses in the manner described in Art. 24. The other force W′

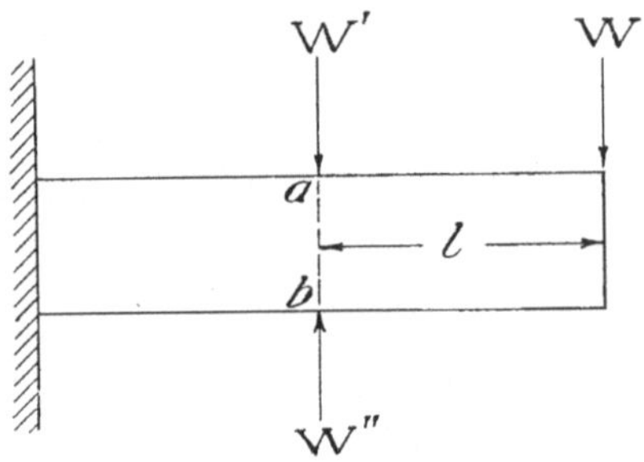

FIG. 27.

produces shearing stress across the section, and the beam must be strong enough to resist both the bending and shearing stresses. If there are several forces acting to the right of ab they may be reduced to a single equivalent resultant force.

In many cases the effect of the bending moment is much greater than that of the shearing force, and the latter may be neglected.

29. *Curves of Bending Moment and Shearing Force.*—Fig. 28 represents a beam *supported* at the ends and carrying a distributed load of varying intensity. Consider any point C in the beam, distant x from A, and let the intensity of loading at the point C be w. For any small length dx in the vicinity of C then we have the load equal to wdx. Let R be the reaction of the support at A. At every point along AB erect an ordinate of length, to some convenient scale, equal to the value of w at that point. Draw a curve through the end of these ordinates, and we have the *load* curve

DHE. Then the area ADHEB represents the total load on the beam, and the resultant of this load acts through the center of figure of the area. The shearing force at C is the resultant of all the forces to the left of C; call this shearing force S. Then

$$S = R - \int_0^x w\,dx. \qquad \text{(a)}$$

If the value of S is calculated for a number of points along the beam and ordinates erected to represent these values, a curve through their ends will give the curve IK, which is the *shearing-force* curve. In the same way, calculate the value of the bending

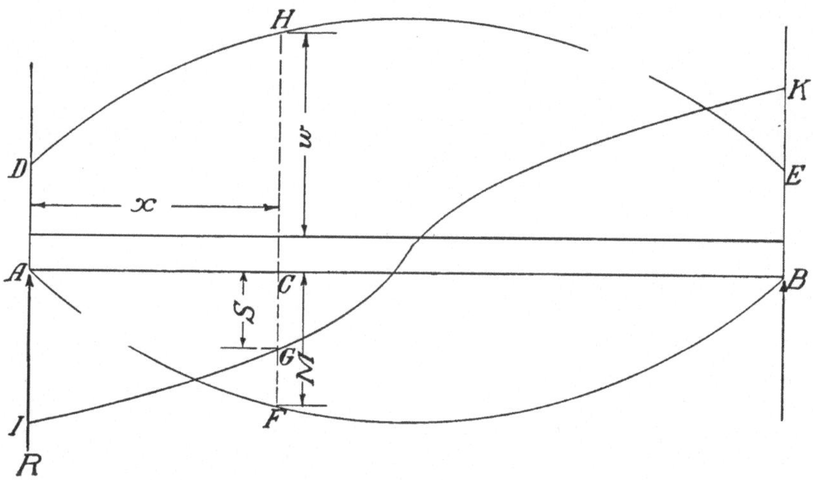

FIG. 28.

moment, M, at C. This moment is the algebraic sum of the moments of all forces to the left of C, and is

$$M = \int_0^x S\,dx. \qquad \text{(b)}$$

Construct the bending moment curve, AFB.

From equations (a) and (b) it is seen that the shearing force at any point is equal to R minus (the area of that part of the load curve between A and the point) and that the bending moment at any point is equal to the area of the shearing-force curve between A and the point. Also from the figure, the bending moment is a maximum at the point at which the shearing force changes its sign. A case of frequent occurrence in practice is that in which w is constant (that is, the load is uniformly distributed), in which case the curve of shearing force is a straight line, and the curve of bending moment is a parabola.

Table 10 (following) shows the greatest shearing force, bending moment, stress and deflection of beams under different loads and methods of support, and Table 11 shows shearing force and bending moment curves.

30. *Beam Fixed or Encastre.*—A beam is said to be fixed, or encastre, when its end is so supported that the direction of the axis of the beam is not changed by the action of the bending forces.

This Fig. 29 represents a beam fixed at one end and loaded with a weight W at the other end.

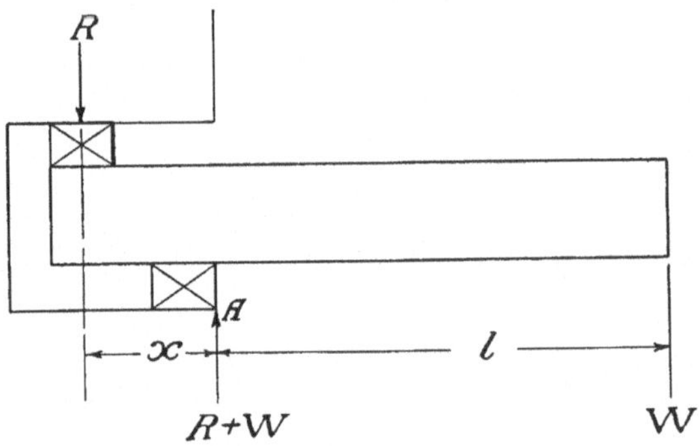

FIG. 29.

Such a beam may be considered as under the action of a couple,

$$Rx = Wl.$$

31. *Economy of Different Sections.*—The resistance of a beam to bending is proportional to its sectional modulus Z. Thus, if we have a number of beams of different cross sections, under the same loads, all having the same value of Z, that one is most economical of material which has the least area. Therefore, the greater the ratio of Z to A the more economical is the form of the beam.

In a beam of circular or rectangular section the top and bottom fibers only are fully strained. At the neutral axis there is no strain, due to bending, so that by removing material from the vicinity of the neutral axis and placing it at the top and bottom edges the beam is made stronger. This gives the I or double-flanged section. In a beam of this section the flanges resist nearly the whole of the stresses due to bending, while the shearing forces are resisted principally by the vertical web.

In a beam of I section, in order that both flanges may carry their due proportion of the result of the load, the following relations must hold:

$$M = f_t Z_t = f_c Z_c.$$

If $f_t = f_c$ then $Z_t = Z_c$, which is the case when the section is symmetrical about the neutral axis. If, however, f_t does not equal f_c, the two flanges will not be under the same stress if $Z_t = Z_c$, and, therefore, in order to have the most economical distribution of material the form of section must be unsymmetrical about the neutral axis.

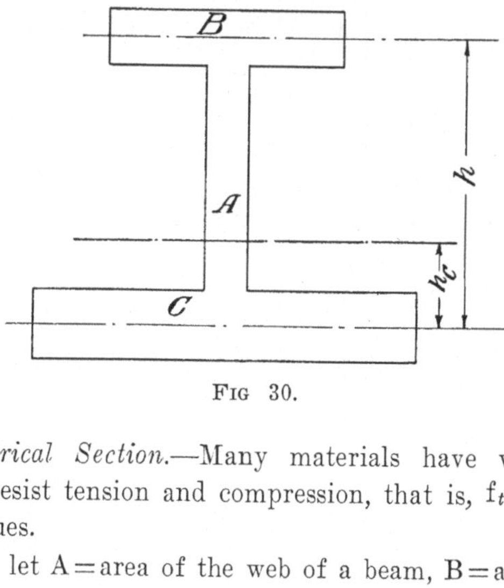

FIG 30.

Unsymmetrical Section.—Many materials have very unequal strength to resist tension and compression, that is, f_t and f_c have different values.

In Fig. 30 let A = area of the web of a beam, B = area of flange in tension, C = area of flange in compression and h = the depth of the beam, from center to center of the flanges. Suppose the area A of the web is small compared to the areas of the flanges. Then, in order that each flange may be stressed in proportion to its strength, we have

$$f_t \times B = f_c \times C. \qquad (a)$$

Now, let the neutral axis of the complete section be at a distance h_c from the center of the lower flange. Assuming that the effect of the web in resisting *bending* is negligible, and calling M the external bending moment, we have, taking moments about the neutral axis,

$$M = f_t B(h - h_c) + f_c C h_c.$$

Substituting from (a)

$$M = f_t B(h - h_c) + f_t B \times h_c,$$
$$\therefore M = f_t Bh$$

or

$$B = \frac{M}{f_t h}; \tag{21}$$

similarly,

$$C = \frac{M}{f_c h}. \tag{22}$$

If S = the total shearing force and f_s the safe shearing stress of the material the necessary area A of the web is given by

$$A = \frac{S}{f_s}. \tag{23}$$

The assumption that the web does not resist any of the bending action may, in some cases, especially when the web is heavy, introduce an error, but this error is on the safe side. In the usual cases of rolled steel sections the error is inappreciable.

TABLE 10.—SHEARING FORCE, BENDING MOMENT, AND DEFLECTION OF BEAMS.

INCH-POUND UNITS.

FIXED AT ONE END.

Case.	Method of loading.	Position of greatest shearing force.	Greatest shearing force.	Position of greatest bending moment.	Greatest bending moment.	Normal stress due to load. f_b	Deflection. δ	Remarks.
1	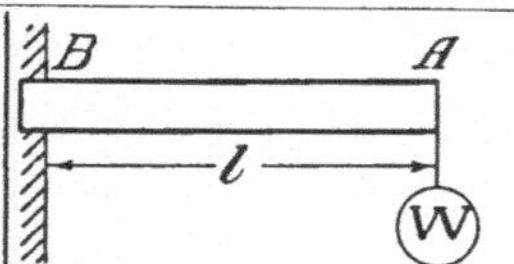	All points between A and B.	W	At B.	Wl	$\frac{Wl}{Z}$	At A $\frac{Wl^3}{3EI}$	Load at free end.
2	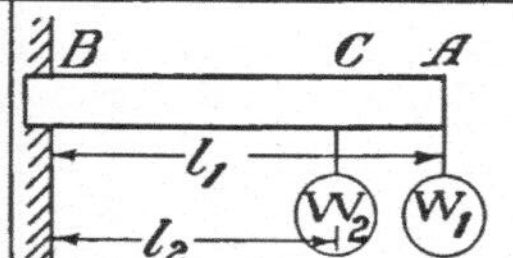	Between C and B. Between A and C.	$W_1 + W_2$ W_1	At B.	$W_1l_1 + W_2l_2$	$\frac{W_1l_1+W_2l_2}{Z}$		More than one load.
3	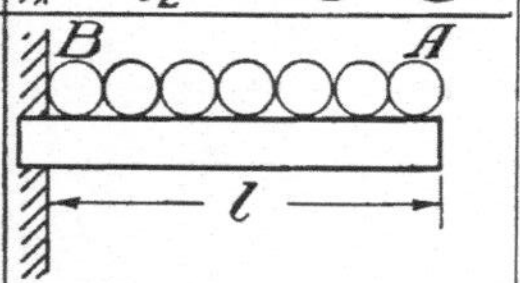	At B.	$wl = Q$	At B.	$\frac{wl^2}{2} = \frac{Ql}{2}$	$\frac{wl^2}{2Z}$	At A $\frac{Ql^3}{8EI}$	Uniform load w lbs. per inch run.
4	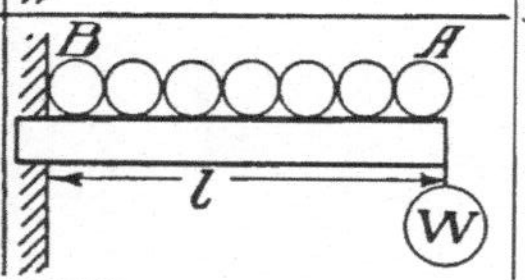	At B.	$W + wl = W + Q$	At B.	$\frac{Wl^2}{2} + Wl = (\frac{1}{2}Q + W)l$	$\frac{wl^2 + 2Wl}{2Z}$	At A $\frac{l^3}{EI}\left(\frac{W}{3} + \frac{Q}{8}\right)$	Uniform load w lbs. per inch run and W at free end.

5

TABLE 10.—SHEARING FORCE, BENDING MOMENT, AND DEFLECTION OF BEAMS (Continued).

INCH-POUND UNITS.

SUPPORTED AT BOTH ENDS.

Case.	Method of loading.	Position of greatest shearing force.	Greatest shearing force.	Position of greatest bending moment.	Greatest bending moment.	Normal stress due to load. f_b	Deflection. δ	Remarks.
5	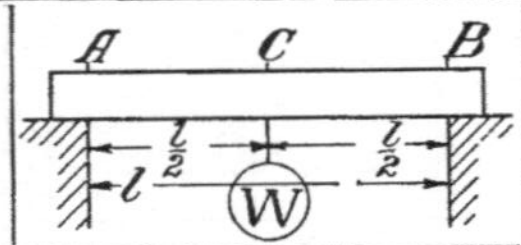	Between A and C. Between B and C.	$\frac{W}{2}$ $-\frac{W}{2}$	At C.	$W\frac{l}{4}$	$\frac{Wl}{4Z}$	At C $\frac{Wl^3}{48EI}$	W at center.
6	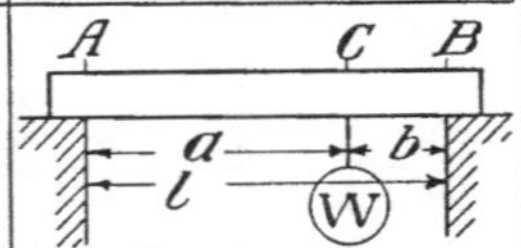	Between A and C. Between B and C.	$\frac{Wb}{l}$ $-\frac{Wa}{l}$	At C.	$W\frac{ab}{l}$	$\frac{Wab}{lZ}$	$\frac{Wa^2b^2}{3lEI}$	W not at center.
7	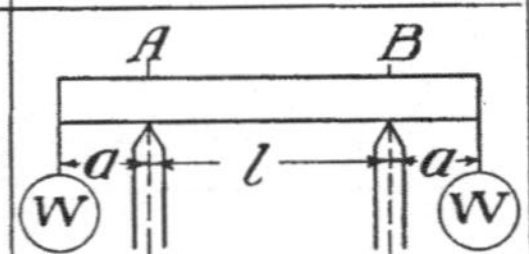	At A or B. Between A and B.	W 0	Between A and B.	Wa	$\frac{Wa}{Z}$	At center $-\frac{Wal^2}{8EI}$ At ends $\frac{W}{EI}\left(\frac{a^3}{3}+\frac{a^2l}{2}\right)$	Equal overhanging loads producing equal couples at free ends.
8	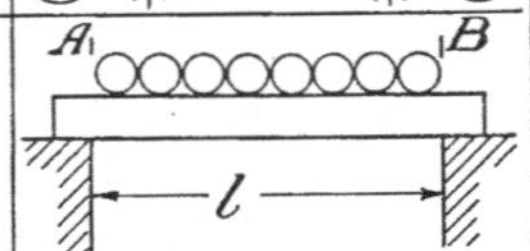	At A or B.	$\frac{wl}{2}=\frac{Q}{2}$	At center.	$\frac{wl^2}{8}=\frac{Ql}{8}$	$\frac{wl^2}{8Z}$	At center $\frac{5l^3Q}{384EI}$	Uniform load, w lbs. per inch span. Wl = Q
9	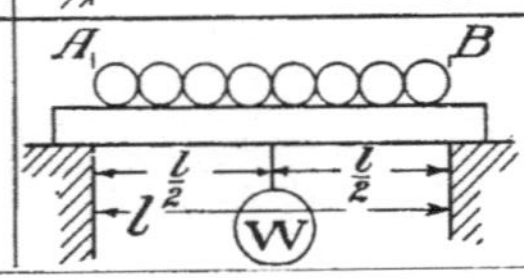	At A or B.	$\frac{W}{2}+\frac{wl}{2}$ $=\frac{1}{2}(W+Q)$	At center.	$(W+\frac{1}{2}Q)\frac{l}{4}$	$\frac{(W+\frac{1}{2}Q)l}{4Z}$	$(W+\frac{5}{8}Q)\frac{l^3}{48EI}$	Uniform load, w lbs. per inch span, and W at center.

TABLE 10.—SHEARING FORCE, BENDING MOMENT, AND DEFLECTION OF BEAMS (Continued).

INCH-POUND UNITS.

FIXED AT BOTH ENDS.

Case.	Method of loading.	Position of greatest shearing force.	Greatest shearing force.	Position of greatest bending moment.	Greatest bending moment.	Normal stress due to load. f_b	Deflection. δ	Remarks.
10	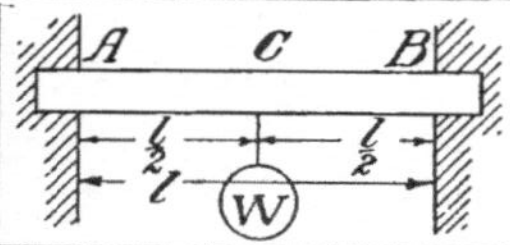			At A, B or C.	$\frac{Wl}{8}$	$\frac{Wl}{8Z}$	$\frac{1}{192} \cdot \frac{Wl^3}{EI}$	W at center.
11	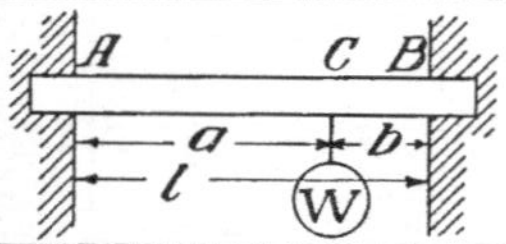			At A. At C. At B.	$\frac{Wab^2}{l^2}$ $-\frac{2Wa^2b^2}{l^3}$ $\frac{Wba^2}{l^2}$			W not at center.
12	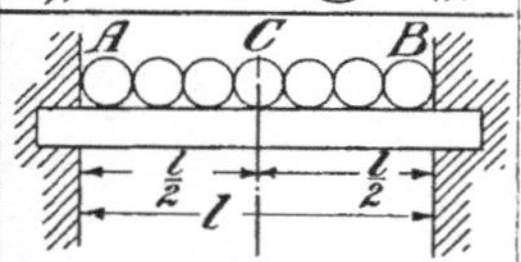			At A or B. At C.	$\frac{wl^2}{12}$ $-\frac{wl^2}{24}$	$\frac{wl^2}{12Z}$		w lbs. per inch of span.

TABLE 11.—DIAGRAMS OF SHEARING FORCE AND BENDING MOMENTS.

Case.	Loading.	Diagram of load. Shearing Force Curve. Bending Moment Curve.	Remarks.
1	Fixed at one end. Single load at free end.	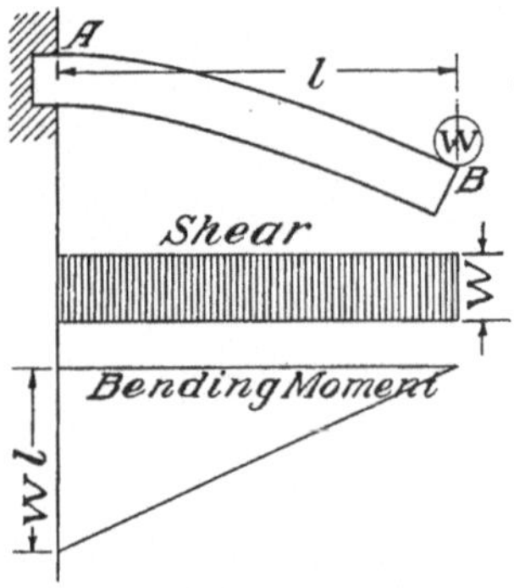	The only moment acting to the right of A is Wl, which is therefore the bending moment at A.
2	Fixed at one end. Two concentrated loads.	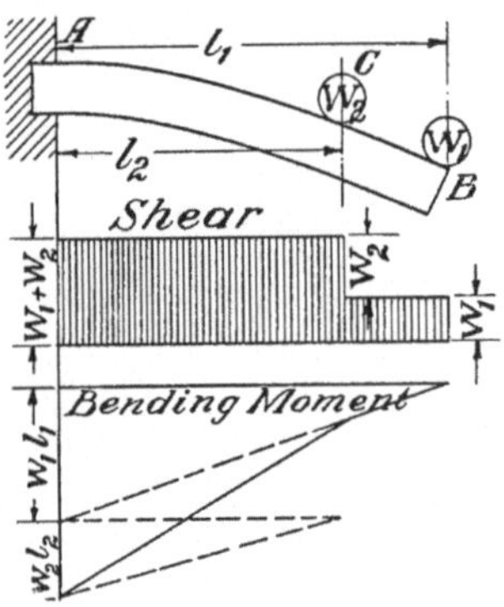	This is simply an example of the combination of two diagrams such as case 1.
3	Uniform load of w lbs. per inch run.	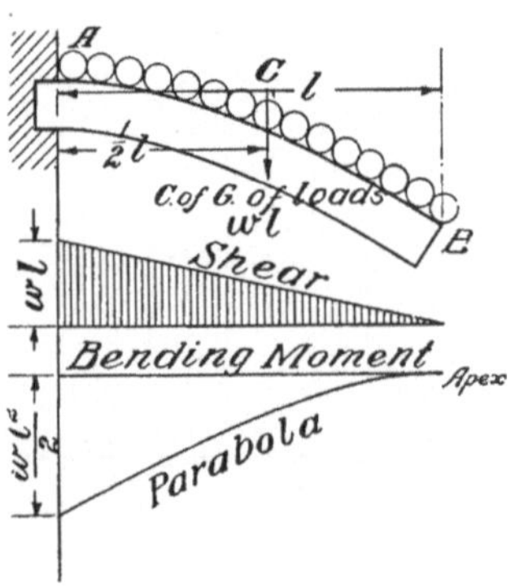	The load being uniformly distributed, its resultant acts at $\frac{1}{2}$ l from A. The total load is wl or Q. ∴ the bending moment at $A = wl \times \frac{1}{2} l = \frac{wl^2}{2}$.

TABLE 11 (Continued).

Case.	Loading.	Diagram of load. Shearing Force Curve. Bending Moment Curve.	Remarks.
4	Supported at both ends. Single central load.	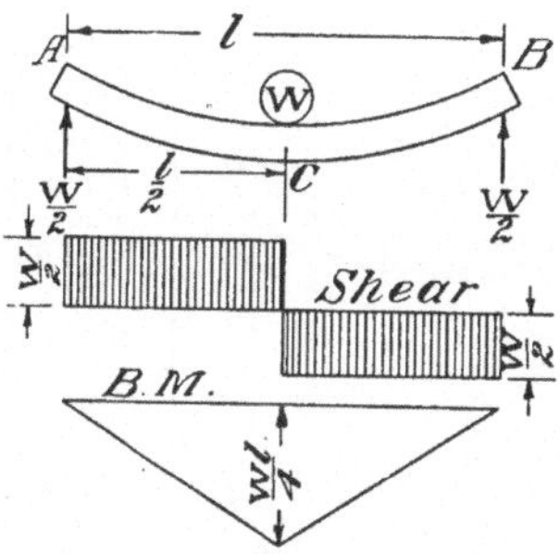	Each support takes ½ the load $= \frac{W}{2}$. The only moment to the right or left of C is $\frac{W}{2} \times \frac{l}{2} = \frac{Wl}{4}$. At any section the moment varies directly as its distance from the support, hence the diagram is triangular. The shearing curve changes sign on opposite sides of the middle section.
5	Supported at both ends. Single load not central.	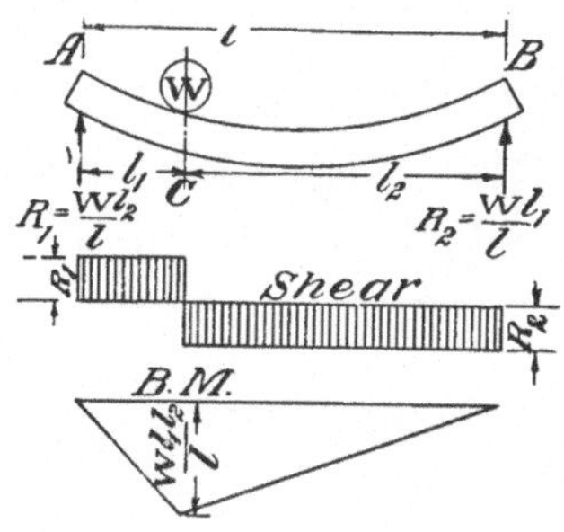	Taking moments about one support gives $R_1 l = W l_2 R_1 = \frac{W \cdot l_2}{l}$. The B. M. at C is B. M. $= R_1 l_1 = \frac{W l_1 l_2}{l}$.
6	Supported at both ends. Two symmetrical equal loads.	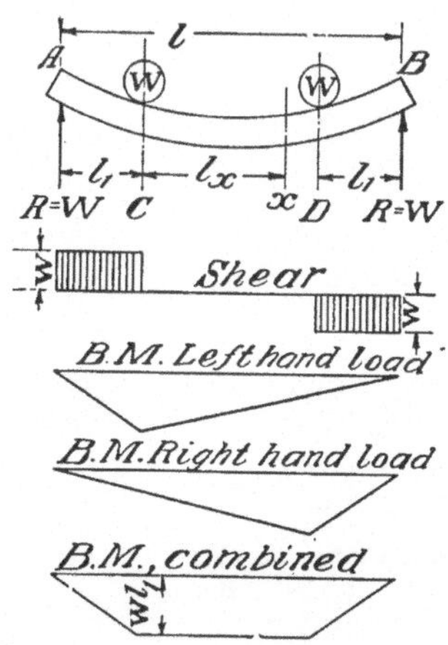	The sum of the forces to the right or left of $x = W - R = 0$, and to the right of D, or to the left of C the sum $= R = W$. The only moment to the right of D is Wl_1, similarly to the left of C. At any section x between the loads and distant l_x from one of them, we have, by taking moments to the left of x, $R(l_1 + l_x) - Wl_x = Rl_1 + Rl_x - Wl_x = Rl_1$ or Wl_1.

TABLE 11 (Continued).

Case.	Loading.	Diagram of load. Shearing Force Curve. Bending Moment Curve.	Remarks.
7	Supported at both ends. Uniformly distributed load, w lbs. per inch run.	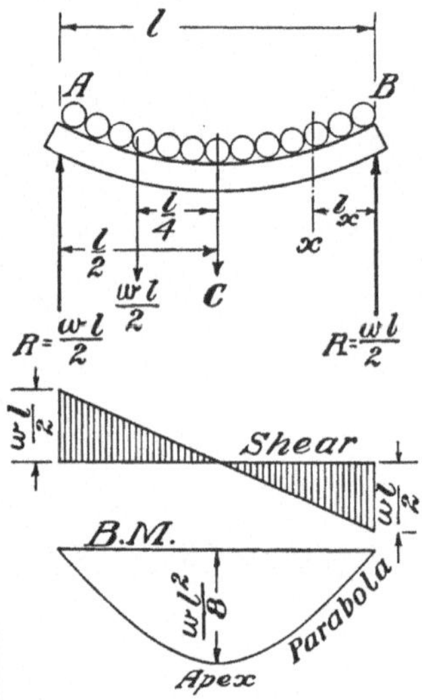	Replace the forces by their resultant. Each support takes one half the load $=\frac{wl}{2}$. The forces acting to the right of C $= \frac{wl}{2} - \frac{wl}{2} = 0$, or shear at middle section in zero. Taking moments to left of C gives $\frac{wl}{2} \times \frac{l}{2} - \frac{wl}{2} \times \frac{l}{4} = \frac{wl^2}{8}$. The $\frac{wl}{2}$ shown midway between C and the support is the resultant of the load on half the beam acting at its center of gravity, $\frac{l}{4}$ from C, or from the support. At any other section x distant l_x from the support, find the B. M. as follows: Take moments to the right of x, $\frac{wl}{2} \times l_x - wl_x \times \frac{l_x}{2}$ $= \frac{wl_x}{2}(l - l_x)$ (a), let $l - l_x = l'_x$, then (a) becomes $\frac{w}{2}(l_x \times l'_x) = \frac{Q}{l}\left(\frac{l_x l'_x}{2}\right)$, the curve is a parabola.
8	Supported at two points equidistant from ends, and uniformly loaded with w lbs. per inch run.	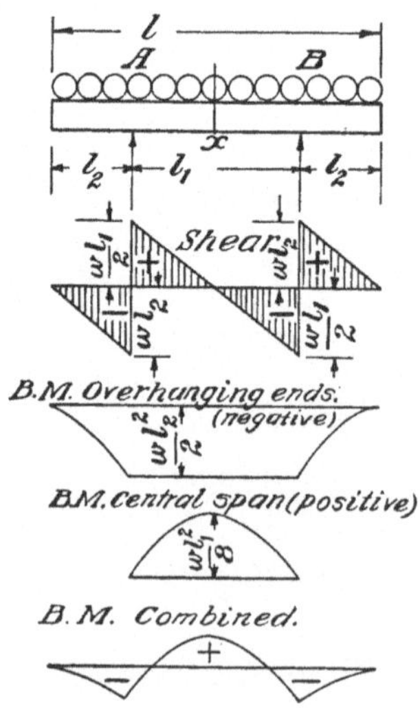	The shearing force curve is a combination of cases 3 and 7. The bending moment is a combination of cases 3 and 6. This case illustrates the importance of giving moments their proper signs. The B.M. will be smallest when M_A (or M_B) $= M_x$, or when $\frac{wl_2^2}{2} = \frac{wl_1^2}{8} - \frac{wl_2^2}{2}$ $wl_2^2 = \frac{wl_1^2}{8}$, $\frac{l_1^2}{l_2^2} = 8$, $l_1 = 2.83 l_2$. But $l_1 + 2l_2 = l$, whence $2.83\,l_2 + 2l_2 = l_1$ or $l = 4.83\,l_2$, or say $l_2 = \frac{1}{5}\,l$ for maximum strength of beam.

TABLE 11 (Continued).

Case.	Loading.	Diagram of load. Shearing Force Curve. Bending Moment Curve.	Remarks.
9	Fixed at both ends. Single central load.	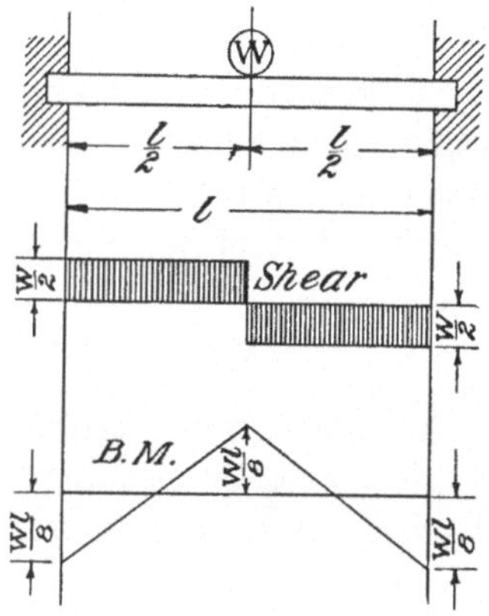	For method of this case see Goodman, *Machines Applied to Engineering*, or Smith, *Strength of Material*.
10	Fixed at both ends. Uniformly loaded with w lbs. per inch.	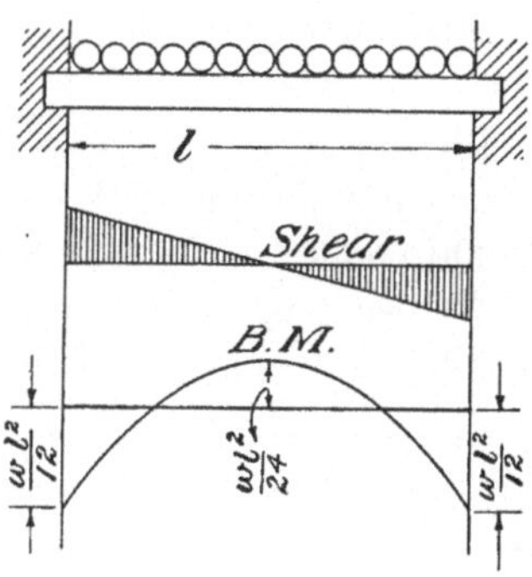	See Goodman or Smith.

Questions and Problems.

Discuss the conditions of simple bending. Show how to deduce the value of the stress f in a beam, in terms of the bending moment M and the modulus of the section, Z.

Explain the limitation of the theory of simple bending. Why may it safely be used in most practical cases?

Define moment of inertia. Given the moment of inertia, I, about an axis through the center of figure of a plane area, show how to find the moment of inertia about a parallel axis distant y from the neutral axis.

Show how to find the position of the neutral axis and the moment of inertia of a complex section. Illustrate by a double-flange section.

Show how to construct curves of shearing force and bending moment.

Deduce expressions for the moment of inertia, square of radius of gyration, and modulus of a rectangle.

Deduce expressions for I, k^2 and Z for a circle about a diameter.

Discuss the economical distribution of material in a beam subject to bending action.

PROBLEMS.

1. A wooden beam of rectangular cross section is 15 feet long and 10 inches wide. If the maximum bending moment is 16.5 foot-tons and the allowed stress is $\frac{1}{2}$ ton per square inch, what must be its depth?

2. What is the radius of the smallest circle into which a cold rod of iron 2 inches diameter may be bent without injury, the stress being limited to 4 tons per square inch? $E=13{,}000$ inch-tons.

3. An I beam, supported at both ends, is 25 feet long; top flange, $3'' \times 2''$; bottom flange, $10'' \times 3''$; web, $12'' \times 1''$; total depth, $17''$. If the working stress is limited to $4\frac{1}{2}$ tons per square inch, find the greatest central load it can carry in addition to its own weight (take weight of beam as 2000 pounds acting at its center).

4. A hollow steel shaft 30 feet long, supported by a bearing at each end carries a load of 4000 pounds at its middle, in addition to its own weight. Outside diameter of shaft$=12''$, inside diameter $=8''$. Find the maximum bending moment and the corresponding fiber stress. A cubic inch of steel weighs 0.287 pound (consider weight of shaft as a uniformly distributed load).

5. A gang plank, 16 feet long, $2\frac{1}{2}$ feet wide, is crowded from end to end by 15 men, weighing 160 pounds each. If the maximum working tensile strength is 2400 pounds per square inch, what must be the thickness of the plank?

6. A plank, 20 feet long, floating in still water, is loaded (1) with 300 pounds at the middle; (2) with 150 pounds at each end; (3) with 150 pounds at points half way from the middle to the ends. Construct the diagrams of shearing force and bending moments.

CHAPTER V.

Directions for Practical Work.

PRACTICAL PROBLEM I.

32. *Directions for Practical Work.*—Data differing from that in the text will be given for the problems.

Calculate problems on loose paper to nearest hundredth reducing this to nearest 32d from table in back of book. If the result falls midway between thirty seconds in table, take the highest 32d. The gauge for accuracy will be the result obtained in hundredths. Note that when results are reduced to nearest 32ds these are used in succeeding steps of problem. Make smooth copies of calculations on interleaved pages. Place important headings in the margin at the left of the page. Make a drawing in pencil of the required views of the problems on cross-section paper (or drawing paper if used) placing all dimensions properly; complete the drawing and make a tracing and two blue prints if there is time. Use a whole sheet of cross-section paper for each problem.

Place the legend, number of views, etc., of each problem according to instructions.

Omit cutting, border and working lines when cross-section paper is used and turn in the sheet without trimming it.

Trim the tracing to the exact size of sheet of cross-section paper.

If drawing paper should be used, secure the sheet on the boards with thumb tacks, and draw cutting, border and working lines of the following dimensions, respectively, $18'' \times 24''$, $16'' \times 22''$ and $15'' \times 21''$.

Each midshipman will use the data corresponding to his own desk number.

Make all lettering, dimension figures, conventional lines, etc., conform to the standards of the drawing course.

PROBLEM I.*

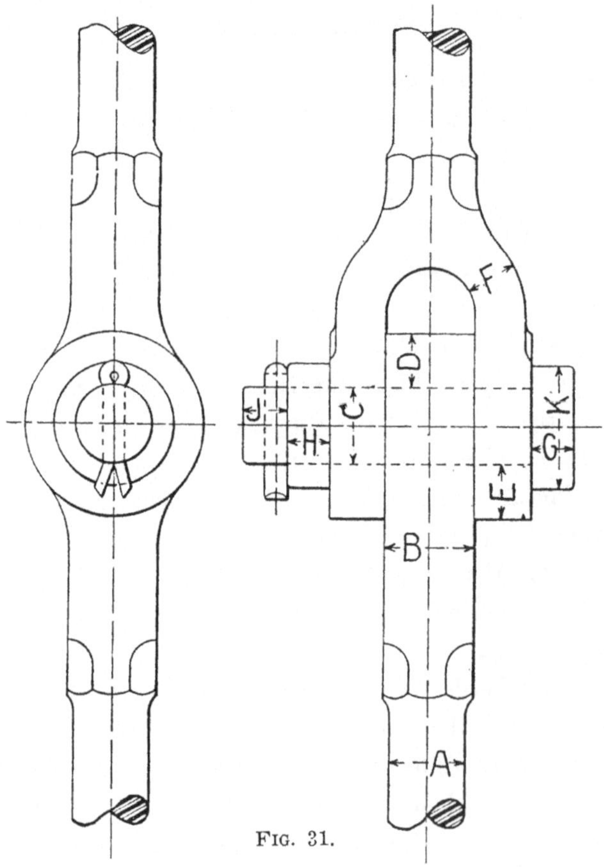

FIG. 31.

33. Design a knuckle joint for a stay rod of a boiler. Area supported 12″×12″. Pressure per gauge, 150 pounds. Material, wrought iron. $f_t = 7200$ pounds per square inch.

DRAW FRONT AND SIDE ELEVATIONS: SCALE, FULL SIZE.

To find the diameter of the rod A, which is in tension, we have:

$P = 12 \times 12 \times 150 = 21{,}600$ pounds. Area of rod $= \dfrac{21{,}600}{7200} = 3.00$ square inches.

* From "Notes on Machine Design."

And A, the diameter, $=1''.95$, or $1\frac{15}{16}''$.

Let B be 1.1A (Unwin), or,

$B=1.1A=1.1\times 1\frac{15}{16}''=2''.13$, or $2\frac{1}{8}''$.

To find the diameter of the pin C, which is calculated to resist bending. The pin is in the condition of a uniformly loaded beam $2\frac{1}{8}''$ long, supported at each end. We have, therefore,

$$M = \frac{wl^2}{8} \text{ and } w = \frac{P}{l}. \quad \text{Hence, } M = \frac{Pl}{8} = \frac{21{,}600\times 2\frac{1}{8}''}{8}$$

$$= \frac{21{,}600\times 17}{8\times 8}.$$

Also, $M=f_t z$ where $z=\frac{\pi d^3}{32}=.0982\ d^3$ and $f_t=7200$ pounds per square inch.

$$\text{Therefore, } \frac{21{,}600\times 17}{8\times 8}=7200\times .0982\ d^3, \text{ or,}$$

$$d = \sqrt[3]{\frac{21{,}600\times 17}{8\times 8\times 7200\times .0982}} = 2''.01, \text{ or } 2''.$$

Then C, or diameter of the pin, is 2″.

To find D, the thickness of the part surrounding the pin. The metal is in the condition of a uniformly loaded beam fixed at both ends.

Fig. 32 shows a perspective view of the eye of the bottom rod and of the pin through it. It is understood that the eye is pulling down upon the pin. The eye will not fail by *shearing,* as might be assumed from the figure. This figure is only intended to show the portion of metal considered as a beam. Two planes shown by dotted rectangles form the ends of the beam where it is "fixed" to the remaining part of the eye. The little beam is drawn separately above, the "fixed" planes at the ends being shown by ruling them with vertical lines.

$$l=\text{length}=C=2''. \quad M=\frac{Pl}{12}=\frac{21{,}600\times 2}{12}=3600.*$$

* This is the bending moment at *ends* of a uniformly loaded, fixed beam.

Also, $M = f_t z$ where $f_t = 7200$ and $z = \frac{bh^2}{6}$. $b = 2\frac{1}{3}''$.

Therefore, $M = \frac{7200 \times h^2 \times 17}{6 \times 8} = 3600$, and

$h = 1''.19$, or $D = 1\frac{3}{16}''$.

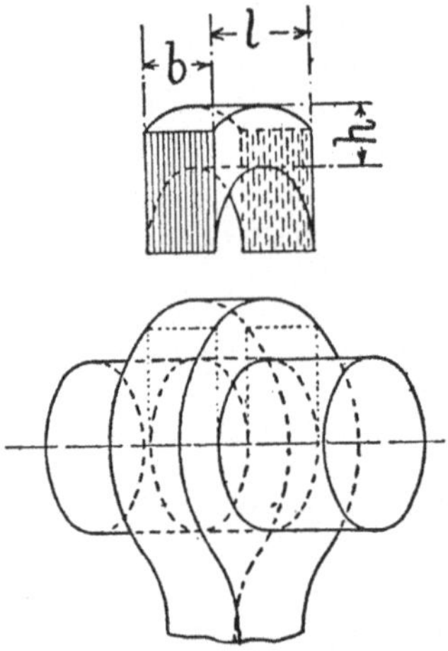

FIG. 32.

The depth E is taken equal to D for symmetry and is thus made amply strong to resist irregular stresses and fitting.

As possibly two-thirds of the stress may come on each arm of the split end, due to irregular fitting, F is made about $\frac{2}{3}$ of $B = 1''.42$, or $1\frac{13}{32}''$.

As one dimension of the cross section at F is the same as one side of the square section at B, the other dimension alone takes the entire variation.

The lower portions of the forked ends are made a little wider than at F to make a good bearing for the pin, or $1\frac{1}{2}''$. Or, from Unwin,

$.75 \times$ diameter of pin $= 1\frac{1}{2}''$.

G, the thickness of the head of the pin, is taken as .5 C (Unwin).

K, the diameter of the head of the pin, is 1.5 $C \simeq 3''$.

H, the collar, has the same dimensions as the head of the pin.

J is taken long enough for a round split pin of a diameter $= \frac{C}{4}$ and an equal amount beyond. This is empirical.

The space above D is taken a little greater than D, or $1\frac{1}{2}''$. The eight-sided portions at top and bottom are empirical and represent the corners of the square head bevelled off to make an eight-sided section. The dimensions of this portion depend upon the work in the blacksmith shop, as a fuller is used to shape the octagon. Make the length that required for use of proper fuller.

The part where the cylinder merges into the octagonal prism must be carefully drawn to show that it is *octagonal* not *hexagonal*. Draw, in pencil only, a plan like that in Fig. 23, draw the octagon and its inscribed and circumscribed circles. On the front elevation, assume some fairly large radius r for the arc of enlargement, ab, of the cylinder, and draw the arc, extending it to g. From g′ on the plan project down to g on this arc. Pass horizontal lines through b and g. The lines bf and gk represent horizontal planes which intersect the surface of revolution generated by the arc abg, in circles. These circles are seen in their true form on the plan as the inscribed and circumscribed circles already drawn. Corresponding points, therefore, can be projected from plan to elevation as b′b, h′h, c′c, i′i, etc. From i and j draw in and jp, the near edges of the octagonal prism. Find by trial arcs of circles to connect hci, idj and jek, as shown.

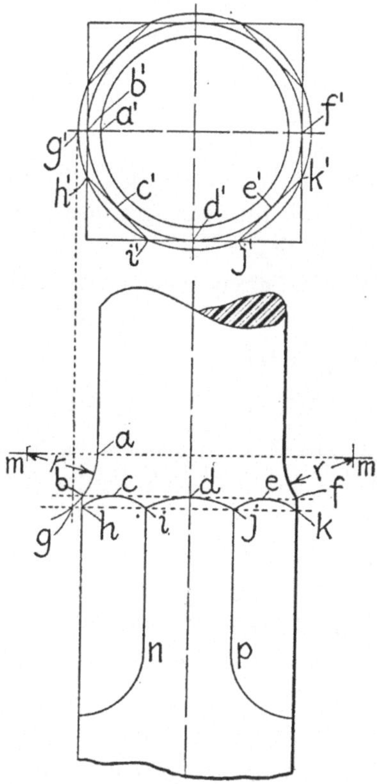

FIG. 33.

The fact that b and f are on a level with c, d, e, and are *not the same points as h and k,* shows the difference between octagon and hexagon. From n and p any arcs may be drawn to the side boundaries.

FIG. 34.

Fig. 34 shows the split pin. This is made of half-round wire bent into an eye as shown. D, the inside diameter of the eye, = d.

$d=\frac{1}{4}C$. After entering the pin the ends are opened out to prevent its falling out of place.

Questions and Problems.

Make a neat sketch of a knuckle joint for a boiler stay, and explain clearly the nature of the stresses to which each part is subjected and the considerations determining the size of each part to resist these stresses.

A rod of circular section expands into an octagonal section. Make a neat sketch on the blackboard showing the construction; plan and front elevation.

Problems.

1. The stay rods connecting the shell to the front head of a boiler are fitted with knuckle joints. The line of the stay makes an angle of 30° with the axis of the boiler. Area of head supported by each stay 15″×12″. Pressure per gauge 160 pounds; $f_t=7500$ pounds. Find the diameter of the stay rod, and diameter of the pin.

Note.—If P is the force acting normally to the head on the area supported, $P \times \sec\theta$ is the resultant force acting in the direction of the axis of the stay.

2. The opening of the forked end in a knuckle of a tie rod which sustains a pull of 24,000 pounds is $2\frac{1}{2}$ inches. Find the necessary diameter of the pin of the knuckle joint. $f_t=7200$ pounds per square inch.

3. In the knuckle joint shown in Fig. 31 find A, B, C and D. Area supported, 14″×14″, pressure per gauge 180 pounds; material, wrought iron; $f_t=7500$ pounds per square inch.

4. In the knuckle joint shown in Fig. 31 find A, B, C and D. Area supported, 10″×10″; pressure per gauge 160 pounds; material wrought iron; $f_t=7500$ pounds per square inch.

5. In the knuckle joint shown in Fig. 31 find A, B, C and D Area supported, 12″×12″; pressure per gauge 200 pounds; material, wrought iron; $f_t=7800$ pounds per square inch.

33. DATA FOR THE DESIGN OF A KNUCKLE JOINT FOR A STAY ROD OF A BOILER.

Problem No.	Desk No. ending in	Boiler pressure.	Area supported.	Material.	Working strength f_t.
1	1	180 lbs. p. g.	14″ x 14″	Wrought iron.	7,500 lbs. / in.
	2	160 " "	10″ x 10″	" "	7,500 lbs. / in.
	3	200 " "	12″ x 12″	" "	7,800 lbs. / in.^{2}
	4	140 " "	14″ x 14″	" "	6,800 lbs. / in.^{2}
	5	180 " "	12″ x 12″	" "	8,500 lbs. / in.^{2}
	6	200 " "	15″ x 15″	Steel.	9,000 lbs. / in.^{2}
	7	220 " "	15″ x 15″	"	10,000 lbs. / in.^{2}
	8	150 " "	″ x 11″	Wrought iron.	6,500 lbs. / in.^{2}
9	9	200 " "	15″ x 15″	Steel.	9,000 lbs. / in.^{2}
10	0	140 " "	14″ x 14″	Wrought iron.	6,800 lbs. / in.^{2}

Use a scale = one-half size for problems 1, 3, 5, 6, 7 and 9; for the remaining problems use scale = full size.

CHAPTER VI.

Bolts. Nuts. Screws.

FORMS OF SCREW THREADS. APPLICATIONS OF BOLTS, STUDS, NUTS, ETC. STRESSES IN BOLTS AND SCREWS. FRICTION AND EFFICIENCY OF SCREWS. CALCULATION OF NUMBER, SIZE AND PITCH OF CYLINDER COVER STUDS.

34. A *screw* is a cylindrical bar on which is formed a helical projection called the *thread.* The screw fits into a corresponding hollow form, which is called the *nut.* The mechanical *pairs* thus

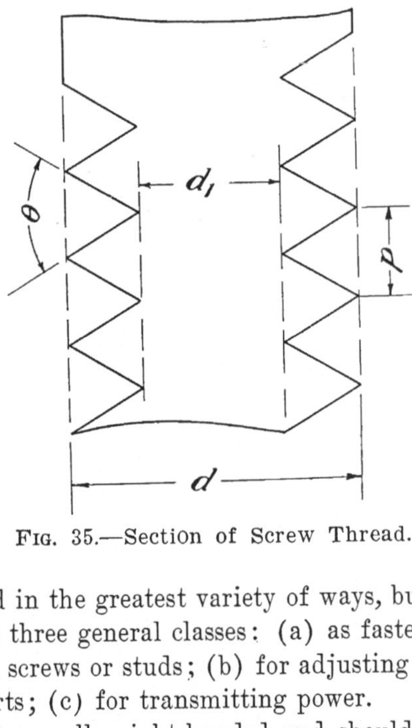

Fig. 35.—Section of Screw Thread.

formed are used in the greatest variety of ways, but these uses may be grouped into three general classes: (a) as fastenings, when they are called bolts, screws or studs; (b) for adjusting the relative positions of two parts; (c) for transmitting power.

The thread is usually right-handed and should always be made so for all purposes, unless, in very special cases, it is necessary to use a left-handed thread.

Bolts and screws used for fastenings are generally subject to tension, the forces acting parallel to the axis of the bolt and normally to the surfaces connected together. When the forces act normally to the axis of the bolt and parallel to the surfaces of the parts fastened together, the bolt is in shear, in the same manner as a rivet. The advantage of a bolt over a rivet in such cases is that the connected parts can be easily taken apart when necessary or convenient. In Fig. 35 d is the *nominal* diameter of the bolt. (This is also usually the diameter of the plain cylindrical part, or diameter of the rod on which the thread is cut.) d_1 is the *effective* diameter, that is, it is the diameter across the roots of the threads. p is the *pitch*: this is the distance the bolt will advance in the direction of its axis for one complete revolution. The pitch is also the reciprocal of the number of threads per inch. θ is the *thread angle.*

35. *Standard Screws.*—No extensive explanation is needed to point out the necessity for complete interchangeability of screws of the same nominal dimensions.

To secure this interchangeability all countries have adopted standard systems of screws, bolts and nuts. Unfortunately, however, the systems of different countries are not like each other. Thus, the United States standard screw has an angle of 60°, while the British has an angle of 55°. There are also other differences, and for some of the nominal sizes the number of threads per inch, or the pitch, is different.

In such a system of standard screws a limited number of diameters is selected, and for each diameter a fixed pitch is determined, while the thread must have an accurately defined form. Every screw must very accurately conform to the standard nominal and effective diameter and pitch, and the thread angle is also important. However, for many sizes the United States nut will go on the British bolt, when the pitch is the same, but the fit between the two is not accurate. The width and depth of the head and nut is also fixed by the nominal diameter of the bolt, thus permitting the use of standard *wrenches.*

The table in the back of the book gives the standard system used in the navy, and generally throughout the United States.

In the absence of the table, or for sizes not tabulated, the pitch and effective diameter may be found from the following formulæ:

$$p=0.24\sqrt{d+0.625}-0.175, \text{ approximately.} \qquad (24)$$

$$d_1=0.91d-0.08, \text{ approximately.} \qquad (25)$$

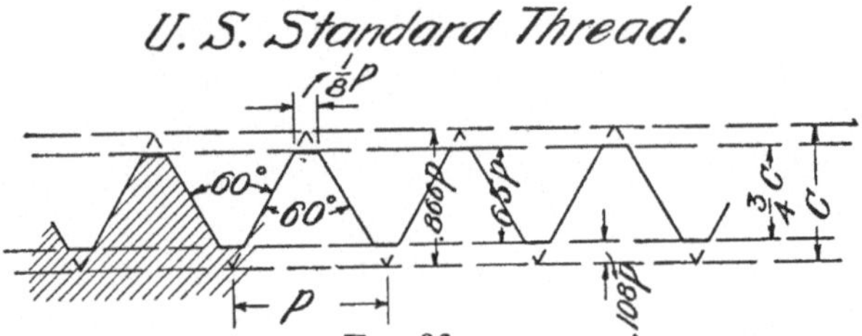

FIG. 36.

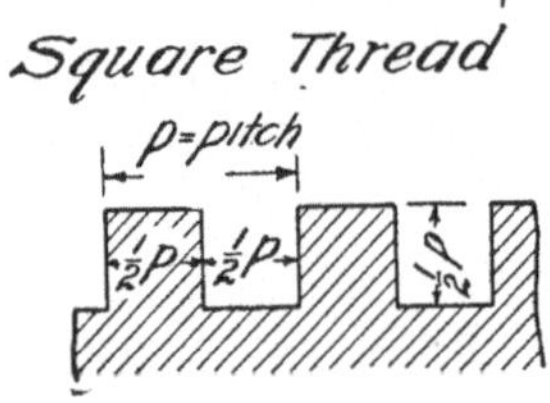

FIG. 37.

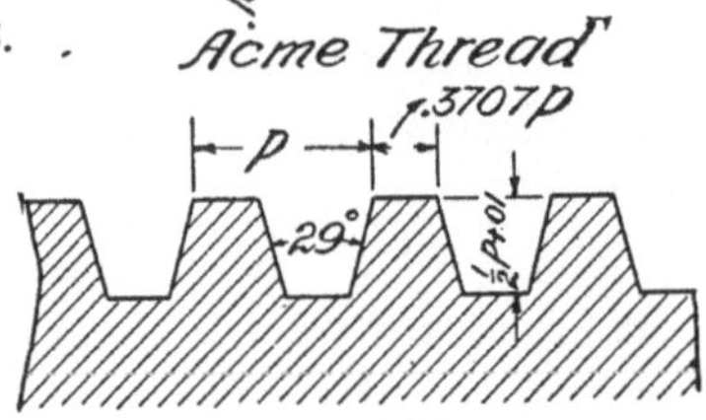

FIG. 38.

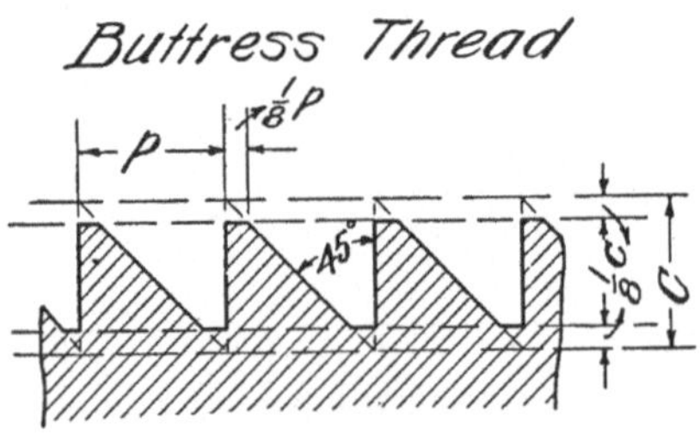

FIG. 39.

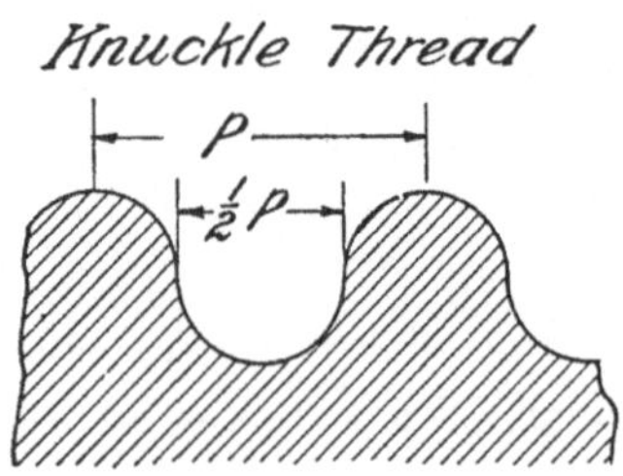

FIG. 40.

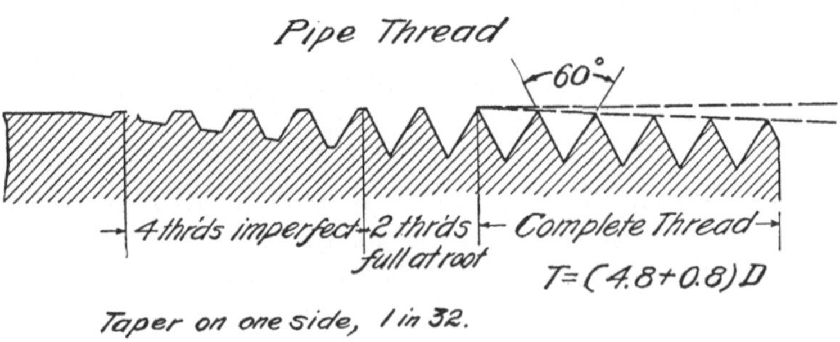

FIG. 41.

36. *Forms and Proportions of Screw Threads.*—Figs. 36-41 show the forms and proportions of threads commonly used in practice. Fig. 36 is the United States standard, or Sellers V thread, used for bolts and nuts. The angle of the thread is 60°. The pitch is determined by the formula given in Art. 35, but in practice is always obtained from a table. One-eighth of the depth is cut off from the top and bottom, forming a flat point, which tends to prevent burring of the thread. The tap for cutting the nuts has a diameter slightly greater than the screw, to allow a small clearance.

Fig. 37 is the *square* thread. The pitch for *standard* square-thread screws being approximately *twice* the pitch of the standard V screws of the same nominal diameter. The pitch is given by the formula

$$p=1/n=0.16d+0.08, \text{ nearly.} \qquad (26)$$

With this form of thread there is no bursting pressure on the nut, as the thrust is parallel to the axis of the screw. These threads cost more to cut than triangular threads, and cannot well be formed by dies. They are very generally used for transmitting power. The depth of the thread of the screw is sometimes made $\frac{19}{40}p$, in which case $\frac{1}{2}p$ is the depth of the thread in the nut, thus giving a slight clearance.

Fig. 38 is the *acme* thread, which is a modified form of square thread, the angle of the thread being 29°. The slight taper thus provided greatly facilitates its being rapidly engaged and disengaged when used with a *split nut,* as in the screw-cutting lathe.

Fig. 40 is the *round* or *knuckle* thread. This makes a very strong screw, and one that will stand much rough usage without damage. It is not suitable for transmitting power.

Fig. 39 is the *trapezoidal* or *buttress* thread. This form combines the important advantage of the square thread, that of having its thrust parallel to the axis of the screw, with the strength of the V thread. The power, however, can only be transmitted efficiently in one direction. It is an excellent form where there is little or no work to be done by it in a reversed direction; as, for example, the screw of a jack, in some form of presses, and for the breech block of large guns.

Fig. 41 is the standard pipe thread, known as the Brigg's system of pipe threads, used in the United States. The following is a synopsis of the standard table, giving the numbers of threads per inch for the various sizes:

$\frac{1}{8}$″ pipe has............	27	threads per inch
$\frac{1}{4}$″ and $\frac{3}{8}$″ pipe has......	18	" " "
$\frac{1}{2}$″ and $\frac{3}{4}$″ pipe has......	14	" " "
1″ to 2″ pipe has.......	$11\frac{1}{2}$	" " "
$2\frac{1}{2}$″ pipe and over has...	8	" " "

The pipe sizes refer to the nominal inside diameters of the pipe.

37. *A Few Practical Applications of Bolts, Studs, Nuts, etc.*— Fig. 42 is the ordinary *stud,* used most commonly in securing the cylinder head to the flange of the cylinder, and other similar connections. The distinction between a bolt and a stud is that the latter is threaded at each end, and one end is permanently screwed into one of the pieces to be connected, and a nut is then screwed on the other end. As indicated in the figure, the width of the flange should be three times the diameter of the stud, and the thickness one and one-half times.

Fig. 43 is a *forcing bolt,* used to break the joint under a cylinder cover, or a follower ring, before lifting with eye bolts.

Fig. 44 illustrates a case where a stud with a nut on each end would have to be used, instead of the ordinary headed bolt.

Fig. 45 illustrates a case in which both a *stud* and a *tap bolt* must be used.

If a stud is used at A it will be impossible to use one at B also, as the angle plate C could not be put on or taken off. Consequently, a screw must be used at B which is put in after the plate is put over the stud at A.

Fig. 46 shows a *flanged* nut, which is sometimes used when the bolt hole is considerably larger than the bolt. This is an expensive nut to manufacture, and in many cases the same purpose may be served by using a washer.

Fig. 47 shows a *flanged cap* nut, used to prevent leakage through the screw threads.

In places where a nut is subject to vibration, as on many parts of moving machinery, it tends to slack back, or even work off its bolt altogether, however perfect its fit might have been originally. One of the most common methods of preventing this is to fit a second nut, screwed down on the first, as shown in Figs. 48-50, to jamb or lock it. Fig. 48 shows the proper method of fitting a lock nut. A little

Ordinary Stud.

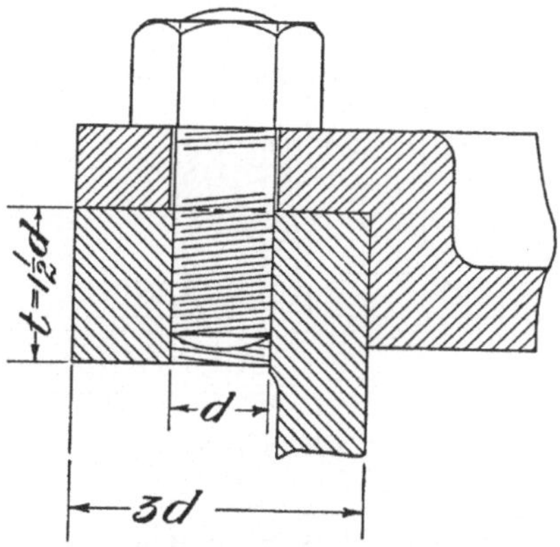

Fig. 42.

Forcing Bolt.

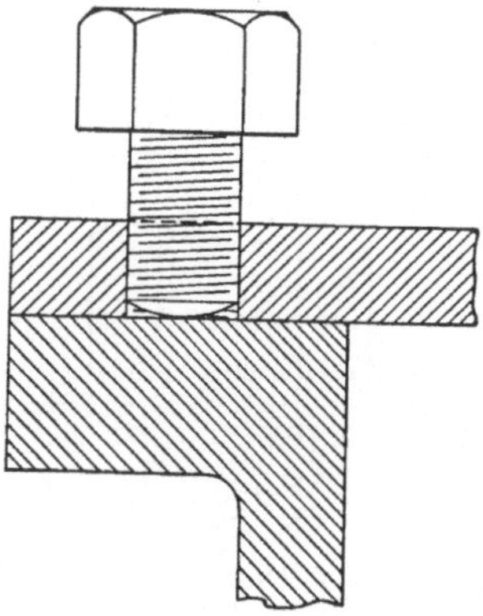

Fig. 43.

Stud Nutted at Each End.

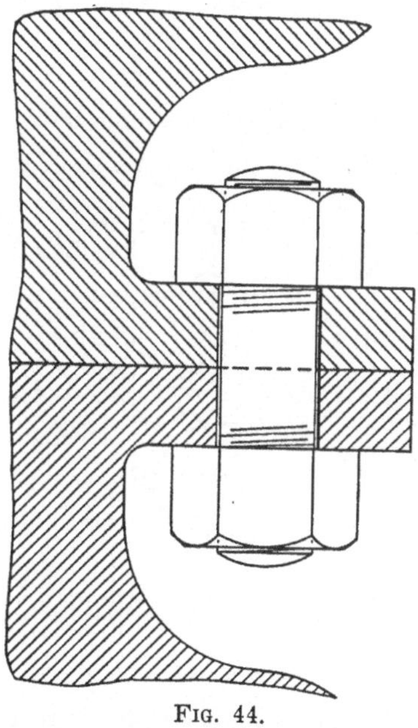

Fig. 44.

Stud and Tap Bolt.

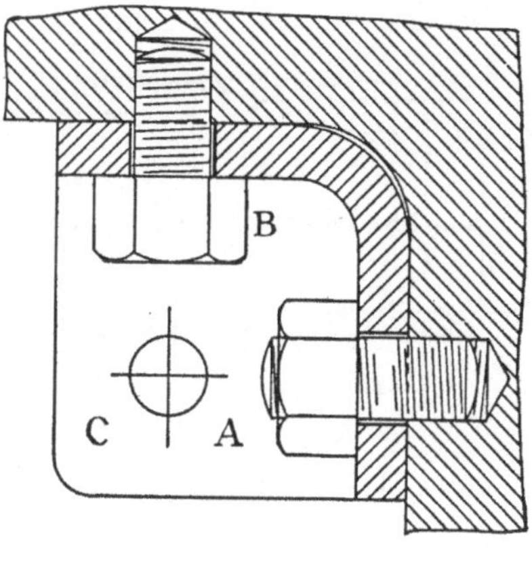

Fig. 45.

consideration will show that the top nut may take the *whole* of the load, and, therefore, this is the proper place for the *thick* nut. The faulty arrangement shown in Fig. 49 is often seen in practice, but

Flanged Nut.

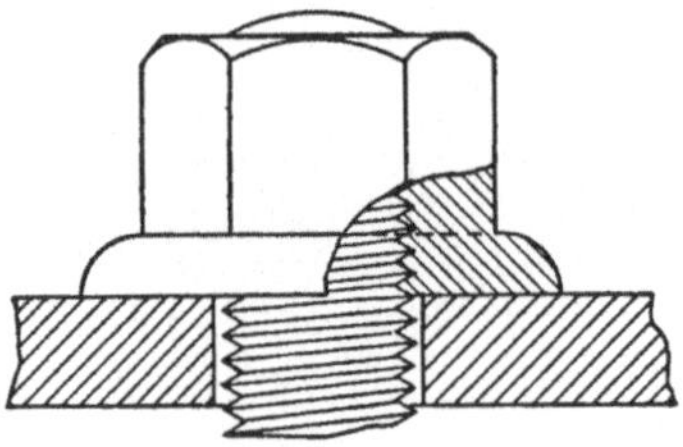

FIG. 46.

Cap Nut.

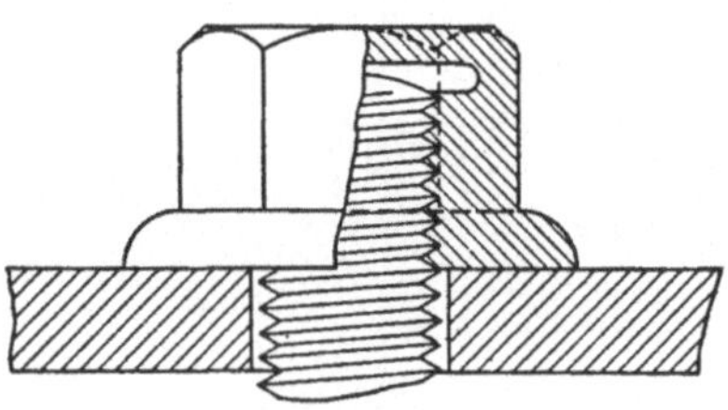

FIG. 47.

it is none the less faulty. This arrangement has come about probably from the fact that most wrenches are too thick to set up the thin nut when it is at the bottom. The obvious way of escaping this difficulty would be to make *both* nuts full thickness, but there may not always be room to do this, and it is rather unsightly in appearance. This has led to the compromise shown in Fig. 48, where both

LOCK NUTS.

Good Practice.

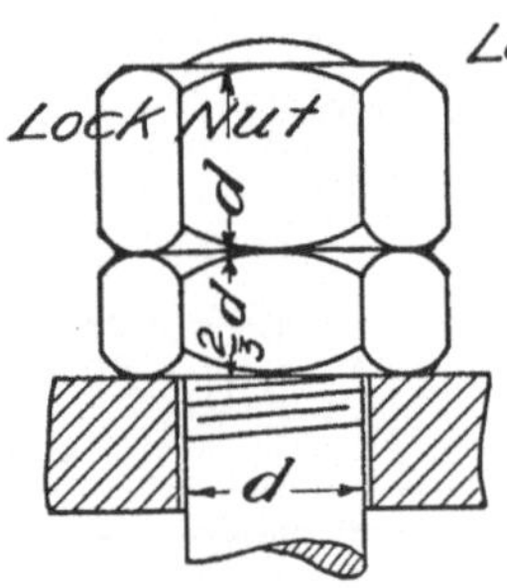

FIG. 48.

Faulty Practice.

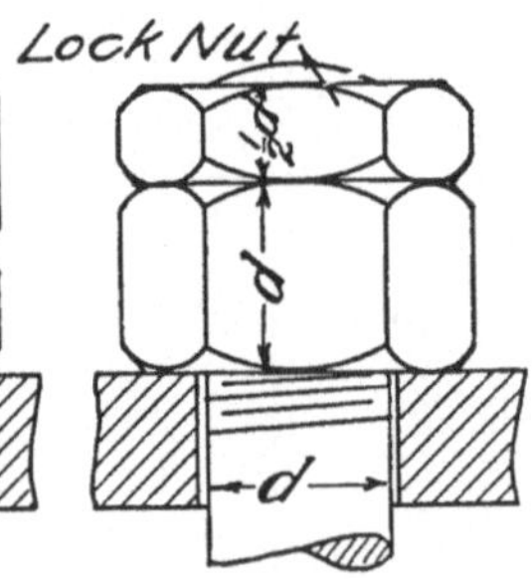

FIG. 49.

Convenient Compromise.

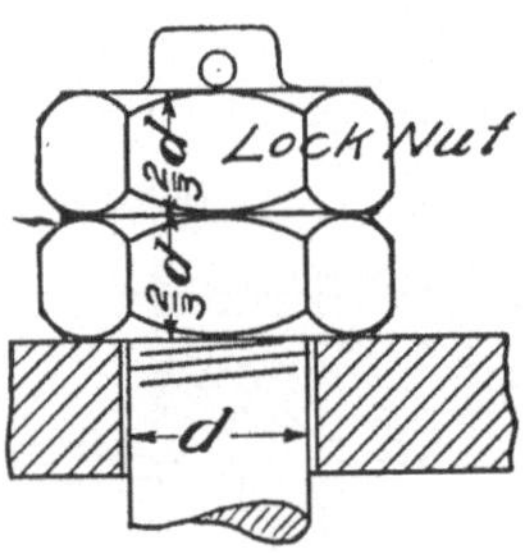

FIG. 50.

nuts are of equal thickness, each from $\frac{2}{3}$ to $\frac{3}{4}$ of the thickness of a full-size nut. In all cases, wherever it can possibly be done, the method of Fig. 48 should be used.

In cases where the jarring is apt to be excessive, the upper nut is still further secured by driving a pin through a hole in the end of the bolt, the upper surface of the nut having a slot into which the pin fits. If, after several readjustments, the nut has to be screwed further down on the bolt, a washer is fitted under the pin, of a thickness just enough to require the pin to be forced in.

Other methods of locking are by means of a set screw through the side of the nut; the use of a locking plate; the use of a collar nut, with a locking ring, or set screw, similar to the method familiar to all, shown on Sheet 4 of the course in mechanical drawing.

38. *Comparison of Square and V Threads.*—If the tensile load on a V-threaded screw acting parallel to its axis be represented by

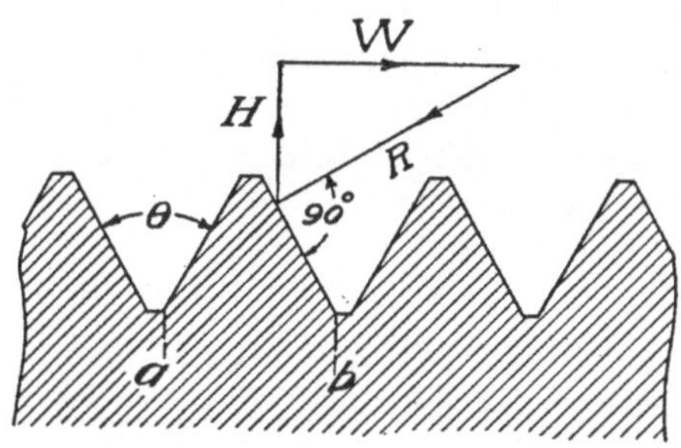

FIG. 52.—Forces Acting on a Screw Thread.

W (Fig. 51), then R will represent the pressure acting normally to the surface of the thread (this is also a measure of the friction), and H will represent the force tending to burst the nut. It is evident, from the triangle of forces, that the larger the angle θ the greater is the bursting pressure, and the larger is the amount of friction. When the angle decreases and the surface of the thread becomes perpendicular to the axis of the bolt, the bursting pressure disappears, and the normal force R becomes equal to W. This is the case of the square-threaded screw, and, therefore, the square thread is used to transmit power. On the other hand, for the same depth of thread in each case, the V thread has about twice the amount of material resisting *shearing* at the root of the thread, ab, as the square thread. Therefore, the V thread is stronger, and is generally used for fastenings, where strength is the primary consideration. It will now be apparent why the buttress thread combines to a considerable extent the special advantages of both the V and the square threads.

39. *Fatigue of Bolts.*—Where bolts are subjected to repeated shocks, or the action of live loads, it is found that they frequently fail by breaking across the threads, even when carrying a less load than they have safely carried before. The cause of such failure is attributed to a slight temporary elongation at the weakest part, the threads, each time the stress occurs. This risk of this failure may be reduced to a minimum by making the bolt of *uniform strength* throughout its length (a) by turning down the body of the bolt to the diameter of the bottom of the threads, or (b) by drilling a hole up from the head through the axis of the bolt to the point at which the threads begin. In either case the object is to make the *effective* area of the bolt the same throughout its length. In case method (b) is used, the size of the hole is obtained as follows:

Let d=nominal diameter of bolt, or diameter of unthreaded part.
d_1=*effective* diameter of bolt, or diameter across roots of threads.
d'=diameter of hole.

Then

$$\frac{\pi d^2}{4} - \frac{\pi d'^2}{4} = \frac{\pi d_1^2}{4}$$

or

$$d' = \sqrt{d^2 - d_1^2}. \qquad (26)$$

40. *Straining Action on Bolts.*—In Fig. 52 a bolt is represented as carrying a load W, without initial tension in the bolt, before the load is applied. Let f be the safe working stress allowed for the material of the bolt, and let d_1=the effective diameter of the bolt. Then

$$W = \frac{\pi d_1^2}{4} f,$$

or

$$d_1 = \frac{\sqrt{4W}}{\pi f}, \qquad (27)$$

from which the size of the bolt can be determined.

Load with no Initial Tension.

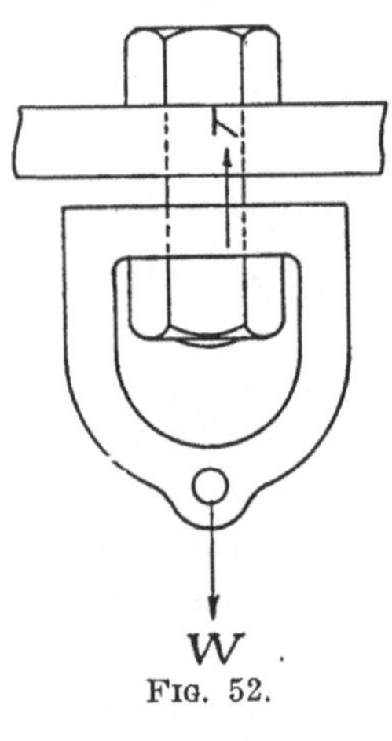

FIG. 52.

In Fig. 53 let C and D represent two rigid flanges which when bolted together form a fluid-tight joint. In order that the joint may remain tight after the load W is applied, the flanges must not separate to the slightest extent. This condition can exist only if the initial tension produced in the bolt by screwing up is equal to or greater than W.* Thus it is seen that the strength required in a bolt is not directly dependent on the load it is to carry, but is also dependent on the strain produced in making the joint.

Fig. 54 represents diagrammatically the case of a joint made with elastic packing, the elasticity of the packing being represented, for clearness of description, by the springs. Then, if the initial thrust

Load with Initial Tension.
Rigid Joint.

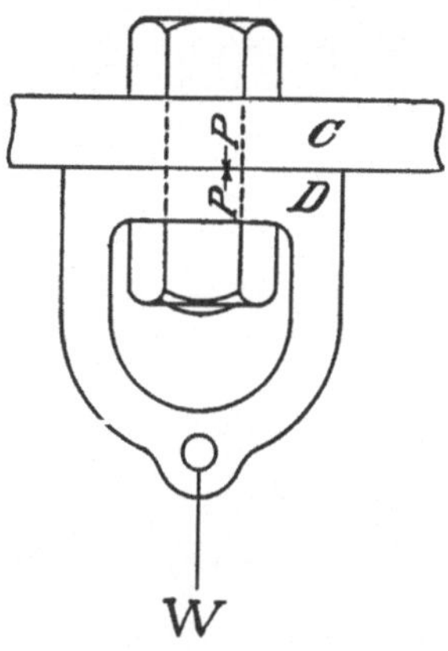

FIG. 53.

Load with Initial Tension.
Elastic Joint.

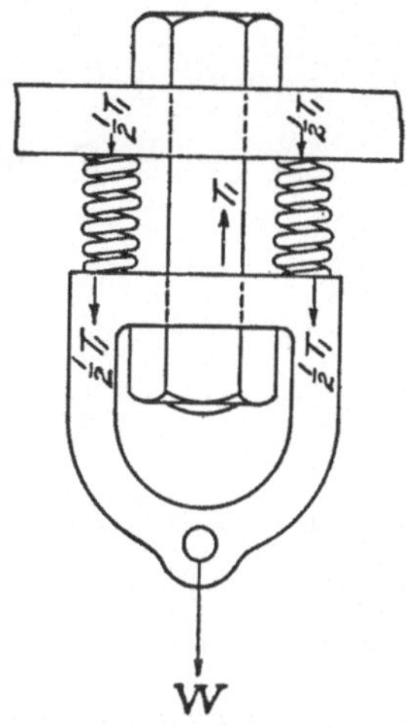

FIG. 54.

of each spring is represented by $\frac{1}{2}T_1$ the initial tension in the bolt will be T_1, due to screwing up the joint.

If we call the extension of the bolt a, and the compression of the springs b per unit of load, the initial compression of each spring is $\frac{1}{2}T_1 \times b$, and the corresponding extension of the bolt is $T_1 \times a$.

If a weight W is applied, the tension in the bolt is increased to a value T between W and $W + T_1$, and the additional tension of the bolt diminishes the compression of the springs. If W is increased until the additional extension of the bolt equals or exceeds the initial compression of the springs, the thrust of the springs dis-

* This statement will be more fully explained a little later.

appears, and the tension of the bolt is that due to the load alone, as represented by Fig. 52. In making a fluid-tight joint, therefore, the bolts must be screwed up tight enough initially to prevent the thrust of the packing being exceeded when the pressure is acting.

When the load is added, the additional extension of the bolt becomes $a(T-T_1)$ and the compression of each spring diminishes to $\frac{1}{2}[bT_1-a(T-T_1)]$.

If we call R the thrust due to *both* springs in this condition, we have

$$bR=bT_1-a(T-T_1),$$

$$R=T_1-\frac{a}{b}(T-T_1);$$

but in this condition $T=W+R$, whence

$$T=W+T_1-\frac{a}{b}(T-T_1);$$

which reduces to

$$bT=Wb+T_1b-aT+aT_1,$$

$$(b+a)T=Wb+(b+a)T_1,$$

$$T=\frac{b}{b+a}W+T_1. \qquad (28)$$

If the springs are easily compressible (corresponding to a joint made with very soft and elastic packing) so that b is large compared to a, the term $\frac{b}{b+a}$ in equation (28) approaches unity, and the tension, when the load is applied, becomes $W+T_1$ approximately. If the compressibility of the springs and the extensibility of the bolts are equal (a case that may be approximated when a very hard gasket is used) b becomes equal to a and the tension in the bolt then becomes $\frac{1}{2}W+T_1$. If the joint is very rigid (as when the joint is made by grinding the flanges to true surfaces, and no gasket is used), b becomes small compared to a and the term $\frac{b}{b+a}$ is a very small fraction, and the tension is not appreciably increased by the application of the load, and is then equal to T_1 approximately.

41. *Friction and Efficiency of Square-Threaded Screws* (modified from Unwin).—Fig 55 represents a square-threaded screw connecting two parts, A and B, whose approach to each other is resisted, either because B carries a load W, or because A and B are the flanges of a joint, the elasticity of the packing between them resisting compression. This resistance is symmetrically disposed, and its resultant is an axial force T producing tension in the body of the screw. Suppose that the screw is tightened by the application of a force F acting at an arm l—for instance, by a wrench on the bolt head.

Let d be the nominal diameter and d_1 the effective diameter of the screw and D the diameter of the collar, or head. Let $\mu=\tan\phi$

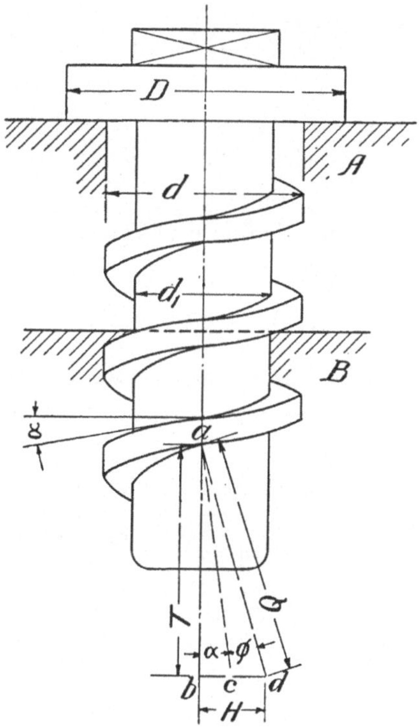

FIG. 55.—Friction of Screw.

be the coefficient of friction between the threads and nut, B; and $\mu'=\tan\phi'$, the coefficient of friction between the collar, or head, and A. If r is the mean frictional radius of the thread, and R the mean frictional radius of the collar, we have:

$$r=\tfrac{1}{2}\left(\frac{d}{2}+\frac{d_1}{2}\right)=\tfrac{1}{4}(d+d_1);$$

and, similarly,

$$R=\tfrac{1}{4}(d+D).$$

Let p be the pitch of the screw, and α the angle of inclination of the thread to the normal, at the radius r. Then $\tan\alpha = \dfrac{p}{2\pi r}$.

If there were no friction the pressure on the thread would act normally to it, or along the direction ac (Fig. 55), which makes an angle α with the axis of the screw. In consequence of the friction, however, the pressure is deflected to an angle ϕ (such that $\tan\phi=\mu$) to the normal to the thread, or it acts along the direction ad, whose angle with the axis is $\alpha+\phi$. We now have the screw in equilibrium under the action of an axial force T along ab; a horizontal force H at radius r along bd due to the force applied to the spanner; and a pressure Q between the threads of the bolt and nut, along ad, making an angle $\alpha+\phi$ with the axis. Then, from the triangle of forces, we have

$$T : H : Q :: ab : bd : ad;$$

$$\text{or } T/H = ab/bd\text{; but } ab/bd = \cot(\alpha+\phi),$$

$$\therefore H = T\tan(\alpha+\phi). \tag{29}$$

The moment of H about the axis is $H\times r$ (r being the arm at which the horizontal component of the friction acts). Therefore, we have:

$$\text{Moment, } M = Hr = Tr\tan(\alpha+\phi). \tag{30}$$

From trigonometry $\tan(\alpha+\phi) = \dfrac{\tan\alpha+\tan\phi}{1-\tan\alpha\tan\phi}$, but $\tan\alpha = p/2\pi r$, and $\tan\phi=\mu$; from (30), $M = Hr = Tr\left(\dfrac{\tan\alpha+\tan\phi}{1-\tan\alpha\tan\phi}\right)$

$$= Tr\left(\frac{\dfrac{p}{2\pi r}+\mu}{1-\dfrac{p\mu}{2\pi r}}\right) = Tr\left(\frac{p+2\pi r\mu}{2\pi r - p\mu}\right). \tag{31}$$

The friction of the collar is $\mu' T$ acting at a radius R, its moment is, therefore,

$$H'R = \mu' TR.$$

The force F applied to the wrench, acting at a leverage l balances the frictional resistances of the screw. Thus we have

$$Fl = Hr + H'R. \tag{32}$$

Efficiency of Screw.—The useful work done in tightening the screw, or driving against a resistance T is Tp per revolution (that is the force T acting through the distance p) and the work expended on the wrench is $F \times 2\pi l$. Therefore, the efficiency

$$E = \frac{\text{useful work}}{\text{total work}} = \frac{Tp}{F2\pi l}. \qquad (33)$$

The value of Fl may be obtained from equation (32).

NOTE.—When the screw is used to lift a weight, as in the screw jack, the expression T in equations (29), (30) and (31) becomes W, the weight to be raised.

If the friction of the collar is eliminated, which may be aecomplished in the case of the screw jack by allowing the weight to revolve with the screw, the expression for efficiency becomes $E = \frac{Tp}{2\pi Hr}$. Since now Fl=Hr (from equation 32), dividing numerator and denominator by $2\pi r$, we get (from equation 29)

$$E = \frac{T\frac{p}{2\pi r}}{H} = \frac{T \tan \alpha}{T \tan(\alpha+\phi)} = \frac{\tan \alpha}{\tan(\alpha+\phi)} \qquad (34)$$

Maximum Efficiency.—From equation (34) it is seen that E becomes zero when $\alpha=0$ and when $\alpha=90-\phi$ and must, therefore, have a maximum value between these limits.

This maximum efficiency occurs when $\alpha=45° - \frac{\phi}{2}$, and we then have

$$\text{Maximum } E = \frac{\tan\left(45° - \frac{\phi}{2}\right)}{\tan\left(45° - \frac{\phi}{2} + \phi\right)} = \frac{\tan\left(45° - \frac{\phi}{2}\right)}{\tan\left(45° + \frac{\phi}{2}\right)}. \qquad (35)$$

The proof of this is long; and the easiest way of arriving at it is to solve equation (34) with several values of α, plot a curve and pick out from it the maximum value.

In case the screw is fitted with a standard size nut, or has a head of standard size (the size being determined by the nominal diameter of the screw) and the thrust of screwing up is taken by the nut, or the head, we have this friction to consider in addition to that of the threads alone. In the case of the nut, or the head, the friction acts at a radius of approximately 1½ times that of the threads.

Then $H'R=\mu'TR=\frac{3}{2}\mu'\,Tr$ (since $R=1\frac{1}{2}\times r$); if $\mu=\mu'=\tan\phi$ this becomes

$$H'R=\tfrac{3}{2}Tr\tan\phi, \qquad (36)$$

and equation 32 reduces to

$$Fl=Tr[\tan(\alpha+\phi)+\tfrac{3}{2}\tan\phi]. \qquad (37)$$

For V threads,

$$M=Pl=Hr=Wr\left(\frac{p+2\pi r\mu\sec\beta}{2\pi r-p\mu\sec\beta}\right) \qquad (38)$$

(see Fig. 56).

$$\text{Efficiency}=\frac{\tan\alpha(1-\mu\tan\alpha\sec\beta)}{\tan\alpha+\mu\sec\beta}. \qquad (39)$$

Comparing these with similar equations for square threads, it is at once evident that considering efficiency and mechanical power, square threads should be used in preference to V threads.

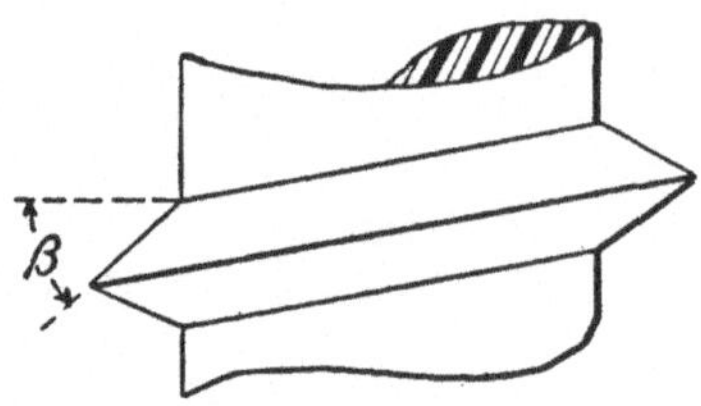

FIG. 56.

42. *The Coefficient of Friction of Screws.*—Experiments have been made to determine the friction between square-threaded screws and nuts of different metals and with different lubricants with the result that no marked differences in the friction were found with different metals or different pressures. The friction varied with different lubricants, as shown in the table.

TABLE 12.—COEFFICIENT OF FRICTION.

	Coefficient $\mu=$		
	Min.	Max.	Mean.
Lard oil	.09	.25	.11
Heavy mineral machinery oil	.11	.19	.14
Heavy oil and graphite	.03	.15	.07

43. *Tensile Stress Produced in Bolts in Making Joints.*—If the resistance in screwing up a joint is that due only to the spring of the packing which is being compressed, the total tension T in the bolt depends only on the moment Fl exerted by the wrench. A series of direct experiments have been made at Sibley College, by which it was found that skilled mechanics would use an effective wrench length of very nearly 15 times the nominal diameter of the bolt, and would graduate the pull on the wrench in proportion to the diameter of the bolt. The average measured tension in the bolt was about

$$T = 16{,}000d, \tag{40}$$

although in some individual cases the tension was much greater than this. Using equation (40) as a basis, the initial stresses produced in bolts of various sizes in making steam-tight joints are as follows:

TABLE 13.—STRESSES IN BOLTS IN MAKING JOINTS.

Nominal diam. of bolt, d........................	5/8″	1″	2″	4″
Total tension, T....................................	10,000	16,000	32,000	64,000
Stress per sq. in., of *effective* area, lbs., f........	49,500	29,100	14,000	6430

In the case of the $\frac{5}{8}$-inch bolt it will be seen that the breaking strength for iron, or the softer varieties of steel, is practically reached and for the 1-inch bolt the stress is dangerously near the yield point of such metals. These experiments are not accepted as absolutely correct, but they show that, for the smaller sizes of bolts, the initial tension produced in making the joint and before any steam pressure is acting, may easily overstrain the bolts. For this reason it is customary in making such joints to use nothing smaller than say $\frac{3}{4}$-inch bolts.

The above-mentioned experiments were conducted by *skilled* workmen. Such a man, it was found, would instinctively select a wrench of a length about 15 times the diameter of the bolt. On board ship it is frequently necessary to have less skillful men make joints, and experience shows that such men have a tendency to use too long a wrench and to exert too great a pull for the size of the bolt. This is a further practical consideration fixing a minimum size of bolt to be used.

In order to provide for the above stresses, and also to allow a margin of safety to provide for shocks such as those produced by water in steam cylinders, as well as the additional stress due to torsion of the bolt, a relatively small working stress is allowed in calculating the sizes of such bolts. In other words, a large factor of safety is used.

In practical designing the engineer has available some standard reference, or pocket book, containing tables to guide him in selecting the safe working stress allowable. In the absence of such a complete table the following may be used:

TABLE 14.—(SPOONER.) SAFE WORKING STRESS FOR BOLTS AND STUDS.

FINISHED JOINTS.

	Values of f.	
	Steel.	Iron.
Bolts and studs 1″ nominal diam. and above	6,000	4,800
" " " 7/8″ " " " under	(4,500 to 3,000)	(3,600 to 2,400)
" " " ordinary marine practice	5,000	4,000
" " " cylinders less than 10″ diam	2,500	2,000

For joints with rough flanges and with comparatively thick gaskets, the values given in the above table (Table 14) may be reduced one-half, to insure safety.

44. *Additional Stress due to Torsion.*—It is shown in Art. 41 that the friction of the threads of a screw produces a moment about the axis of the screw, whose arm is r (the mean radius between the outside and the root of the thread) and whose *force* is H; or a moment Hr. Now this kind of moment is called a *twisting* or *torsional* moment and produces internal stresses in the body of the screw in a manner somewhat similar to that of a beam subject to a bending moment; the stress produced being in this case, however, in the nature of *shearing*. As in the case of a beam we have the general formula $M=fZ$, so now, in the case of a twisting action, we have

$$M=f_sZ_p, \tag{41}$$

in which Z_p may be called the *polar* modulus of section, being the polar moment of inertia divided by the distance of the most distant fiber from the axis.

In the case of the screw the twisting moment produced in its body is the moment applied by the wrench minus the moment of friction of the head (or nut). That is, the friction of the head absorbs a part of the total effort, leaving only the remainder effective in producing twisting in the body of the screw. Then $M=Hr$ (see eq. 30); the polar moment of inertia of a solid circular section is $I_p=\pi D^4/32$ and $Z_p=I_p/D/2=\pi D^3/16=0.196D^3$. In this case D is the effective diameter of the screw, or d_1 of Fig. 53. Substituting the above values in (41) we have

$$f_s=M/Z_p=Hr/0.196d_1^3. \tag{42}$$

The usual proportions of square threads are such that $d_1=0.8d$ (approximately), hence

$$r=\tfrac{1}{4}(d+d_1)=\frac{1.8d}{4}=.45d \tag{42}$$

so becomes

$$f_s=\frac{H\times.45d}{0.196\times.512d^3}=\frac{4.4H}{d^2}\ \text{(nearly)},$$

but

$$H=T\tan(\alpha+\phi),$$

see eq. (30); thus

$$f_s=\frac{4.4T\tan(\alpha+\phi)}{d^2}. \tag{43}$$

The total tension is T and we have, then,

$$T=f_t\times\frac{\pi d_1^2}{4}=f_t\times\frac{22\times.64d^2}{7\times4}=.506f_td^2;$$

whence

$$f_t=\frac{2T}{d^2}\ \text{(nearly)}, \tag{44}$$

therefore,

$$\frac{f_s}{t}=\frac{\dfrac{4.4T\tan(\alpha+\phi)}{d^2}}{\dfrac{2T}{d^2}}=2.2\tan(\alpha+\phi). \tag{45}$$

An average value of α may be taken as about 3°, and a value of μ of .12, corresponding to a value of ϕ of about 7°, or $\alpha+\phi=10°$; and $\tan(\alpha+\phi)=0.176$, substituting this value in (45), gives

$$\frac{f_s}{f_t}=.387\ \text{or say 0.4 (nearly)}. \tag{46}$$

Thus we see that in tightening up a bolt or a screw, there is a shearing or twisting stress produced in addition to the direct tension, of a value of about $\frac{4}{10}$ of the direct tension.

These two stresses must be combined to find the resulting stress in the bolt. Now the *maximum* principal stress is found from the equation

$$f_{max}=\tfrac{1}{2}f_t+\tfrac{1}{2}\sqrt{4f_s^2+f_t^2} \qquad (47)$$

(from Smith, "Strength of Material," with the symbols changed to accord with the method used in this book). Substituting in equation (47) the value of $f_s=0.4f_t$, we get

$$f_{max}=1.14f_t. \qquad (48)$$

In other words *the resultant stress due to the combined tension and twisting increases the stress by 14% for the ordinary square-threaded screw.* In the case of the standard V thread the increase of the stress is about 16% to 17%.

45. *Steam-Tight Joints.*—In making steam-tight joints, such as the joint under the cylinder head or valve chest cover, the studs or bolts must be made strong enough to carry the load due to the steam pressure acting on the area of the cover and also the initial tension produced in setting up the nuts in making the joint. They must also be spaced near enough together to prevent leakage under the flange between them, owing to the flange being sprung or opened by the action of the steam pressure. When studs are used the width of the cylinder flange should be 3d, where d is the nominal diameter of the stud. When bolts are used the width of the flange should be greater than this. The pitch to be used (that is, the distance from center to center of stud or bolt), depends on the steam pressure used, and also to some extent on the thickness of the flanges. The following gives a good rough approximation for the pitch:

For H. P. cylinders, pitch=3.5d.
I. P. " " =4.5d.
L. P. " " =5.5d.

In general, pitch$=\sqrt{\dfrac{100t}{P}}$

where P=pressure in pounds per square inch.
t=thickness of cover in *sixteenths* of an inch.

The thickness of the flanges are *at least equal to the diameter of the stud or bolt,* and if possible they should be from $1\frac{1}{4}$ to $1\frac{1}{2}$ times d. As pointed out in Art. 43, the use of a bolt or stud of less diameter than $\frac{3}{4}$ inch is to be avoided, so it is evident that most joints of steam pipes and small cylinders will have a large excess of strength, as far as the steam pressure alone is concerned.

Example.—Calculate the size, number and pitch of the steel studs for a cylinder cover: diameter of cylinder, 24 inches; steam pressure, 180 pounds per square inch.

From the pressure this is evidently a H. P. cylinder, and, therefore, we may pitch the studs about $3\frac{1}{2}$d. Then the diameter of the pitch circle will be about 24″+3d, since the least width of flange should be 3d. As a first approximation, assume a stud diameter of 1 inch, then the diameter of the pitch circle of the studs is 27 inches and number of studs$=\frac{27\times\pi}{3.5}=24.2$, say 24. (It is customary to use an even number of studs for symmetry.)

It is usual to assume that the steam pressure acts on a circle whose diameter is the pitch circle; that is, it is assumed that the steam pressure extends somewhat beyond the inner edge of the cylinder flange. Now, equating the total strength of the studs to the total load on the cover, we have

$$24\times\frac{d\pi_1^{\,2}}{4}\times f=\frac{\pi 27^2}{4}\times 180.$$

From Art. 43 (Table 14) we find that a safe value for f is 5000 pounds per square inch. Therefore $d_1^{\,2}=\frac{27^2\times 180}{24\times 5000}$ whence $d_1=1.045$. This is the effective diameter.

Referring to the table in the back of the book, we find the corresponding nominal diameter to be between $1\frac{1}{8}$ and $1\frac{1}{4}$ inches, therefore, we will use 24 $1\frac{1}{4}$-inch studs. The width of the flange will then be $3\frac{3}{4}$ inches and the diameter of the pitch circle $27\frac{3}{4}$ inches and the actual pitch$=\frac{\pi\times 27.75}{24}=3.567$ (about).

Then

Ans. $d=1\frac{1}{4}''$; N°$=24$; $p=3.567$ (about).

Questions and Problems.

Make a neat sketch of the U. S. standard screw thread, showing the nominal diameter, the effective diameter, the pitch of the thread, the angle of the thread, the amount of flattening at the point and root of the thread. Explain the importance of a system of standard screw threads (Arts. 34, 35, 36).

Sketch the various forms of screw threads commonly used, and explain in general the relative uses and advantages of the several forms. Discuss the use of lock nuts (Arts. 36, 37).

Explain clearly why the square-threaded screw is better to use for transmitting power than the V thread. What is the fatigue of bolts due to? How may the risk of failure in such cases be minimized (Arts. 38, 39).

Discuss the straining action of bolts due to setting up on a joint with and without elastic packing. Illustrate by supposing the flanges separated by springs. Deduce an expression for the thrust in the bolts in terms of the initial thrust, the load, the compression of the elastic material and the extension of the bolt (Art. 40).

A square-threaded bolt is tightened by a force F applied with a leverage l, producing an axial force T in the bolt. Calling r and R the mean frictional radii of the thread and head respectively, μ the coefficient of friction, p the pitch of the screw and α the angle of inclination of the thread; deduce an expression for the moment Fl in terms of, T, r, p, μ and α. Let $R=1\frac{1}{2}r$. Also deduce an expression for the efficiency (Art. 41).

Discuss the stresses produced in bolts in making joints and show, in a general way, why it is inadvisable to use a bolt less than about $\frac{3}{4}$ inch diameter (Art. 42).

Deduce an approximate relation between the shearing stress and the tensile stress in a square-threaded screw. Make the following assumptions: $I_p=\pi D^4/32$; $d_1=0.8d$; total direct thrust$=T$; $\alpha=3°$; $\mu=.12$ (Art. 44).

Problems.

1. A man in setting up on a square-threaded tap bolt 1 inch diameter exerts a force of 40 pounds on the end of a wrench 15 inches long. Assuming that the friction of the bolt head acts at a radius $R=1.4r$; diameter across roots of thread$=0.68$ inch; pitch $=0.24$ inch; coefficient of friction$=0.125$; find the total thrust produced in the body of the bolt and the efficiency.

2. A man in setting up on a square-threaded tap bolt 2 inches diameter, exerts a force of 80 pounds on the end of a wrench 30 inches long. Diameter at bottom of thread$=1.63$ inches; pitch$=0.40$ inch; coefficient of friction$=0.125$. Assuming that the friction of the bolt head acts at a radius $R=1.4r$, find the total thrust produced in the body of the bolt, and the efficiency.

3. Two men in setting up on a square-threaded tap bolt 4 inches diameter exert each a force of 80 pounds on the end of a wrench 60 inches long. Diameter at bottom of thread $=3.33$ inches; pitch $=.72$ inch; coefficient of friction $=0.125$. Assuming that the friction on the bolt head acts at a radius $R=1.4r$, find the total thrust produced in the body of the bolt, and the efficiency.

4. A man in setting up on a square-threaded tap bolt 2 inches diameter exerts a force of 80 pounds on the end of a wrench 30 inches long. Diameter of bolt at bottom of thread $=1.63$ inches; pitch of thread $=0.40$ inch; coefficient of friction $=0.125$. Assume that the friction of the bolt head acts at a radius of $R=1.4r$ $[r=\frac{1}{4}(d+d_1)]$. Find the tensional and shearing stresses produced in the bolt. From these find the maximum stress produced.

5. The diameter of a steam cylinder is 30 inches; steam pressure $=180$ pounds per gauge. Find the number, size and pitch of the studs for the cylinder cover (steel studs).

6. The diameter of a steam cylinder is 32 inches; steam pressure $=200$ pounds per gauge. Find the number, size and pitch of the studs (steel) for the cylinder cover.

NOTE.—In this case use a stud diameter of $1\frac{1}{2}$ inches for first approximation.

7. The diameter of a steam cylinder is 74 inches; working steam pressure $=50$ pounds per gauge. Find the number, size and pitch of the studs (steel) for the cylinder cover.

CHAPTER VII.

The Screw Jack.

46. *Practical Problem II.**—Design a screw jack to raise a load of fifteen tons. Height of lift 12 inches. The frame is to be of cast iron; the screw, head, nut and band of wrought iron. The head of the screw is to have two $1\frac{1}{4}$-inch holes at right angles to each other for the turning bar. There is to be a cast-iron swivel plate on top of the head. The screw must not overhaul.

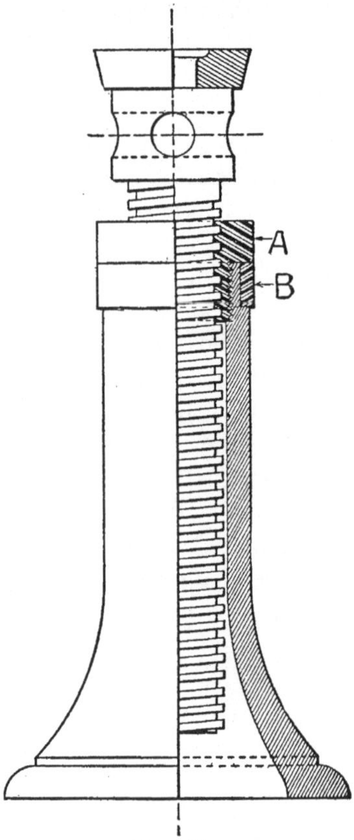

Fig. 57.

1. *To Find the Diameter of the Screw.*—This is calculated for the bottom of the thread (as the threads do not assist in resisting the compression due to the load).

Let A = area of cross section of screw;

W = total load = 15×2240 = 33,600 pounds;

M = Pl = turning moment required to raise W.

This moment will induce a torsional stress $f_a = \frac{yM}{I} = \frac{M}{Z}$, which must be combined with the pure compressive stress $f_c = \frac{W}{A}$, and this combined stress for wrought iron must not exceed 10,000 pounds per square inch (assuming that the screw is frequently loaded and unloaded, and not subject to shock or reversal of stress).

* From "Notes on Machine Design."

In the preliminary calculations, we will assume that an increase of 20% in the area (over that required for pure compression alone) will be a sufficient allowance for the effect of the torsional stress f_a,* or $A=\frac{33,600}{10,000}\times\frac{1.2}{1}=4.032$ square inches, and d_1 (effective diameter) $=2.27$, or $2\frac{9}{32}''$, and A becomes 4.08 square inches. (Fig. 57 shows a half-sectional view of the screw jack with all parts assembled.)

2. *To Find the Dimensions of the Threads.*—For a screw which will not overhaul, the pitch angle must be less than the angle of friction, *i. e.*, $\tan\alpha<\tan\phi$.

Threads will be made square for the best transmission of power and for durability. To be safe against overhauling with the materials used and good lubrication, μ must not be given a greater value than 0.10.

$$\therefore \phi=\tan^{-1}\frac{0.10}{1}=\tan^{-1}\tfrac{1}{10}=5°45', \text{ and } \alpha<\phi.$$

From $\tan\alpha=\frac{p}{2\pi r}$, substituting proper values (see Note 1) and solving we get $p=0.84$, which is the maximum value to satisfy the conditions that the screw will not overhaul.

The value of p thus found is a check only, to insure that the screw will not overhaul. Practically, with a pitch so great, the pull on the turning bar would be excessive; to reduce the pull on the bar the pitch is reduced and its value is found as follows:

From equation 25, Art. 35, $d_1=0.91d-0.08$ (approximately). By substituting the value of d_1 in this equation we get $2\frac{9}{32}=.91d-.08$ or d, the nominal diameter $=2\frac{19}{32}$ inches. This value of d is only approximate, but is close enough to be used in the calculation of the pitch. From Art. 36

$$\begin{aligned} p&=.16d+0.08\\ &=.16\times2\tfrac{19}{32}+0.08=''496 \text{ or } \tfrac{1}{2}''. \end{aligned}$$

To Find the Nominal Diameter.—For *square* threads, consider the depth of the thread $=\frac{1}{2}$ pitch, then

$$d=d_1+p \;\therefore\; d=2\tfrac{9}{32}''+\tfrac{1}{2}''=2\tfrac{25}{32}''.†$$

* See Art. 44.

† For buttress threads, the angle of the thread is 45° and the depth of the thread is $\frac{3}{4}p$, after $\frac{1}{8}p$ has been taken off at point and root of the thread (see Fig. 39). That is to say $d=d_1+\frac{3}{2}p$. Be careful to make the flat part of buttress threads the bearing surface.

3. *To Check the Calculations for Cross Section of Screw.*—The moment required to raise load W is equal to $M=Pl=Hr=Wr\left(\frac{p+2\pi r\mu}{2\pi r-p\mu}\right)$ (see Art. 41), which equals

$$33{,}600\times 1\tfrac{17}{64}\left(\frac{\frac{1}{2}+2\times\pi\times 1\frac{17}{64}\times\frac{1}{10}}{2\pi\times 1\frac{17}{64}-\frac{1}{2}\times\frac{1}{10}}\right)=6970 \text{ inch-pounds.}$$

This is the moment which produces the torsional stress

$$f_a=\frac{M}{Z}; \text{ and } Z=\frac{\pi d_1^3}{16} \therefore f_a=\frac{6970\times 16}{\pi\times(2\frac{9}{32})^3}=3000$$

pounds per square inch, and this torsional stress must be combined with the pure compressive stress,

$$f_c=\frac{W}{A}=\frac{33{,}600}{4.08}=8230 \text{ pounds per square inch,}$$

in order to obtain the combined stress f on the screw which is given by the formula

$$* f=.35f_c+.65\sqrt{f_c^2+4f_a^2}=.35\times 8230+.65\sqrt{\overline{8230}^2+4\times\overline{3000}^2}$$
$$=9500 \text{ pounds.}$$

As this is well within the limit of safety, the area of cross section of screw previously calculated will be taken.

4. *To Find the Dimensions of the Nut.*—The height of the nut is determined from the equation $W=K\times n\times\frac{\pi}{4}\times(d^2-d_1^2)$, in which K is the maximum allowable pressure in pounds per square inch on bearing surface, and its value depends on the speed.

For a rubbing velocity of less than 50 feet per minute (and it is fair to assume that the screw will never exceed this limit), the value for wrought iron is K=2500. Then the number of threads required in the nut will be

$$n=\frac{W}{K\times\frac{\pi}{4}(d^2-d_1^2)}=\frac{33{,}600}{2500\times\frac{\pi}{4}[(2\frac{25}{32})^2-(2\frac{9}{32})^2]}=6.76.$$

Height of nut $=p\times n=\frac{1}{2}\times 6.76=3.38''$, or $3\frac{3}{8}''$ (a short calculation will show that the direct shearing stress on the threads will come well within the limit allowed for wrought iron, with ample allowance for wear).

The nut is screwed into the inside of the frame with a pipe thread, which need not be calculated, as the nut is screwed hard up against

* This is a modified form of equation (47). The constants .35 and .65 are derived from a consideration of Poisson's ratio, and their deduction will be given in a later chapter.

the shoulders a and b and against the band at c (Fig. 59). B is a wrought-iron band shrunk on the frame to prevent the nut and frame from spreading.

The dimensions are empirical, and must be decided upon for each size of jack. For this size of jack the band is $\frac{1}{4}$ inch thick and $1\frac{1}{2}$ inches long. The outside diameter of band = the outside diameter of the nut = the outside diameter of frame + $\frac{1}{2}$ the thickness of band.

5. *To Find the Dimensions of the Head of the Screw.*—The area remaining after the holes are taken out is made equal to the area

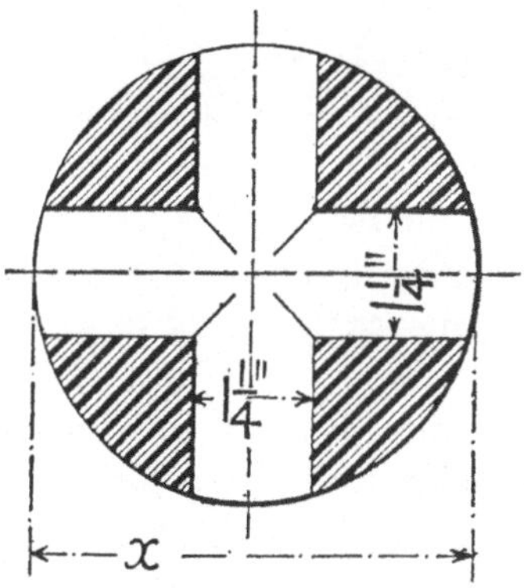

FIG. 58.

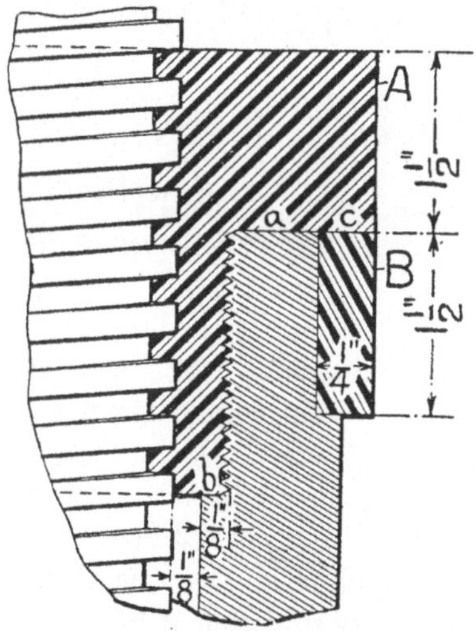

FIG. 59.

of the screw at bottom of thread, as it must sustain about the same stress.

Fig. 58 is a horizontal section taken at the middle of the holes.

Let x = diameter required.

A = area of screw, at bottom of thread.

d = diameter of hole in the head; then

$\frac{\pi x^2}{4} = A + [2(x-d)d + d^2] = A + (2dx - d^2)$ since two lengths d across, less the central portion $d \times d$, are removed.

In this case, substituting proper values,

$$\frac{\pi}{4} x^2 = 4.08 + (1\tfrac{1}{4} \times 2x - 1\tfrac{1}{4} \times 1\tfrac{1}{4}).$$

Solving, we find

$$x = 4''.04, \text{ or } 4\tfrac{1}{32}''.$$

The depth of the head below the holes is calculated on the assumption that the end of the screw tends to punch upward in to the head, and that the resistance to this is one of pure shear (which is on the safe side, as there is an additional resistance to crushing offered by the small centrally projecting portions in the vicinity of the holes).

Let t, Fig. 60, = thickness required such that

$$\pi d_1 t f_s = W,$$

where f_s is the maximum shearing stress for wrought iron = 7500. Solving, we find

$$t = \frac{W}{\pi d_1 f_s} = \frac{33{,}600}{\frac{22}{7} \times 2\frac{9}{32} \times 7500} = .625, \text{ or } \tfrac{5}{8}''.$$

It is customary to make the distance from the top of the hole to the top of the head the same as that below $= \frac{5}{8}$ inch,

$$\therefore \text{ total depth of head} = 2 \times \tfrac{5}{8} + 1\tfrac{1}{4} = 2\tfrac{1}{2}''.$$

6. *Details of the Swivel Plate.*—The central pin is a part of the screw head, and for this size of jack is taken as $1\frac{1}{2}$ inches diameter. It stops just below the level of the top of the swivel plate, the latter being countersunk. It is arranged to turn by dropping it into place and then slightly upsetting the central pin over the top of the plate. Height of swivel plate = 1 inch; upper diameter = 5 inches; and lower diameter = $4\frac{1}{2}$ inches. Depth of counter bore, $\frac{3}{16}$ inch; diameter of counter bore, 2 inches. These dimensions vary according to the size of jack. By calculation, the bearing surface may be worked out and compared with the crushing strength of cast iron. It is ample in this case.

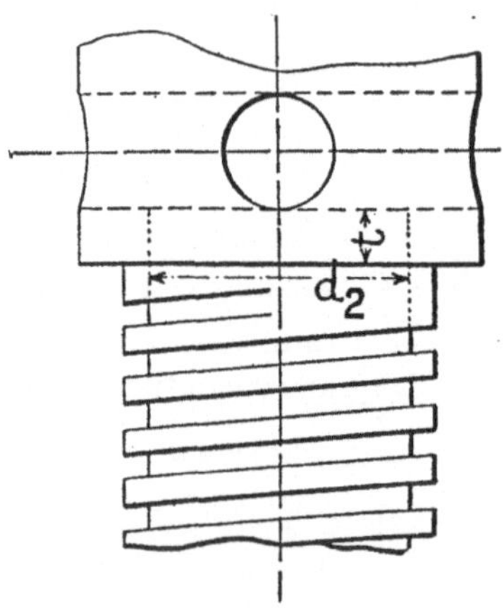

FIG. 60.

7. *Details of the Frame.*—The diameter of the inside of the frame is that of the outside of the screw plus $\frac{1}{8}$ inch to $\frac{1}{16}$ inch clearance on each side. In this case it is $2\frac{25}{32}'' + \frac{1}{4}'' = 3\frac{1}{32}''$. (Area corresponding = 7.2 square inches). The outside diameter is found as follows:

Consider the crushing strength of the cast iron of the frame equal to that of the material of the screw. Then, for equal strength:

Area of screw at bottom of thread=area of outside diameter of frame *minus* area of inside diameter; or

Area of outside diameter of frame=area of inside diameter +area of screw.

Or 7″.2+4″.08=11.28 square inches, and

Outside diameter=3″.79, or $3\frac{25}{32}$″.

Thickness of frame=$\frac{1}{2}$ (outside diameter−inside diameter) =$\frac{12}{32}$″.

This is for strength only, and if used it would be necessary to increase the thickness at the top of the frame for the threaded portion of the nut; also, in order to avoid the danger of breaking accidentally from rough usage, and to insure a better casting the thickness is increased to $\frac{5}{8}$ inch.

The outer diameter of frame is then=$3\frac{1}{32}$″+$1\frac{1}{4}$″=$4\frac{9}{32}$″.

The diameter of the base of the frame is made about $2\frac{1}{2}$ times the above or $10\frac{23}{32}$ for a good spread to prevent tipping, and for a larger surface. The base is made thicker so as to press on a greater quantity of material. It is finished in some such manner as shown in sketch. The bottom of the frame must be at least $\frac{1}{2}$ inch below the bottom of the screw when at its lowest position in order to protect the screw.

The inside diameter of the frame at the upper end is counterbored $\frac{1}{8}$ inch all round for a length of 2 inches and threaded for the nut, terminating in a square shoulder against which the bottom of the nut is screwed up hard. Also the outside diameter at the top of the frame is turned down $\frac{1}{8}$ inch all round (over a length of $1\frac{1}{2}$ inches) to a square shoulder on which the band rests.

NOTE 1.—From $\tan\alpha=\frac{p}{2\pi r}$ we have $p=2\pi r\tan\alpha$.

Remembering that for square threads, depth of thread $\frac{19}{40}p=\frac{p}{2}$ nearly, and that r_1=radius at bottom of thread, and $r=r_1+\frac{p}{4}$,

$$p=2\pi\left(r_1+\frac{p}{4}\right),\ \tan\alpha$$

and substituting

$$p=\frac{2\pi r_1\tan\alpha}{1-\frac{\pi}{2}\tan\alpha}.$$

Questions and Problems for Recitation on the Screw Jack.

Make a neat sketch of a screw jack (elevation), half in section.

Deduce an expression for pitch in terms of the mean pitch angle and the effective diameter of the screw.

Deduce an expression for the turning moment in terms of W, r, p and μ for a screw raising a weight W.

Make the calculation referred to in Art. 46, 4 (a short calculation, etc.).

Problems.

1. Find the diameter of the screw head required for a screw jack, weight to be raised=15 tons. Effective diameter of screw=$2\frac{1}{4}$ inches. Two through holes in head (at 90°) 1 inch diameter.

2. Find the diameter of screw head required for a screw jack. Weight to be lifted=15 tons. Effective diameter of screw, $2\frac{1}{4}$ inches. Three through holes (at 60°) 1 inch diameter.

3. Design the thickness of frame for screw jack to lift 20 tons on a screw $2\frac{3}{4}$ inch outside diameter, $\frac{1}{2}$ inch pitch.

4. Find the depth of screw head required for screw jack; weight to be raised=20 tons; effective diameter of screw, $2\frac{1}{2}$ inches; $f_s=8000$; two 1-inch holes at right angles.

5. Determine the total length of stock steel required to make the screw jack of the problem illustrated in the text, to raise the load 1 foot.

48. *Sequence of Calculations for the Screw Jack:*

1. Find the effective diameter d_1 of the screw, using the given percentage of increase to allow for torsion.

2. Find approximately the diameter d over the outside of the threads.

3. Find the pitch p and depth of the threads, being careful to make the pitch angle α less than the friction angle ϕ to prevent overhauling.

4. Find the nominal diameter.

5. Check the calculations for cross-sectional area, by finding the moment of the friction, and from it the torsional stress. Combine this with the pure compressive stress by means of formula (47) to be certain that the value of f thus found falls within the value of the combined f given in the data.

6. Find the dimensions of the nut.

47. DATA FOR THE DESIGN OF THE SCREW JACK.

Problem No.	Desk No. ends in.	Capacity W. Tons.	Material. Frame.	Material. Screw, head and band.	Material. Nut.	Material. Swivel plate.	Kind of thread.	Combined f for screw, less than	$f_s =$	Bronze $f_s =$	Cast iron $f_c =$	K =	% increase for torsion.	Height of lift.	Diam. of holes in head.
1	1	20	Cast iron.	Wrought iron.	Wrought iron.	Cast iron.	Square.	10,000	7,500	....	....	2,500	20	18″	1½″
2	2	20	Cast steel.	Forged steel.	Forged steel.	Cast steel.	"	20,000	12,00	....	....	2,700	20	16″	1½″
3	3	20	Cast iron.	Wrought iron.	Wrought iron.	Cast iron.	Buttress.	12,00	8,200	....	....	2,500	20	18″	1½″
4	4	20	Cast steel.	Forged steel.	Forged steel.	Cast steel.	"	18,00	13,60	....	....	2,700	15	16″	1½″
5	5	10	Cast iron.	"	Phos. bronze.	Cast iron.	Square.	18,500	11,00	7,000	10,000	2,700	20	10″	1″
6	6	10	"	"	"	"	Buttress.	18,500	11,000	7,000	....	2,600	20	10″	1″
7	7	5	"	"	"	"	Square.	17,000	17,000	10,000	....	2,700	20	10″	¾″
8	8	5	"	"	"	Cast steel.	Buttress.	17,500	10,60	7,200	....	2,650	20	10″	¾″
9	9	20	Cast steel.	"	Forged steel.	"	"	18,00	13,60	....	....	2,700	15	16″	1½″
10	0	10	Cast iron.	"	Phos. bronze.	Cast iron.	Square.	18,500	11,00	7,000	10,000	2,700	20	10″	1″

NOTE.—Each jack has two through holes in the head. Assume $\mu = 0.10$.

7. Check the height of the nut to prevent stripping of threads.

8. Find the dimensions of the head.

9. Calculate the inside and outside diameters of the frame, and its thickness.

10. Lay down the jack on the drawing board, working out the minor details, such as the swivel plate, the reinforcing band, etc., as the work progresses.

NOTE.—Instead of making a complete half section it is suggested that a "break" be made in the frame a short distance below the bottom of the nut, and only the part above the break shown in half section. Continue the outline of the screw below the break by broken lines, so as to be able to record its length, which is an important dimension, on the drawing. The object of this is to avoid the tedious work of actually drawing the threads over the whole length of the screw, and to reduce the amount of cross hatching.

The legend reads:

SCREW JACK.
Capacity, —— tons.
Designed by (*name*), Midshipman, First Class.
Date.

CHAPTER VIII.

Compound Stresses.

BENDING WITH TENSION OR COMPRESSION. COMBINED STRESSES. STRUTS AND COLUMNS. EULER'S FORMULA. EMPIRICAL FORMULÆ, RANKINE'S OR GORDON'S. STRAIGHT LINE FORMULA.

49. *Bending Combined with Tension or Compression.*—Fig. 61 represents the stresses produced in a bar under a load P acting normally to the cross section AB and parallel to the axis of the bar, but to one side of the axis and at a distance x from it. We may introduce two equal and opposite forces P′ and P″, each equal to the original force P, acting against each other along the axis, without disturbing the conditions. We thus have the bar subject to the direct compressive force P″ and a couple whose moment $= P \times x$. Let A be the area of the cross section of the bar. We then have a stress distributed over the section $= \frac{P}{A}$ due to the direct thrust and a stress $\pm \frac{M}{Z} = \pm \frac{Px}{Z}$ at the edges due to the couple. See equation (18), Art. 24. Adding these, the total stresses on the edges of the bar are

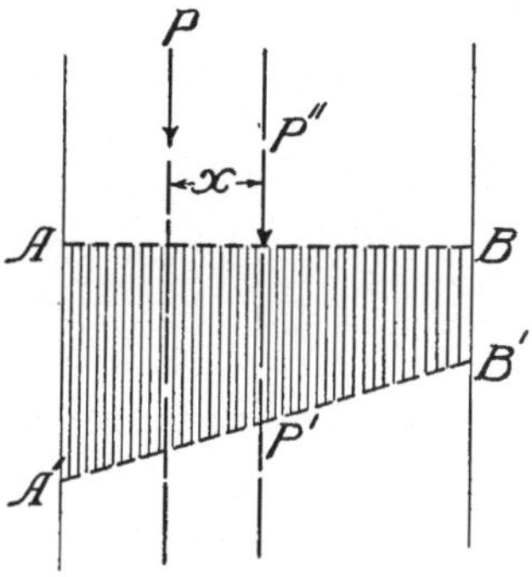

FIG. 61.

Action of Eccentric Load.

$$f = \frac{P}{A} \pm \frac{Px}{Z}. \tag{49}$$

The greater stress is of the same kind as P; the other may be the same, or opposite, according to the value of x.

The distribution of the stress is shown by the shaded part ABB′A′ of Fig. 61, AA′ and BB′ representing the stresses along the edges.

If the bar is of a circular section, diameter d, we have by substituting in equation (49)

$$f = P/\frac{\pi d^2}{4} \pm Px/\frac{\pi d^3}{32} = \frac{4P}{\pi d^2}\left(1 \pm \frac{8x}{d}\right) \qquad (50)$$

If of rectangular section, side h in plane of bending, and b at right angles, equation (49), reduces to

$$f = \frac{P}{bh} \pm \frac{Px}{\frac{1}{6}bh^2} = \frac{P}{bh}\left(1 \pm \frac{6x}{h}\right) \qquad (51)$$

From equation (50) it is seen that the stress along one side becomes zero and along the other side is doubled when $\frac{8x}{d} = 1$ or when $x = \frac{d}{8}$.

Similarly for the rectangular section the stress along one side is zero and is doubled along the other when $\frac{6x}{h} = 1$ or $x = \frac{h}{6}$.

It will be negative, or of opposite sign to P along one edge when the value of x is smaller than the above ratios.

In attempting to find the dimensions of a section to sustain a given eccentric load P, with an assumed allowable value for the safe working stress f, from equation (49) it will be found that a troublesome cubic equation results. In such cases the practical way is to assume a trial section and check this for P and f. In the case of a solid circular or square section, where but one dimension is unknown, equation (49) can be solved.

Example.—A crane has a swing of 30 inches, and has the cross section shown by the figure (Fig. 62). The safe working stress of the material (for compression) is 9000 pounds per square inch. Find the load W the crane can safely lift.

The section being symmetrical, the neutral axis will be at a distance of 2 inches from the edge of either flange. The moment of inertia I is

$$\tfrac{1}{12}[2 \times (4)^3 - 1.5 \times (3)^3] = 7.3.$$

$$\therefore Z = I/y = 7.3/2 = 3.65. \quad A = 2 \times 4 - 1.5 \times 3 = 3.5.$$

Now

$$f = \frac{W}{A} + \frac{Wx}{Z}, \text{ equation (49)},$$

$$f = W\left(\frac{1}{A} + \frac{x}{Z}\right),$$

whence

$$W = \frac{fAZ}{Z + Ax},$$

$$W = \frac{9000 \times 3.5 \times 3.65}{3.65 + 3.5 \times 30} = \frac{9000 \times 3.5 \times 3.65}{108.65 \times 1} = 1060 \text{ lbs.}$$

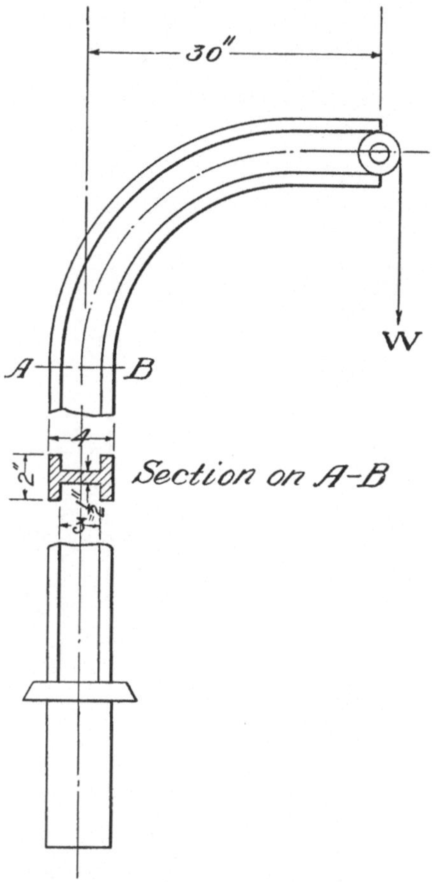

FIG. 62.—Small Crane.

This method is used also for the design of large hooks, the frames of machines, such as punching machines, etc.

50. *Combination of Stresses in General.*—Let the elementary particle shown in Fig. 63 have a parallelopipedal form, the length of the top and bottom sides being B and the ends A, and the thickness being unity. Assume the ends A to be subject to a unit tensile (or compressive) stress f_t and at the same time to the unit shearing stress f_s. This shearing stress tends to rotate the particle, but the tendency to rotate produces a shearing stress along the sides, which balances that along the ends. The area of each side is $B \times \text{unity} = B$. Let f_s' be the unit shearing stress produced along the sides. Then, since equilibrium is sustained, the moment of the forces along the ends balances that of the forces along the sides; therefore, f_sA (the total force at each end) $\times B$ (the arm of the couple) = the moment, or

$$f_sA \times B = f_s'B \times A \therefore f_s = f_s',$$

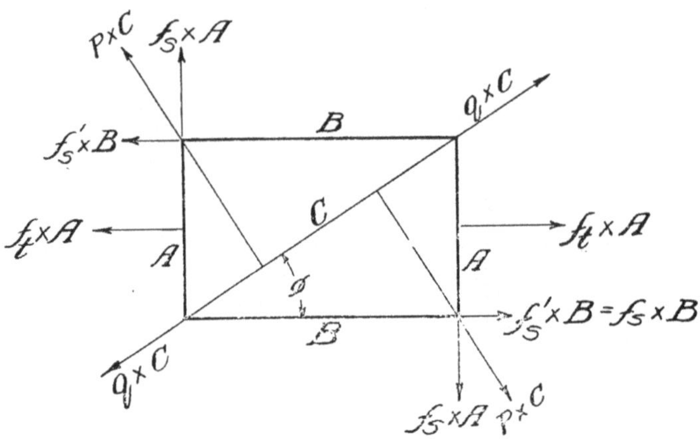

FIG. 63.—Compound Stress.

or a shearing stress at the ends involves a shearing stress of equal intensity at the sides of the particle. Let C be a diagonal of the particle, making the angle ϕ with the top and bottom sides. This diagonal section is subject to an aggregate shearing force $q \times C$ (q being the intensity of shearing stress along C), and to an aggregate tensile force $p \times C$ (p being the intensity of the tensile stress). Resolving all the forces acting into directions along and parallel to C, we have

$$q \times C = f_tA \cos\phi + f_sB \cos\phi - f_sA \sin\phi, \qquad (52)$$

$$p \times C = f_tA \sin\phi + f_sB \sin\phi + f_sA \cos\phi : \qquad (53)$$

from which we get

$$q = f_t \frac{A}{C} \cos\phi + f_s \frac{B}{C} \cos\phi - f_s \frac{A}{C} \sin\phi,$$

but

$$A/C = \sin\phi, \text{ and } \frac{B}{C} = \cos\phi;$$

hence

$$q = f_t \sin\phi \cos\phi + f_s \cos^2\phi - f_s \sin^2\phi, \qquad (54)$$

$$p = f_t \sin^2\phi + f_s \sin\phi \cos\phi + f_s \sin\phi \cos\phi. \qquad (55)$$

These equations reduce to

$$q = \tfrac{1}{2} f_t \sin 2\phi + f_s \cos 2\phi, \qquad (56)$$

$$p = \tfrac{1}{2} f_t (1 - \cos 2\phi) + f_s \sin 2\phi. \qquad (57)$$

Solving for maxima

$$\frac{dq}{d(2\phi)} = 0 = \tfrac{1}{2} f_t \cos 2\phi - f_s \sin 2\phi,$$

from which

$$\tan 2\phi = \frac{\frac{1}{2} f_t}{f_s} \text{ for max. } q; \qquad (58)$$

similarly, from (57)

$$\frac{dp}{d(2\phi)} = 0 = \tfrac{1}{2} f_t \sin 2\phi + f_s \cos 2\phi;$$

whence

$$\tan 2\phi = -\frac{f_s}{\frac{1}{2} f_t} \text{ for max. } p. \qquad (59)$$

When

$$\tan 2\phi = \frac{\frac{1}{2} f_t}{f_s},$$

$$\sin 2\phi = \frac{\frac{1}{2} f_t}{\pm\sqrt{\frac{1}{4} f_t^2 + f_s^2}}$$

and

$$\cos 2\phi = \frac{f_s}{\pm\sqrt{\frac{1}{4} f_t^2 + f_s^2}}$$

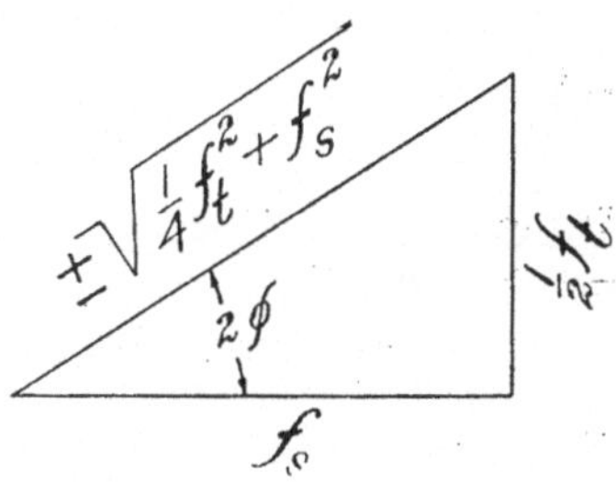

Substituting these values in (56) we have

$$q_{max} = \tfrac{1}{2} f_t \frac{\frac{1}{2} f_t}{\pm\sqrt{\frac{1}{4} f_t^2 + f_s^2}} + f_s \frac{f_s}{\pm\sqrt{\frac{1}{4} f_t^2 + f_s^2}},$$

which reduces to

$$q_{max} = \sqrt{\tfrac{1}{4} f_t^2 + f_s^2}. \qquad (60)$$

Similarly, by substituting the values from (59) in (57), we get

$$* \; p_{max} = \tfrac{1}{2}f_t + \sqrt{\tfrac{1}{4}f_t^2 + f_s^2}, \qquad (61)$$

$$* \; p_{min} = \tfrac{1}{2}f_t - \sqrt{\tfrac{1}{4}f_t^2 + f_s^2}. \qquad (62)$$

51. *Struts or Columns.*—A short compression member, whose ratio of length to lateral dimensions is not greater than 4 or 5 will fail, when overloaded, by direct crushing, if loaded centrally. The effect of eccentric loading of such short struts is discussed in Art. 49. When the length of the column, in proportion to its lateral dimensions, becomes great, failure will occur by lateral bending, unless the column is absolutely straight, with the load acting exactly along the axis of the center of figure of the section, and the material is absolutely homogeneous. These conditions are impossible to obtain. Therefore, the nature of the stresses are those of a bar subject to both bending and compression at the same time.

The first approximately satisfactory rules for the resistance of long thin columns were obtained by Euler, and these are still used where the ratio of length to least lateral dimension is about 30. The formulæ obtained by Euler are based on the elastic theory of mate-

* In the screw-jack problem a formula for combined stress was given in the form

$$f = .35f_t + .65\sqrt{4f_s^2 + f_t^2} \text{ (Art. 46, 3.).}$$

This form of the formula is obtained as follows:

From equations (61) and (62) the principal stresses are

$$p_1 = \tfrac{1}{2}f_t + \tfrac{1}{2}\sqrt{4f_s^2 + f_t^2}$$

and

$$p_2 = \tfrac{1}{2}f_t - \tfrac{1}{2}\sqrt{4f_s^2 + f_t^2}.$$

The deformation produced by these stresses is $\frac{p_1}{E} - \lambda \frac{p_2}{E}$ where $\lambda = \frac{1}{m}$ = Poisson's ratio.

The equivalent normal stress, that is, the single linear stress that would produce the same deformation as that actually produced by the combined stress is

$$p_e = p_1 - \lambda p_2 = \tfrac{1}{2}f_t + \tfrac{1}{2}\sqrt{4f_s^2 + f_t^2} - \tfrac{1}{2}\lambda f_t + \tfrac{1}{2}\lambda\sqrt{4f_s^2 + f_t^2}$$

$$= \tfrac{1}{2}f_t\,(1 - \lambda) + \tfrac{1}{2}\,(1 + \lambda)\sqrt{4f_s^2 + f_t^2}.$$

When $\lambda = .25$ this reduces to

$$p_e = \tfrac{3}{8}f_t + \tfrac{5}{8}\sqrt{4f_s^2 + f_t^2}. \qquad (63)$$

When $\lambda = .30$ we get

$$p_e = .35f_t + .65\sqrt{4f_s^2 + f_t^2}. \qquad (64)$$

rials, and their deduction is given in Smith's "Strength of Material."

*Euler's Formula.**—Let E=the coefficient of elasticity of the material; A=area of section of the column; I=the moment of inertia for bending; k=the least radius of gyration of the section such that $k^2=I/A$; L=the actual length of the column; l=the length of arc of the curved bar measured between two points of contrary flexure; W=the greatest load consistent with stability; p=W/A=the corresponding greatest stress per unit of sectional area; $\pi^2=10$. Then

$$W=\pi^2\frac{IE}{l^2}=10\frac{IE}{l^2}\ \text{(nearly)}, \tag{65}$$

and

$$p=\frac{W}{A}=\frac{10IE}{Al^2}=\frac{10Ek^2}{l^2}. \tag{66}$$

Equations (65) and (66) are Euler's general equations. The ratio of l to L has different values, depending on the way in which the ends of the strut or column are held. The following tabular diagram shows the usual cases, together with the corresponding modifications of formulæ (65) and (66).

The feet of the columns in Fig. 64 are shown bolted down to emphasize the importance of having the ends *rigidly* fixed. If the feet are flat flanges merely *resting* on the supporting surfaces, the columns are little better than those with rounded ends.

In the formulæ given in connection with Fig. 64, W is the *ultimate* load the column can carry without failure, and the greatest safe working load is obtained by dividing W by a factor of safety N. The safe working load is, then, $\frac{W}{N}$, and the safe working stress f must not exceed $\frac{p}{N}$. The value of f must not exceed the value of the safe crushing resistance of the material f_c.

The commonly used values of the factor of safety are:

Cast Iron.	Wrought Iron. Mild Steel.	Timber.
N=8.	N=5.	N=10.

When there is doubt that the load acts in the axis of the column, the above values of N should be increased by one-half.

* For the deduction of Euler's formula, see Smith, "Strength of Material."

DIAGRAM OF STRUT LOADING AND EULER'S FORMULÆ.

CASE 1.	CASE 2.	CASE 3.	CASE 4.
One end fixed, other end free.	Both ends rounded or pivoted.	One end rounded or pivoted, other end fixed.	Both ends fixed.
$l = 2L$	$l = L$	$l = \frac{L}{\sqrt{2}}$	$l = \frac{L}{2}$
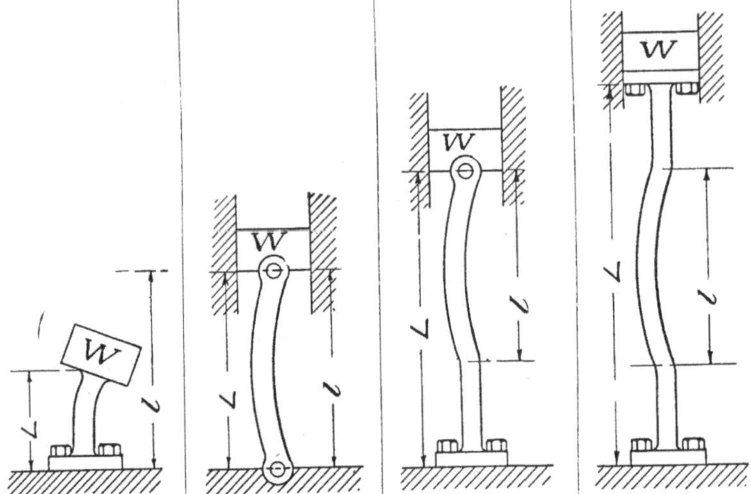			
$W = \frac{2.5EI}{L^2}$	$W = \frac{10EI}{L^2}$	$W = \frac{20EI}{L^2}$	$W = \frac{40EI}{L^2}$
$p = \frac{2.5Ek^2}{L^2}$	$p = \frac{10Ek^2}{L^2}$	$p = \frac{20Ek^2}{L^2}$	$p = \frac{40Ek^2}{L^2}$

FIG. 64.

52. *Limit of the Application of Euler's Formulæ.*—The principal defect in Euler's formulæ is that they neglect the direct compressive stress in the material, and give only the load that will cause failure by bending. Even this load is given only approximately, and the load calculated by them is too great, because the bending moment increases more rapidly than the formulæ assume, due to the fact that after bending once starts, the load no longer acts in the neutral axis, because the neutral axis shifts when bending starts (see Art. 49). However, in very long columns in which the compressive stress is small compared to the stress due to bending, the error is not serious. Therefore, they are used for such very long columns when the *virtual* length l is not less than 30 times the least lateral dimension for wrought iron or steel; or is not less than 12 times for cast iron or wood.

If applied to very short, or moderately short, columns or struts, whose ratio of length to lateral dimension is below about 10, they give results for the load which are absurdly too high. Thus, for example, take an iron strut, 4 inches diameter and 40 inches long, with rounded ends. Then

$$p=\frac{10Ek^2}{L^2},$$

$$I=\frac{\pi}{64}d^4,$$

$$k^2=I/A=\frac{\pi}{64}d^4/\frac{\pi}{4}d^2=\tfrac{1}{16}d^2=1,$$

$$p=\frac{10\times 29{,}000{,}000\times 1}{1600}=181{,}000 \text{ pounds per square inch.}$$

This result is absurd, as the value is far higher than the crushing strength of the material.

53. *Empirical Formulæ for Struts or Columns of Moderate Length.*—When a strut is short, that is, when the ratio of length to least lateral dimension is about 5 or less, its strength, or rather the load it can support, is obtained by the relation, $W_c=A\times f_c$, where W_c = the load; A = the area of cross-section; and f_c = direct crushing resistance of the material. When the column is long, that is, when the ratio of length to least lateral dimension is about 30 or more, Euler's formulæ give satisfactory working approximations. When the ratio of length to lateral dimension is between the limits of 5 and 30 approximately, neither Euler's nor the crushing formula give sufficiently accurate results. Various empirical formulæ have been proposed to cover such cases. Of these the best known is one devised by Gordon and made more general by Rankine. It is an interpolation formula, which gives results agreeing with the crushing formula when the strut is very short, and with Euler's formula when it is very long, and at the same time gives results agreeing with actual experiments for lengths between very short and very long.

54. *Rankine's Formula.*—For a short strut, where buckling is impossible, we have

$$W_c=A\times f_c. \tag{67}$$

For a *long* column, with the ends free, Euler's formula is

$$W_b=\pi^2EI/L^2=\pi^2EA(k/L)^2. \tag{68}$$

Now, if W_a is the *ultimate* load for any length L and cross-section A, we may assume an equation

$$\frac{1}{W_a} = \frac{1}{W_b} + \frac{1}{W_c}. \tag{69}$$

If the strut is very short this equation will evidently give a value of W_a, which will approximate to W_c, because $\frac{1}{W_b}$ becomes so small as to be negligible, since a small value of L in equation (68) will make W_b large. Consequently, $W_a = W_c$ very nearly. Again, if the strut is long, $\frac{1}{W_c}$ becomes negligibly small in comparison with $\frac{1}{W_b}$ and $W_a = W_b$ very nearly. Now, since the change in W_a caused by changing L is continuous, for a fixed value of A, it is reasonable to assume that equation (69) will give a proper value of W_a for *any* length of strut.

Then, for a strut with free ends we have

$$W_a = \frac{1}{\frac{1}{W_c} + \frac{1}{W_b}} = \frac{1}{\frac{1}{f_c.A} + \frac{L^2}{\pi^2 EI}}$$

$$= \frac{f_c.A}{1 + \frac{f_c.L^2}{\pi^2 Ek^2}} = \frac{f_c.A}{1 + a\left(\frac{L^2}{k^2}\right)}, \tag{70}$$

where $a = \frac{f_c}{\pi^2 E}$ and is, therefore, a constant for any given material. The symbols have the same meaning as given in Art. 51.

TABLE 15.—VALUES OF f_c AND a FOR USE WITH RANKINE'S FORMULA.

Material.	f_c = lbs. per square inch.	a = Case 1. One end fixed, other end free.	Case 2. Both ends pivoted.	Case 3. One end fixed, other end pivoted.	Case 4. Both ends fixed.
Cast iron	80,000	$\frac{1}{400}$	$\frac{1}{1600}$	$\frac{1}{3200}$	$\frac{1}{6400}$
Wrought iron	36,000	$\frac{1}{2250}$	$\frac{1}{9000}$	$\frac{1}{18000}$	$\frac{1}{36000}$
Mild steel	55,000	$\frac{1}{1375}$	$\frac{1}{5500}$	$\frac{1}{11000}$	$\frac{1}{22000}$
Hard steel	70,000	$\frac{1}{1250}$	$\frac{1}{5000}$	$\frac{1}{10000}$	$\frac{1}{20000}$
Timber	7,200	$\frac{1}{188}$	$\frac{1}{750}$	$\frac{1}{1500}$	$\frac{1}{3000}$

The value of the constant varies for the different methods of end fixing; thus for a column *fixed* at both ends (Case 4 of Fig. 64), the value is a/4; one end fixed and the other end free it is 4a (Case 1 of Fig. 64); one end fixed, the other end pivoted it is a/2 (Case 3 of Fig. 64).

Table 15 gives the values of f_c and a commonly used.

NOTE.—The values of a in this table are to be used *as given,* in equation (70), since the differences due to the different kinds of end supporting have been applied to the constant. Also in equation (70) L is the *actual* length of the column from center to center.

Equation (70) gives the breaking *load* for a column, and to obtain the breaking or buckling *stress,* W is divided by A, the area of cross section. Thus we have for the breaking stress p

$$p=\frac{W}{A}=\frac{f_c}{1+a\left(\frac{L^2}{k^2}\right)}. \tag{71}$$

The *safe working* stress is, then, p/N, where N is the factor of safety. In the case of connecting rods, the load being a "vibrating" one, the value of N is between 8 and 10.

The practical application of the above formulæ is more fully shown in the solution of the connecting rod problem.

55. *Straight Line Strut Formula.*—Experiments on full-size struts and columns show that, unless the most elaborate precautions are taken to insure dead accuracy of loading, and unless the material is perfectly homogeneous, the results may easily be from 10% to 15% out when compared with the results as calculated from such formulæ as Rankine's, or Gordon's, or similar ones. Even when every possible precaution is taken the results are often 5% or more out. For these reasons many engineers prefer the use of a straight line formula of the form

$$p=M-N\frac{l}{d}, \tag{72}$$

where p=buckling load in pounds per square inch.

M=a constant depending on the strength of the material.

N=a constant depending on the form of section and the elasticity of the material.

l=the *equivalent length* of the strut.

d=the *least* lateral dimension of the strut.

The following table gives the values of M and N.

TABLE 16.—VALUES OF M AND N FOR STRAIGHT-LINE FORMULA.

Material.	Form of section.	$\frac{l}{d}$ not to exceed	M.	N.
Wrought iron.	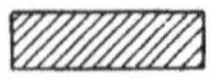	40	47,000	825
		40	47,000	900
		40	47,000	775
	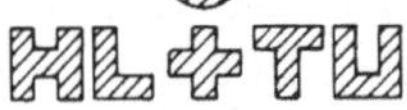	30	47,000	1,070
Mild steel.	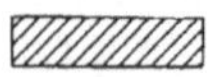	30	71,000	1.570
		30	71,000	1,700
		30	73,000	1,430
		30	71,000	1,870
Hard steel.		30	114,000	3,200
		30	114,000	3.130
		30	114,000	2,700
		30	114,000	3,500
Soft cast iron.	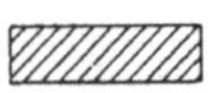	15	90,000	4,100
		15	90,000	4,700
		15	90,000	3,900
		15	90,000	5,000
Hard cast iron.		15	140,000	6,600
		15	140,000	7,000
		15	140,000	6,100
		15	140,000	8,000
Timber.	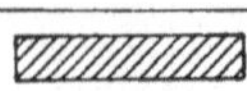	10	8,000	470
		10	8,000	500

The straight line formula is very commonly used for columns and in bridge designing, etc. For piston rods and connecting rods, however, the Bureau of Steam Engineering uses Rankine's formulæ, or a modification of them, as will be given under the practical connecting rod problems.

QUESTIONS AND PROBLEMS.

Discuss the action of a compressive load when its line of action is parallel to, but not in, the axis of the load. Deduce equations for

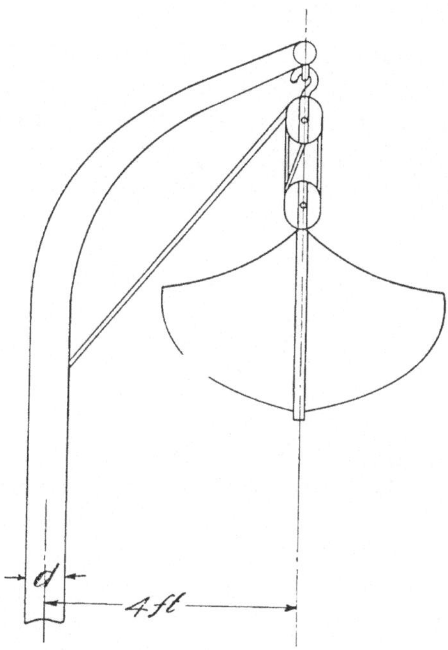

FIG. 65.

the stresses produced (1) for a circular section; (2) for a rectangular section.

Show how to find the load a simple crane, such as a boat davit, can safely carry.

A piece of material is under the action of a tensile and at the same time a shearing force. Show how to find the resulting maximum shearing and tensile stresses produced.

Given Euler's formula $p = \frac{10Ek^2}{l^2}$, explain the meaning of the symbols. Make sketches showing the four usual cases of end supporting and show how the above formula is modified in each case.

What is the serious defect of Euler's formula? Show how Gordon's or Rankine's formula is deduced given Euler's formula $W=\frac{10EI}{L^2}$.

Problems.

1. A boat weighing 8 tons is supported by two davits as shown in the sketch. f=6000. Find the diameter of the vertical part of the davit.

2. A hollow cast-iron column, fixed at each end, is 20 feet high, and has a mean diameter of 12 inches. It is to carry a load of 100 tons. Factor of safety 8. What is the thickness of the metal? Use Rankine's formula.

3. A solid mild steel column, with round ends, is 6 inches in diameter and 37 feet long. What load will it bear? Solve by Euler's and by Rankine's formulæ and compare the results.

4. A tie bar of rectangular section, 2 inches wide and 1 inch thick, carries a tensile load of 10 tons, which acts at a distance of $\frac{1}{10}$ inch from the axis of the bar, in the direction of the thickness, but is in the center of the width. Find the extreme stresses.

5. A short cast-iron strut, 8 inches external diameter with the metal 1 inch thick, carries a load of 20 tons which acts $1\frac{3}{4}$ inches from the axis of the strut. Find the extreme stresses. At what distance from the axis must the load act to cause tension along one side of the strut.

6. Find the principal stresses and their directions in a beam at a point where the unit normal stress is 400 pounds per square inch and the unit shearing stress is 250 pounds per square inch.

7. A round steel propeller shaft is subject to a normal stress, due to bending, of 5000 pounds per square inch, and to a shearing stress, due to torsion, of 8000 pounds per square inch. Poisson's ratio=.25. Find the equivalent stress.

CHAPTER IX.

COTTERS.

56. *Cottered Joint.*—Fig. 66 represents the *cottered joint* very often used to connect two rods A and B which are required to transmit a force of tension or compression in the direction of their length. The end of rod B is made in the form of a socket into which is fitted the end of rod A and the two are held firmly together by driving in the *cotter* C. The clearance at G, H and I is a very important point to be considered in designing such a joint as is shown in Fig. 66. It is obvious that since the cotter is to draw the rod A into the socket there must be clearance at H and I, and also an equal amount at G. The usual amount allowed is from $\frac{1}{16}$ inch to $\frac{1}{8}$ inch.

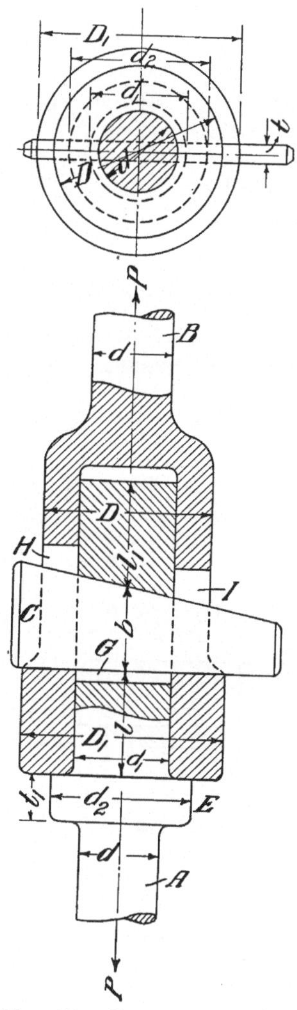

FIG. 66.—Cottered Joint.

The taper of the cotter is also important. It will be shown later that the total taper should not *exceed* 1 in 7, but in practice, especially if no provision is made for locking the cotter, the taper is usually much less than this; about from $\frac{1}{4}$ inch to $\frac{1}{2}$ inch to the foot being common practice.

The material of the cotter is practically always steel, whatever may be the material of the rods.

57. *Method of Proportioning the Parts of a Cottered Joint for Uniform Strength.*

Let f_t = safe working strength in tension of the rods, pounds per square inch.

f_s = safe working strength of the rods in shear, pounds per square inch (f_s = from $0.7f_t$ to $0.8f_t$ for steel, or $0.5f_t$ for iron).

f_c = safe working crushing strength of the bearing surfaces, pounds per square inch.

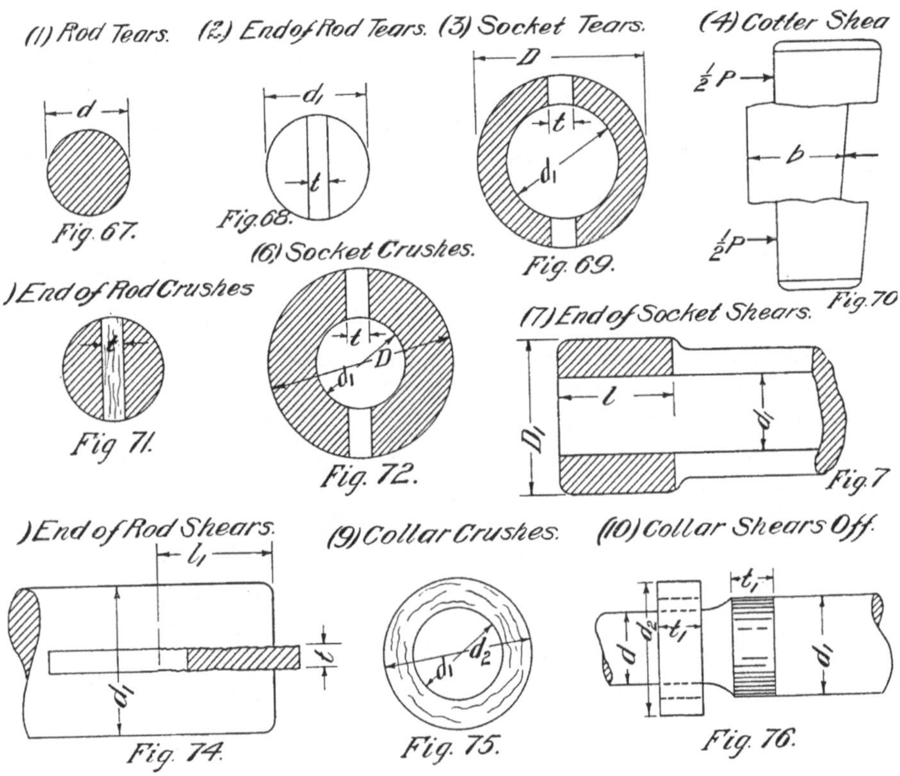

ILLUSTRATING FAILURE OF COTTER JOINTS.

f_s' = safe working shearing stress of the cotter, pounds per square inch.

P = safe working load of the joint.

The above figures, 67 to 76, illustrate the various ways in which the several parts of the joint may fail either in tension or compression. When the load acts alternately in compression and tension, the safe working stresses are only one-half those for either tension or compression alone.

In a properly designed joint of this kind the resistance to failure in each of the above illustrated ways should be equal, so that every part has the same strength. The diameter of the main body of the rods A and B is first found and the proportions of the parts of the joints are found in terms of d.

Resistance to Tension.—(1) The rods may fail by tearing at A or B (Fig. 67). We then have

$$P = \frac{\pi d^2}{4} f_t, \tag{73}$$

or

$$d = \sqrt{\frac{4P}{\pi f_t}}. \tag{74}$$

(2) Failure may occur by the end of the rod tearing across the cotter hole (Fig. 68). Then

$$P = \left(\frac{\pi d_1^2}{4} - d_1 t\right) f_t; \tag{75}$$

equating (73) and (75) we have

$$\frac{\pi d^2}{4} = \frac{\pi d_1^2}{4} - d_1 t.$$

But it is usual to make $t = \frac{1}{4} d_1$,

$$\therefore \frac{\pi d^2}{4} = \frac{\pi d_1^2}{4} - \frac{d_1^2}{4};$$

whence

$$\pi d^2 = (\pi - 1) d_1^2,$$

$$d_1 = d\sqrt{\frac{\pi}{\pi - 1}} = d\sqrt{\frac{3.1416}{2.1416}},$$

$$d_1 = 1.21d. \tag{76}$$

To find the value of t in terms of d we have

$$t = \frac{d_1}{4} = \frac{1.21d}{4} = 0.3d. \tag{77}$$

(3) The socket may tear across the cotter hole (Fig. 69). Then

$$P = \left[\frac{\pi(D^2 - d_1^2)}{4} - (D - d_1)t\right] f_t. \tag{78}$$

Equating to equation (73)

$$\frac{\pi d^2}{4} f_t = \left[\frac{\pi}{4}(D^2 - d_1^2) - (D - d_1)t\right] f_t. \tag{79}$$

Substituting the values of d_1 and t

$$.785d^2 = .785(D^2 - 1.464d^2) - (D - 1.21d) \times .3d.$$

Simplifying and transposing,

$$785D^2 - 300Dd - 1571d^2 = 0.$$
$$D^2 - .382Dd - 2d^2 = 0.$$

Completing the square,

$$D^2 - .382Dd + .0365d^2 = 2.0365d^2,$$
$$D - .191d = 1.428d,$$
$$D = 1.62d. \qquad (80)$$

For additional stiffness and for convenience it is usual to increase this somewhat, say

$$D = 1.75d. \qquad (81)$$

(4) The cotter may fail by double shearing (Fig. 70).

The mean breadth of the cotter is b. Then

$$P = 2btf_s';$$

equating to equation (73),

$$2btf_s' = \frac{\pi d^2}{4} f_t,$$
$$2b \times .3d \times f_s' = .785d^2 f_t,$$
$$b = 1.31d \frac{f_t}{f_s'}. \qquad (83)$$

When all parts of the joint are of either wrought iron or steel, the breadth of the cotter is usually 1.6d.

(5) The bearing surface of the cotter in the rod end may fail by crushing (Fig. 71).

$$P = d_1 t f_c. \qquad (84)$$

This equation is to be used as a check on the values of d_1 and t already found. If the value of P obtained from equation (84) works out less than the load for which the joint is being designed, then d_1 and t must be increased, keeping the proportion, $t = \frac{d_1}{4}$.

(6) The bearing surfaces of the cotter in the socket may fail by crushing (Fig. 72).

$$P = (D_1 - d_1) t f_c. \qquad (85)$$

Equating this to the crushing of the rod end,

$$(D_1 - d_1)tf_c = d_1 tf_c,$$
$$D_1 - d_1 = d_1,$$
$$D_1 = 2d_1 = 2 \times 1.21d = 2.42d. \qquad (86)$$

(7) The cotter may shear through the end of the socket (Fig. 73).

$$P = 2[l(D_1 - d_1)f_s], \qquad (87)$$
$$= 2[l(2.42d - 1.21d)f_s] = 2.42ldf_s;$$

equating to equation (73),

$$2.42ldf_s = \frac{\pi}{4} d^2 f_t;$$

the shearing strength of wrought iron parallel to the fibers is about $.5f_t$; then

$$l = .65d. \qquad (88)$$

If the rods are of steel f_s may be taken equal to $.8f_t$ when

$$l = .43d. \qquad (89)$$

In practice it is usual to increase these values to:

$$\text{for wrought iron, } l = .75d. \qquad (90)$$
$$\text{for steel, } l = .5d. \qquad (91)$$

(8) The cotter may shear through the end of the rod (Fig. 74).

$$P = 2l_1 d_1 f_s, \qquad (92)$$
$$= 2.42 l_1 d f_s,$$

which is exactly the same as the case above. So that

$$l_1 = l. \qquad (93)$$

When the joint is in compression (9) it may fail by crushing the collar E (Figs. 66 and 75).

$$P = \frac{\pi}{4}(d_2^2 - d_1^2)f_c. \qquad (94)$$

In order to keep equal strength, the bearing surface on the collar must equal that of the cotter on the end of the rod, so

$$\frac{\pi}{4}(d_2^2 - d_1^2)f_c = d_1 tf_c,$$
$$\frac{\pi}{4}(d_2^2 - d_1^2) = \frac{d_1^2}{4} \text{ since } t = \tfrac{1}{4}d_1,$$
$$3.1416d_2^2 = 4.1416d_1^2,$$
$$d_2 = 1.15d_1 = 1.15 \times 1.21d = 1.4d. \qquad (95)$$

But in practice this is usually increased to

$$d_2 = 1.5d. \qquad (96)$$

(10) The collar may be sheared off the rod (Fig. 76).

$$P = \pi d_1 t_1 f_s; \qquad (97)$$

equating to equation (73),

$$\pi d_1 t_1 f_s = \frac{\pi}{4} d^2 f_t,$$

$$\text{but } d_1 = 1.21d \text{ and } f_s = 0.5f_t,$$

substituting

$$t_1 = .413d, \text{ say } 0.42d. \qquad (98)$$

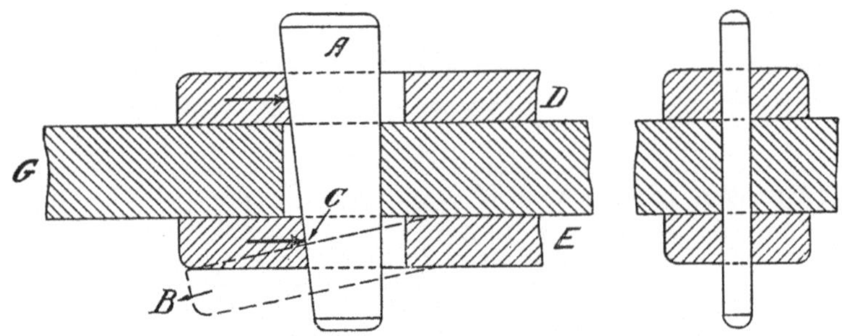

FIG. 77.—Effect of Cotter Without Gib.

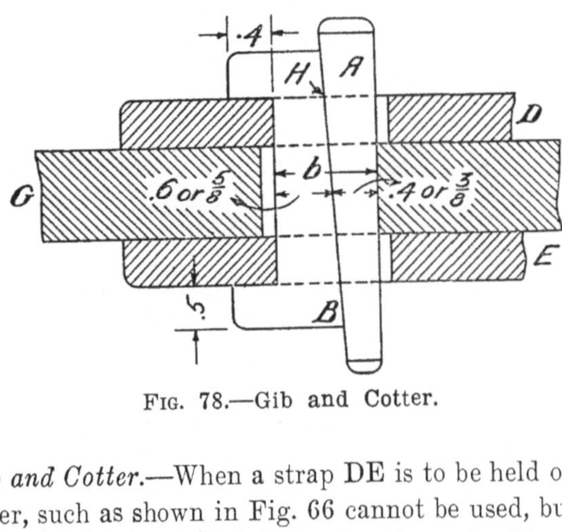

FIG. 78.—Gib and Cotter.

58. *Gib and Cotter.*—When a strap DE is to be held on a rod G a plain cotter, such as shown in Fig. 66 cannot be used, but a *gib* and *cotter* (sometimes called *gib* and *key*), as shown by Fig. 78, must be fitted. If the cotter without a gib were used, the friction between the strap and cotter at the point C, Fig. 77, would cause the former to be sprung away as shown by the broken lines at B. The use of the gib H, Fig. 78, prevents this action and is the method in common use for connecting the strap to the end of the connecting rod for many slow-moving engines.

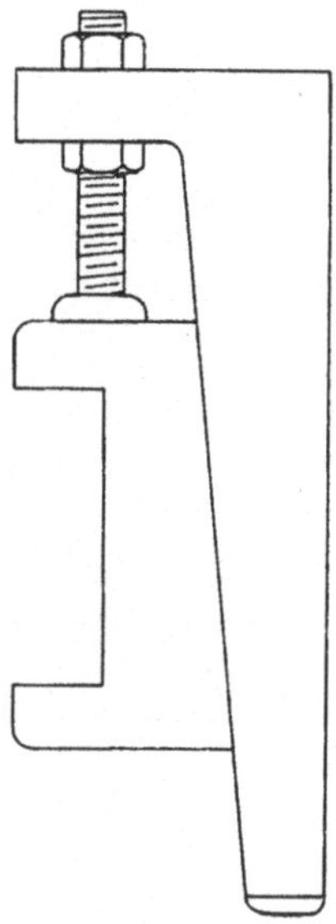

FIG. 79.—Method of Locking the Key.

It should be noted that when the gib and cotter are used the holes in the strap and rod are parallel, the taper being given to the back of the gib, as shown at AB, Fig. 78. The dimensions shown in Fig. 78 refer to the total breadth b of the gib and key, and are the usual proportions given the parts, having been found satisfactory in practice. In order to prevent the key slacking back while at work, it must be locked fast in some manner. Fig. 79 shows one method of doing this, the screw shown being rigidly attached to the gib. Another simple method is to fit a set screw through the body of the rod bearing directly on the side of the key.

59. *The Taper of the Cotter.*

Let H = total horizontal force necessary to drive the cotter home when the connected parts are under a load P.

α = angle between the edges of the cotter, in other words, the taper of the cotter in degrees.

ϕ = angle of repose of the materials, or the *friction* angle.

It is then evident from Fig. 80 that

$$H = H_1 + H_2 = P[\tan(\phi+\alpha) + \tan\phi]. \tag{99}$$

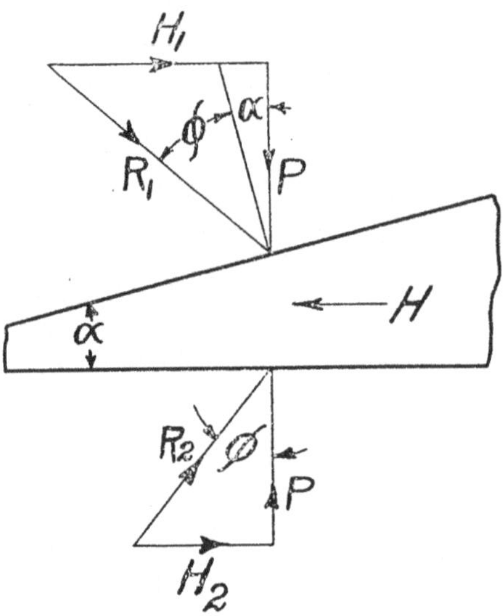

FIG. 80.—Force to Drive Cotter Home.

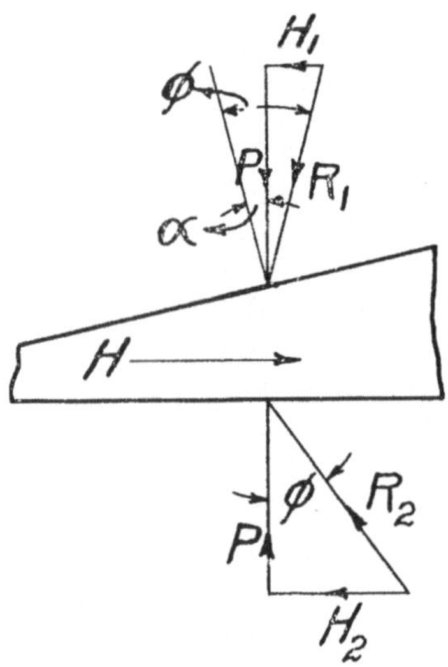

FIG. 81.—Force to Drive Cotter Out.

This is the force necessary to tighten up the cotter under the load.

Similarly from Fig. 81, the force necessary to loosen the cotter is

$$H = H_1 + H_2 = P[\tan(\phi-\alpha) + \tan\phi], \tag{100}$$

but the cotter will just slip back without any additional force when H = 0; then

$$P[\tan(\phi-\alpha)+\tan\phi]=0,$$
$$\tan(\phi-\alpha)=-\tan\phi,$$

or

$$\frac{\tan\phi-\tan\alpha}{1+\tan\phi\tan\alpha}=-\tan\phi,$$
$$\tan\phi-\tan\alpha=-\tan\phi-\tan^2\phi\tan\alpha,$$
$$2\tan\phi=\tan\alpha(1-\tan^2\phi),$$
$$\tan\alpha=\frac{2\tan\phi}{1-\tan^2\phi}=\tan 2\phi,$$

or

$$\alpha=2\phi. \qquad (101)$$

Thus it is seen that when the angle of the taper of the cotter is equal to twice the friction angle the cotter will be just on the point of working loose; and to prevent this occurring α must be less than 2ϕ.

For slightly greasy metal the value of ϕ is taken at $4\frac{1}{2}°$, therefore, α must be less than 9°, which corresponds to a taper of 1 in 7. In practice, as already stated in Art. 56, the taper is made from $\frac{1}{4}$ inch to $\frac{1}{2}$ inch to the foot for safety. This corresponds to a taper of from 1 in 48 to 1 in 24.

Questions and Problems.

Make a neat sketch of a socket and cotter joint for connecting two round rods. Explain the various ways in which the joint may fail under the action of a tensile or compressive load.

Show how to find the proportions of a socket and cotter joint in terms of d the diameter of the rods, to resist both tension and compression.

Make a neat sketch showing the use of a gib and key. Explain why it is necessary to use a gib. In a gib and key connection, show a method of locking the key.

Find the greatest taper that can be given a cotter so that it will just not slide out of place when the load is acting; given the friction angle for slightly greasy metal as $4\frac{1}{2}°$.

PROBLEMS.

1. Two rods are to be connected together by a socket and cotter joint. The load is a steady pull of 10 tons. Design the joint, all parts to be of mild steel. Use Table 8 for the values of the working stresses.

2. A long pump rod is made of two parts connected by a socket and cotter joint. The load is an alternate tensile and compressive one of 5 tons in each direction. The rod is so guided that buckling may be neglected. Material, rods, phosphor bronze; cotter, mild steel. Use working stresses given in Table 8. Design the joint.

CHAPTER X.

THE THEORY OF THE CONNECTING ROD.*

60. The connecting rod of a steam engine is subjected to rapidly alternating, and nearly equal, tensions and compressions. Theoretically it may fail in one of three ways:

1st. As a "long column" in compression, by bending out the line of its axis, *i. e.*, "buckling."

2d. As a "tie rod" by pulling apart under tension.

3d. As a "short column" in compression, by the metal breaking down or "crushing."

Practically we need consider the rod only as a long column in compression, because the formula for strength against failure by buckling gives a larger section than for either of the other two cases.

61. *Gordon's Formula.*—The formula for long columns, known as Gordon's formula, has been given in a great variety of forms. One of the most general forms for a column "free at both ends" has been used in the Bureau of Steam Engineering for the *tabulation* of the designs of the connecting rods of all the U. S. naval vessels.

It can also be used for *designing* rods, or special formulæ deduced from it for special cases.

It is

$$W = \frac{\frac{C}{N} \times A}{1 + \frac{C}{\pi^2 E} \times \frac{l^2}{k^2}}. \qquad (102)$$

An intelligent use of it requires an understanding of the terms and the values given to them under different circumstances.

A, the area, and k, the radius of gyration, of the cross section of the rod are the unknown factors whose values are to be found by the application of the formula for one or more places along the length of the rod.

* From "Notes on Machine Design."

The Bureau of Steam Engineering uses the following method to determine the load W the rod is designed to bear: It is supposed that on starting the engine, before the receiver spaces have been filled with steam, the piston of the H. P. cylinder may be subjected for one or two strokes to full boiler pressure on the upper surface, while the lower surface is under atmospheric pressure only. R, the thrust on the H. P. piston rod, is then the area of the H. P. piston $\times$ boiler pressure. W, the thrust on the connecting rod, is greater than R by a small percentage. The difference is a maximum when the crank is at right angles to the axis of the cylinder and the connecting rod in consequence makes its greatest angle with that axis. This angle is of course $\sin^{-1}\frac{\text{crank length}}{\text{connecting rod length}}$ (see Fig. 82) and with the proportions there given is $\sin^{-1}\frac{24''}{96''}=\sin^{-1}\frac{1}{4}$ $=14° 29'$. At the moment considered, the cross-head pin is acted

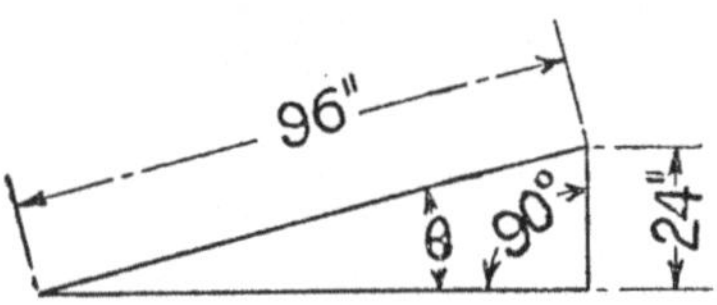

FIG. 82.

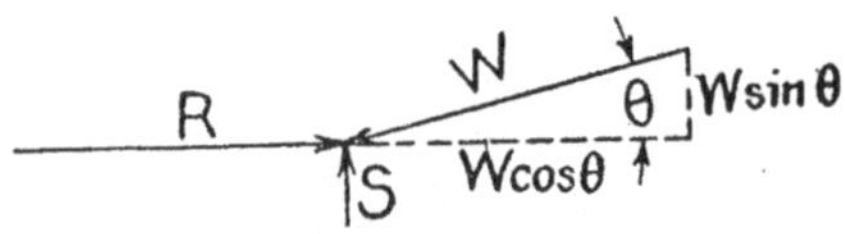

FIG. 83.

on by three forces, W, the thrust of the connecting rod; R, the thrust of the piston rod; and S, the resistance of the guides (see Fig. 83). Neglecting friction, *i. e.*, taking S perpendicular to R and resolving forces horizontally, we have $W\cos\theta=R$, or $W=R\sec\theta$.

In marine practice the crank is very frequently one-fourth the length of the connecting rod. In all such cases

$$W=R\sec 14° 29'=R\times 1.033.$$

This calculation applies only to the H. P. cylinder, but in nearly all cases the rods are made equal and interchangeable.

The factor $\frac{C}{N}$ is often written as one symbol, f, and is defined as the "safe working fiber stress." However, since C also appears separately, the older method of using C, the "ultimate strength" of the material, and providing a "factor of safety" N is adhered to as the best in this case.

The ultimate strength of the metal used is largely a matter of cost, and constantly changes with improved methods of manufacture. For slow-running or small-sized engines, where saving in weight is not of prime importance, wrought iron, on account of its cheapness is still used. For quick-running engines the tendency has been towards the best attainable material. This tendency is well shown by the specifications of our naval vessels here briefly tabulated.

Name of material.	Ult. Str.	Vessels.
Wrought iron	36,000	Puritan and Miantonomoh.
Mild steel	55,000	Boston to Dolphin.
Steel	65,000	Maine (old) to Iowa.
Oil-tempered steel	80,000	Kearsarge to Illinois and destroyers.
Oil-tempered nickel steel..	95,000	Maine (new) and all later vessels.

The first steel used was of very low carbon, 0.20%, for the sake of great ductility and consequent safety against fracture from sudden shock.

The process of oil-tempering has enabled steel of 0.30 to 0.40% carbon to be used, with its attendant great strength and without loss in ductility. It is requisite, however, that the thickness of the metal tempered should nowwhere exceed 2½ inches to three inches, and, consequently, *for large rods oil tempering necessitates a hollow section.*

The presence of a small percentage of nickel in the steel, makes it possible to reach an ultimate strength of 95,000 pounds. Without it, it is barely possible to get 80,000 pounds.

The factor of safety must be a large one by reason of the "vibrating load" on the rod. Its value usually lies between 8 and 10. Recently the Bureau of Steam Engineering has used values of 8.5 and 8.9.

π is, of course, 3.1416. π^2, in Gordon's formula, is, however, taken as 10 for simplicity.

E is the modulus of elasticity of the metal of the rod. For all iron and steel rods its value is taken as 30,000,000 pounds.

l, the length of the column when the formula is used to determine the section at the middle of the rod is the length "from center to center."

62. *The Rod Considered as a Column.*—The strength of a column to resist buckling is known to depend largely on the nature of the "end fastenings." The connection of the rod to the cross head at one end and to the crank at the other is of such a kind as to allow perfect freedom of adjustment in the "*plane of motion of the rod,*" while very little, if any, is permitted in that *at right angles to the plane of motion.*

In theory, therefore, the bending of the rod in the two planes should be treated as two different cases. The formula we have assumed applies directly to buckling in the plane of motion of the rod. It can be applied also to determine the strength against buckling in the other plane by substituting $\frac{l}{2}$ for l, on the ground that the strength of a column "fixed at both ends" is equal to that of one of half its length "free at both ends." One other change must be made. The radius of gyration employed is always that for the section of the rod about an axis in the plane of the section *perpendicular to the plane in which bending takes place.*

Since all other factors are unaltered, if the value of W in the second application of the formula is not to change, the alteration in k must counterbalance that in l. If k changes to $\frac{k}{2}$ this condition is evidently fulfilled. Therefore, *to give equal strength against both kinds of buckling the radius of gyration of the section about an axis in the plane of motion must equal half the radius of gyration about an axis perpendicular to the plane of motion.*

For any circular, or flattened circular section, solid or hollow, likely to be used for a connecting rod, the two radii of gyration will be so nearly equal that we shall have a great excess of strength against bending sidewise.

For a rectangular section the two radii are proportional to the sides to which they are parallel, in other words, to the sides perpendicular to the axes considered. If it is taken as a rule that the thickness of the section at right angles to the plane of motion shall equal or exceed half that in the plane of motion, bending sidewise will again be provided against.

Thus no double application of the formula is ordinarily required and the only k considered is that for bending in the plane of motion of the rod.

63. *Tapering the Rod.*—The rod does not need to be of equal section throughout its length. Were it really a column pure and simple it could be made to taper in both directions from the middle.

The section at any point along the rod at a distance x from one end can be obtained as follows: The piece of rod of length x forms a column "fixed at one end" (*i. e.*, fixed or made fast by the molecular forces which bind it to the other part of the rod) and "free at the other." Such a column is equivalent to one of twice its length "free at both ends." The section at a distance x from one end is found, therefore, by putting 2x in place of l in our formula.

If the sections at a great number of points are determined the resulting column will be one swelling gradually at the middle like an ornamental architectural column. Rods were at first so designed. More frequently the sections at two points near the ends were determined and the rod made to taper uniformly from the middle in both directions. Many slow-moving rods are still so made.

With the increase of speed of engines it has been found desirable to increase the strength of the rods near the crank end. This is due to the additional bending stresses set up by the "transverse inertia" of the connecting rod itself. These stresses increase rapidly with the speed of the engine and are a maximum in the part of the rod near the crank end where the cross motion is greatest.

The tapering of the rod at the crank end was first given up, leaving that portion of the rod of parallel section. Finally the practice has come in of carrying on the taper, or rate of increase, which is determined by the sections at the cross-head neck and the middle of the rod, to the crank end. This practice will be followed in the succeeding pages.

64. *The Rod Considered as a Tie Rod.*—The formula for the strength of a tie rod or short column is $W' = \frac{C}{N} \times A$. This expression is the numerator in the formula for the strength of the rod as a column, equation (102), and we note also that the denominator is *necessarily greater than unity.* W' will, therefore, always exceed W, and the rod designed for strength as a column will be amply strong as a tie rod.

CHAPTER XI.

PRACTICAL PROBLEM III.*

TO DESIGN A CONNECTING ROD WITH STRAP END FOR A SLOW-MOVING ENGINE.

65. *Specifications and Data.*—Compound engine. Diameter of H. P. cylinder, 30 inches. Stroke, 36 inches. Boiler pressure, 80 pounds. Connecting rod of wrought iron. Length between centers, 90 inches. Dimensions of crank pin, $8\frac{1}{2}'' \times 10''$. Assumed ratio of length to diameter of connecting rod, 15. Assume a solid circular section for the rod: $C=36,000$, $N=9$, $E=30,000,000$.

Gordon's formula is

$$W=\frac{\frac{C}{N}\times A}{1+\frac{C}{\pi^2 E}\times\frac{l^2}{k^2}}.$$

Since k^2 for a solid circular section equals $\frac{d^2}{16}$ the formula can be written

$$W=\frac{\frac{C}{N}\times\frac{\pi d^2}{4}}{1+\frac{16C}{\pi^2 E}\times\frac{l^2}{d^2}},$$

or (at middle of rod)

$$d=\frac{\sqrt{W\left(1+\frac{16C}{\pi^2 E}\times\frac{l^2}{d^2}\right)}}{\sqrt{\frac{C}{N}\times\frac{\pi}{4}}}. \qquad (103)\dagger$$

* From "Notes on Machine Design."

† In equation (103) the factor $\frac{l^2}{d^2}$, is often written r^2, where r is the assumed ratio of length to diameter. The factor $\frac{16C}{\pi^2 E}$ is also written as 4a and $\sqrt{\frac{C}{N}\times\frac{\pi}{4}}$, as b and 103 becomes

$$d=\frac{\sqrt{W(1+4ar^2)}}{b} \qquad (104)$$

When formula (104) is used the values of a and b must be first calculated for the material to be used. This simplified formula is convenient only when a number of different rods of the *same* material are to be designed.

The formula (103) is solved by assuming a value of $\frac{l}{d}$ from a knowledge of similar designs; having found the value of d, the true value

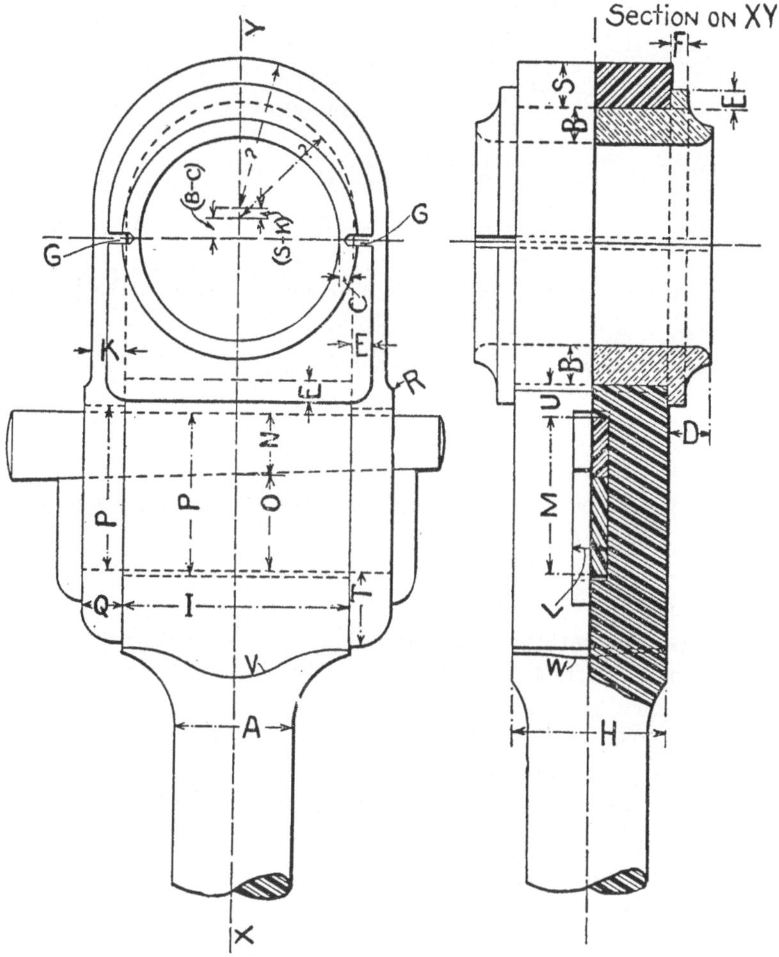

FIG. 84.

of the ratio $\frac{l}{d}$ can be found and the result must agree reasonably well with the ratio assumed. If not found to agree, a new trial value of $\frac{l}{d}$ is taken and a second solution made.

Application of formula to find the middle diameter of the connecting rod.

R, the thrust on the piston rod, is, by the Bureau's method, $706.9 \times 80 = 56{,}552$ pounds. $\theta = \sin^{-1} \frac{1}{5} = 11° \ 31'$. $\sec\theta = 1.021$.*

$W = R \sec\theta = 56{,}552 \times 1.021 = 57{,}700$ pounds.

Substituting in (103),

$$d = \frac{\sqrt{57{,}700\left(1 + \frac{16 \times 36{,}000}{10 \times 30{,}000{,}000} \times 15^2\right)}}{\sqrt{\frac{36{,}000}{9} \times \frac{\pi}{4}}} = 5.12.$$

This would make the ratio $\frac{l}{d} = 17.7$, or nearly 18, instead of 15 as taken. A second application of the formula, with $\frac{l}{d} = 18$, gives

$$d = \frac{\sqrt{57{,}700\left(1 + \frac{16 \times 36{,}000}{10 \times 30{,}000{,}000} \times 18^2\right)}}{\sqrt{\frac{36{,}000}{9} \times \frac{\pi}{4}}} = 5''.46 \text{ or } 5\tfrac{15}{32}''.$$

The diameter of the rod at the cross-head end may be taken as $.9d = .9 \times 5\frac{15}{32} = 4\frac{15}{16}''$.

The diameter of the rod at the crank-pin end may be taken as $1.1d = 6\frac{1}{32}''$.

Brasses (empirical dimensions, Unwin).—The dimensions of the crank pin, obtained from other calculations, are $8\frac{1}{2}$ inches diameter and 10 inches long. The unit for calculating the brasses is found from $t = .08d + \frac{1}{8}''$, where d is the diameter of the pin.

Then unit $= t = .08 \times 8''.5 + \frac{1}{8}'' = ''.80$.

B, the thickness at the bottom, $= 2t = 1''.6 = 1\frac{19}{32}''$.

C, the thickness at the edges, $= .7t = ''.56 = \frac{9}{16}''$.

D, the overhang, $= 2\frac{1}{2}t = 2.5 \times ''.8 = 2''$.

E, the width of the flange to hold the brass in the strap, $= t = \frac{13}{16}''$.

F, the thickness of the same, $= t = \frac{13}{16}''$.

* It is difficult to pick accurately from the omnimeter the secants of small angles. This can be avoided by using Fig. 82 and working out the problem geometrically. In this example, replace 96″ by 90″ and 24″ by 18″, then

$$\sec\theta = \frac{90}{\sqrt{90^2 - 18^2}} = 1.02.$$

At G, the brasses are planed away at the edges, as shown, in order to reduce the labor of fitting the brasses.

The contour of the line joining the end of the overhang and the end of the flange is empirical. No. 5 is generally used in naval designs, as it is amply strong and is lighter than all except No. 1,

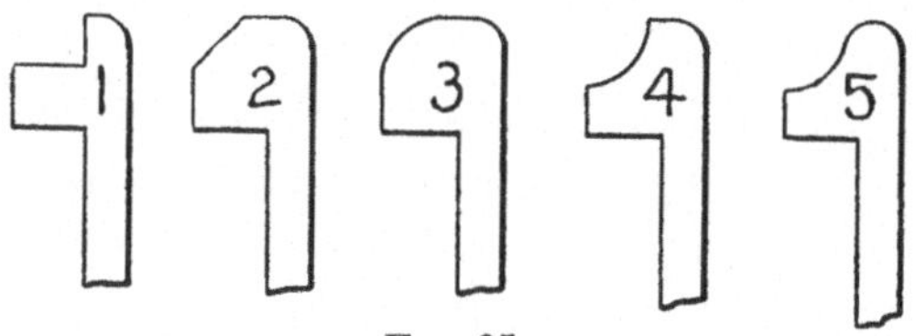

FIG. 85.

which is seldom used. In this case the curves at outside and inside are about as drawn in No. 4.

Stub End and Strap.

NOTE.—In calculating the dimensions of the stub end, strap, etc., f_t is taken as 4000 and f_s as 3000 pounds per square inch.

H, the width of the strap, = the width of the stub end = length of crank pin $-2\times$overhang $=10''-4''=6''$.

I, the breadth of the stub end, = diameter of pin $+2\times$thickness of the brasses at the edges $=8\frac{1}{2}+2\times\frac{9}{16}''=9\frac{5}{8}''$.

The thickness of the strap is found by calculating the dimensions to resist a tensile stress. Then,

K, the thickness of the strap, $\times$width$\times f_t=\frac{1}{2}$W, or, $K\times6\times4000=\frac{1}{2}\times57{,}700$. Whence $K=1''.20$, or $1\frac{3}{16}''$, nearly.

If the calculated value of K is ever less than the value taken for E, then either K must be increased or E diminished, so that K is greater than E by at least $\frac{1}{16}''$. It is usually better to increase K.

Key and Gib.—These are given a thickness of about one-fourth the width of the strap.

L, the thickness of the key and gib, $=\frac{6}{4}''=1\frac{1}{2}''$.

Then, since the gib and key need be calculated only for shear,

M (breadth of key and gib) $\times$ L (thickness) $\times f_s=\frac{1}{2}$W, or, $M\times1\frac{1}{2}''\times3000=\frac{1}{2}\times57{,}700$, and $M=6''.41$, or, $6\frac{13}{32}''$.

From Fig. 78:

N = breadth of key $=\frac{3}{8}M=\frac{3}{8}\times6\frac{13}{32}''=2''.40$, or, $2\frac{13}{32}''$.

O = breadth of gib $=M-N=6\frac{13}{32}''-2\frac{13}{32}''=4''$.

The taper of the key is $\frac{1}{2}$ inch to the foot.

The key is made of such length that the small end projects from $\frac{1}{2}$ inch to 1 inch and the large end from 1 inch to 3 inches past the ends of the gib, depending on the size of the connecting rod.

The thickness and length of the hooked ends of the gib are, as shown in Fig. 78, viz, 0.5b and 0.4b, where b is the combined breadth of key and gib.

It is necessary for this size of connecting rod to allow $\frac{1}{4}$ inch in the length of the slots in the stub end and strap for drawing down the key in adjusting brasses. So P, the length of the slots in this case, is $6\frac{1}{2}'' + \frac{1}{4}'' = 6\frac{3}{4}''$. The dimensions of the slot are, then, $6\frac{3}{4}'' \times 1\frac{1}{2}''$.

Strap.—Since one-fourth of the strap has been cut out for the key way, it is necessary to thicken it at this point in order to keep the necessary strength. Hence

Q, the thickness of the ends of the strap, $= \frac{4}{3} \times 1\frac{3}{16}'' = 1''.58$, or $1\frac{19}{32}''$.

This enlargement of the thickness begins a short distance above the key at R, and is carried to the ends.

Crown of Strap.—To allow for extra bending stresses, the thickness of the crown, at S, is made 1.2 times the thickness of the strap, or

$$S = 1.2 \times 1\frac{3}{16}'' = 1''.42, \text{ or } 1\frac{13}{32}''.$$

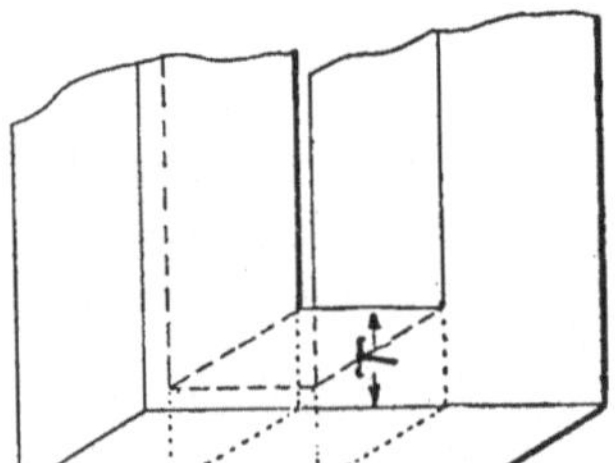

FIG. 86.

The distance T of the key way from the end of the strap is calculated for shearing, the metal being in double shear. Hence

$$3000 \times T \times 2 \times 1\frac{19}{32}'' = \frac{1}{2} \times 57{,}700,$$

$$\text{and } T = 3''.02, \text{ or } 3\frac{1}{32}''.$$

The distance U of the key way from the end of the stub end is calculated in the same way:

$$3000 \times 2U \times 9\frac{5}{8}'' = 57{,}700, \text{ and } U = 1''.$$

In case the value of U in this calculation ever falls below $\frac{1}{2}$ inch it should, nevertheless, be made at least $\frac{1}{2}$ inch to provide for shock and for slight flaws in the material.

PRACTICAL PROBLEM III.

To find the curves of intersection of the rectangular stub end with the surface of revolution formed by sweeping the assumed curve of the enlargement of the rod around the central axis: The curve at a is assumed. Make a front elevation b of the stub end. Pass cylinders with centers at c longitudinally through the rod. These will cut the surface swept by a in circles concentric with the axis, the projections of which on d will be straight lines. They also

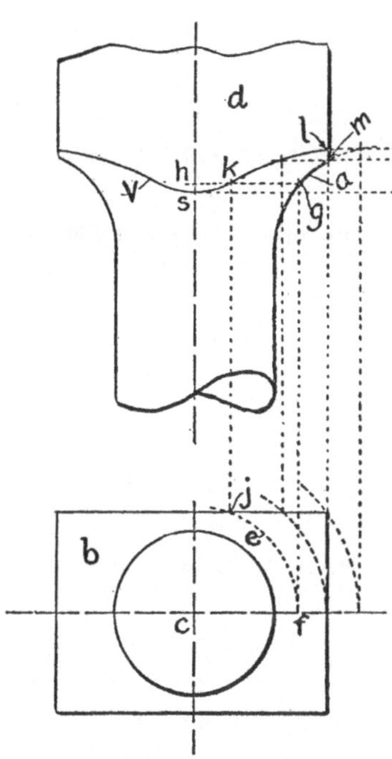

FIG. 87.

cut the sides of the rectangle. The intersections of these projections give the points of the curves V and W.

The cylinder e cuts the curved figure at a circle whose radius is cf. Project f to g and draw gh, the projection of the circle cut by the cylinder. e also cuts the rectangle at the point j. Project j to k and one point is determined. l and m are projected to n and o for the top and bottom of the curve, W, in the side projection. The remaining points of W are found by using the vertical side of the rectangle (front elevation) in the same manner.

QUESTIONS AND PROBLEMS.

State the three ways in which a connecting rod may fail. Which one is it necessary to consider, and why? In the formula

$$W=\frac{\frac{C}{N}\times A}{1+\frac{C}{\pi^2}\times\frac{l^2}{k^2}}.$$

explain the meaning of the several terms and show how the maximum load W is obtained.

Explain why a connecting rod of circular section is stronger as regards bending in the direction parallel to the shaft than in the direction perpendicular to it. What considerations govern the taper to be given a connecting rod?

Make a neat sketch of the stub end of a strap connecting rod, showing in detail all the features, including the provisions to compensate for the wear of the brasses.

PROBLEMS.

1. Find the diameter at the neck of a connecting rod from the following data: Diameter H. P. cylinder, 35 inches; stroke, 38 inches; boiler pressure, 100 pounds; length between centers, 95 inches. Assume a trial ratio of length to diameter of 15. Solid wrought-iron rod.

2. A connecting rod is subject to a maximum load of 60,000 pounds; the width of the strap is 7 inches. Find the dimensions of the key and gib and of the strap. See Fig. 84, P, O, N, L, K, Q, M, S and U. Assume I=10 inches; $f_t=4000$; $f_s=3000$.

3. From the following data find the diameter at the middle and at the neck of a connecting rod. Diameter H. P. cylinder=12 inches; stroke=16 inches; boiler pressure, 100 pounds per gauge; length between centers, 40 inches; C=40,000; N=8.6; E=30,000,000; $f_t=5000$; $f_s=4000$. Assume l/d=15. Solid circular section.

4. A connecting rod is subject to a maximum load of 11,550 pounds. The width of the strap is 3 inches. Find the dimensions of the key and gib and of the strap. See Fig. 84, P, O, N, L, K, Q, M, S and U. Assume I=4⅛ inches; $f_t=5000$; $f_s=4000$.

66. DATA FOR THE DESIGN OF A CONNECTING ROD.

Prob. No.	Desk No. ends in	Diam. H. P. cylinder.	Stroke.	Boiler pressure.	Material.	Length center to center.	Crank pin diam. x length.	Ratio length to diam. of rod assumed.	C.	N.	E.	f	f
1	1	24″	30″	120 lbs.	Wrought iron.	82″	9″ x 11″	15	36,000	9.0	30,000,000	4,000	3,000
2	2	12″	16″	100 lbs.	" "	40″	3½″ x 5″	15	40,000	8.6	30,000,000	5,000	4,000
3	3	40″	72″	50 lbs.	" "	12 ft.	8″ x 12″	15	36,000	9.0	30,000,000	4000	3,000
4	4	18″	42″	150 lbs.	Forged steel.	7 ft.	8″ x 10″	15	55,000	10.0	30,000,000	8,000	6,000
5	5	20″	36″	120 lbs.	"	6 ft.	7½″ x 9½″	15	65,000	9.0	30,000,000	8500	6,500
6	6	10″	15″	150 lbs.	Steel.	3 ft.	3″ x 5″	15	70,000	8.0	29,000,000	9,000	6,800
7	7	36″	48″	80 lbs.	Wrought iron.	9 ft.	10″ x 12½″	15	34,000	9.2	30,000,000	4,400	3,200
8	8	20″	36″	180 lbs.	Steel.	6 ft. 6″	8″ x 10½″	15	65,000	8.9	30, 000	10,000	7,500
9	9	18″	42″	150 lbs.	Forged steel.	7 ft.	8″ x 10″	15	55,000	10.0	30,000,000	8,000	6,000
10	0	20″	36″	120 lbs.	" "	6 ft.	7½″ x 9½″	15	65,000	9.0	30, 000	8500	6,500

All rods are to have a circular section. Use Gordon's formula for the diameter of the rod.

Sequence of Calculations for the Connecting-Rod Problem.

I.

1. Find the maximum load on the rod.

2. Find the middle diameter of the rod using the assumed ratio of length to diameter.

3. Check the ratio of length to diameter from 2 and if necessary repeat the calculation.

4. Find the diameter at the neck of the rod.

5. Find the unit t for the calculations for the brasses.

6. Calculate the dimensions of the brasses, B, C, D, E and F of Fig. 84.

II.

7. Calculate H and I, Fig. 84.

8. Calculate the thickness of the strap, K, Fig. 84.

9. Calculate L, M, N and O, Fig. 84, and the length and depth of the hooked end of the gib, Fig. 78.

10. Find P, Fig. 84.

III.

11. Calculate in order Q, S, T and U, Fig. 84.

CHAPTER XII.

Practical Problem IV.*

67. *To Design a Connecting Rod for a Naval Vessel. Specifications.*—A method of designing a connecting rod of a type much used at present for naval vessels is illustrated in detail in the following pages by giving the calculations for a rod for the battleship Alabama, as an example.

Before beginning the design, certain main dimensions and specifications are decided upon from general considerations of the type of engine and from the design of the adjacent parts.

For the Alabama they are as follows:

Name of ship,	: Alabama.
Number of screws,	: 2.
Type of engines,	: Vertical triple expansion.
I. H. P.,	: 10,000.
Diameter of cylinders,	: $33\frac{1}{2}''$, $51''$, $78''$.
Stroke,	: $48''$.
Revolutions,	: 120 per minute.
Boiler pressure,	: 180 pounds.
Material of connecting rod,	: Forged nickel-steel; oil-tempered.
Ultimate strength of material,	: 80,000 pounds per square inch.
Factor of safety for rod,	: 8.5.
Length between centers,	: $96''$.
Dimensions of crank pin,	: $14\frac{3}{4}''$ diameter $\times 17''$.
Dimensions of cross-head pin,	: $9\frac{3}{4}''$ diameter $\times 14''$. $5''$ bore hole.
Depth of fork from center of cross-head pin,	: $19''$.
Cross-head pin faced to,	: $8\frac{7}{8}''$.

68. *The Load the Rod Must Bear.*—A general idea of the type of rod to be designed and the names of the parts can be gained from Fig. 88. The sections to be first determined are those at aa and bb. For them Gordon's formula is used, and as a preliminary the value of W must be determined from the specifications. Thus

R, the thrust on piston rod = area $\times$ pressure = 8814×180 $=158,612$ pounds, or 158,600.

* From "Notes on Machine Design."

$\theta = \sin^{-1} \frac{24}{96} = \sin^{-1} \frac{1}{4} = 14° \ 29'$. Sec $\theta = 1.033$. Whence $W = R \sec \theta = 158{,}600 \times 1.033 = 163{,}800$ pounds.

68a. *Application of Gordon's Formula.*—C and N are given in the "specifications" as 80,000 pounds and 8.5 respectively. $E = 30{,}000{,}000$ pounds. $\pi^2 = 10$. Thus we have

$$W = 163{,}800 = \frac{\frac{80.000}{8.5} \times A}{1 + \frac{80{,}000}{10 \times 30{,}000{,}000} \times \frac{96^2}{K^2}}. \qquad (105)$$

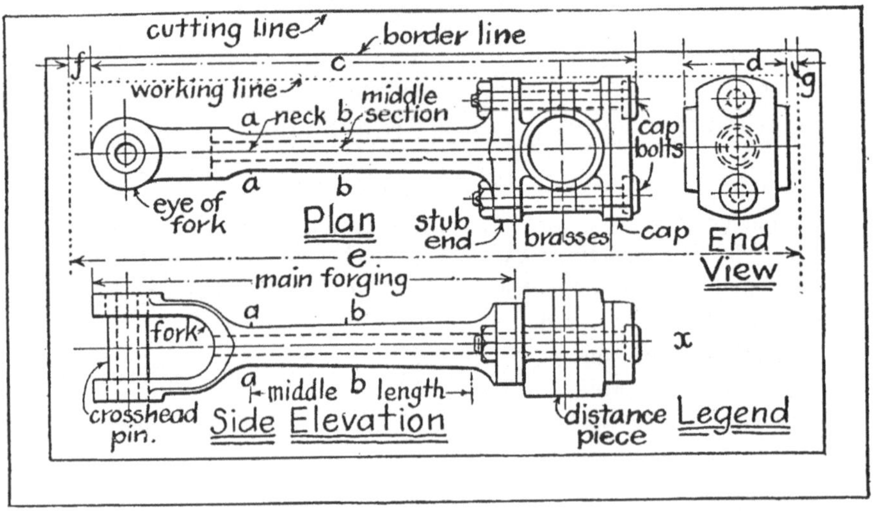

FIG. 88.

A and K^2, our unknowns, are not independent. If made to depend on common variables (as d_1 and d_2 below) the equation is of high degree and not easily managed. It is best to solve simply by trial and error.

Since we are confined to hollow sections (b or d in Fig. 89) because of the oil-tempering, our first trial will be a circular section of 8 inches diameter with a 5-inch bore hole. The diameter of the bore is usually from .5 to .6 of the outside diameter. We will calculate proper values for A and k from these dimensions and substitute in the right-hand member of the equation. The value of W found, will then be compared with the value required, 163,800 pounds.

Calling the diameter, area, and radius of gyration of the 8-inch circle, d_1, A_1 and k_1, and those of the 5-inch circle, d_2, A_2 and k_2, then $A = A_1 - A_2$.

Referring to the table of areas of circles, we find $A = 50.27 - 19.64 = 30.63$ square inches.

The value of k^2 in terms of d_1 and d_2 can be derived from the fact that the moment of inertia of the hollow section is equal to the moment of inertia of the solid 8-inch circle—the moment of inertia of the 5-inch circle. Expressed mathematically, $I = I_1 - I_2$, or $k^2A = k_1^2A_1 - k_2^2A_2$. Then $k^2 = \frac{k_1^2A_1 - k_2^2A_2}{A_1 - A_2}$.

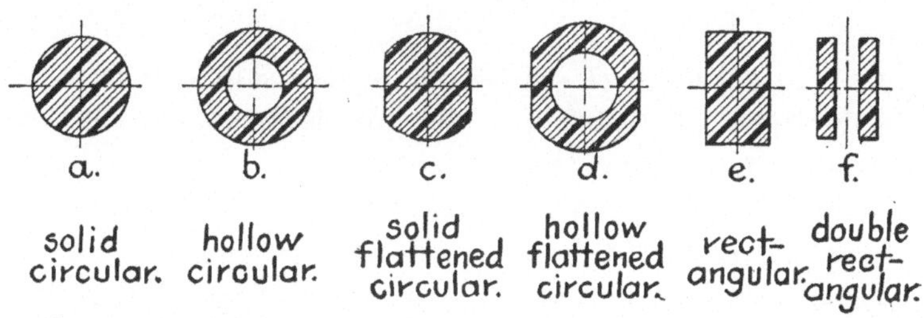

USUAL SECTIONS OF CONNECTING RODS.

FIG. 89.

But, $k_1 = \frac{d_1}{4}$, $k_2 = \frac{d_2}{4}$, $A_1 = \frac{\pi}{4} d_1^2$, and $A_2 = \frac{\pi}{4} d_2^2$. Making these substitutions,

$$k^2 = \frac{d_1^4 - d_2^4}{16(d_1^2 - d_2^2)} = \frac{d_1^2 + d_2^2}{16} = \frac{d_1^2}{16} + \frac{d_2^2}{16}.$$

A table of values of $\frac{d^2}{16}$ has been calculated and is given in the back of the book.

From it we find $k^2 = 4.000 + 1.563 = 5.563$. Then,

$$W = \frac{\frac{80,000}{8.5} \times 30.63}{1 + \frac{80,000}{10 \times 30,000,000} \times \frac{96^2}{5.563}} = \frac{9,410 \times 30.63}{1 + \frac{2.4574}{5.563}}$$

$$= \frac{288,200}{1.442} = 199,900 \text{ pounds.}$$

As this value exceeds 163,800 pounds by a considerable amount we make a second trial reducing each diameter by $\frac{1}{2}$ inch. Then

$d_1=7\frac{1}{2}''$, $d_2=4\frac{1}{2}''$, $A=44.18-15.90=28.28$, $k^2=3.516+1.266$ $=4.782$, and $W=\dfrac{9410\times 28.28}{1+\dfrac{2.4574}{4.782}}=\dfrac{266,100}{1.514}=175,800$ pounds.

This value is almost exactly two-thirds of the way from 199,900 to 163,800 pounds. Reducing again by $\frac{1}{4}$ inch we may expect a very close result. Thus, $d_1=7\frac{1}{4}''$, $d^2=4\frac{1}{4}''$, $A=41.28-14.19=27.09$, $k^2=3.285+1.129=4.414$, and $W=\dfrac{9,410\times 27.09}{1+\dfrac{2.4574}{4.414}}=\dfrac{254,900}{1.557}=$ 163,700 pounds, a solution as close as need be.

The diameter of rod and bore hole are, then, $7\frac{1}{4}$inches and $4\frac{1}{4}$ inches.

69. *Section at the Neck.*—The neck is assumed to be at about one-third the length of the rod from the cross-head end, say 30 inches. The true position of the neck will be determined later. The section of the rod at 30 inches from the free end is theoretically the middle section of a rod 60 inches long. The bore hole, $4\frac{1}{4}$ inches in diameter, is of course retained and at first trial the outside diameter may be supposed to decrease about $\frac{1}{2}$ inch. Then, $d_1=6\frac{3}{4}''$, $d^2=4\frac{1}{4}''$, $A=21.59$, $K^2=3.977$, from which

$$W=\frac{9410\times 21.59}{1+\dfrac{80,000}{10\times 30,000,000}\times\dfrac{60^2}{3.977}}=\frac{203,200}{1.241}=163,700 \text{ pounds,}$$

a very good result, so that no further trial is needed.

The value 203,200 pounds, is the strength of the rod as a tie rod, since it is the strength of the smallest section.

70. *The Fork.*—The fork is faced at the sides to a thickness greater than the diameter at the neck by from 5 to 25%. In this case we may choose 8 inches (A in Fig. 90). The width between the jaws is $\frac{1}{8}$ inch or $\frac{1}{4}$ inch greater than the specified length of the cross-head pin. This allows a clearance, or room for a slight motion of the rod. The distance B (Fig. 90) is, therefore, $14\frac{1}{4}$ inches. The distance C must be such that the fork of the rod may clear by a large margin (1 inch to 6 inches) the cap and bolts of the cross-head brasses in every position that the rod may take relative to the cross head. It must be obtained from the drawing of the cross head, as shown in Fig. 91, or must be given in the data prepared by a previous designer. In this case C is 19 inches. The radius D is, of course, $\frac{1}{2}$ B.

The thickness E must be such that the section of the fork is from 70 to 100% of the rod at the neck. Large rods require the larger amount as the tempering of the thick metal (over $2\frac{1}{2}$ inches),

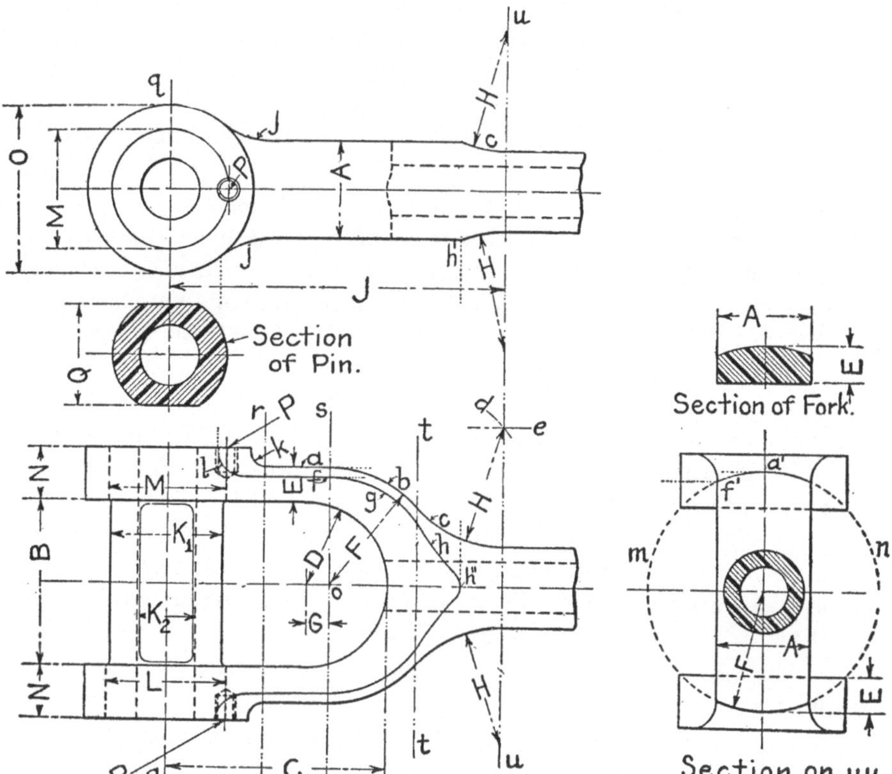

FIG. 90.

is imperfect. It is supposed that much more than half the load may come on one side of the fork from defective alignment. Taking 95% in this case we have $.95 \times 21.63 = 20.55$ square inches. Since the width of the section is 8 inches the average height must be $\frac{20.55}{8}$ $= 2''57 = 2\frac{9}{16}''$. If we take $E = 3''$ it will allow amply for the rounding down of the corners of the section when the outer surface of the fork is turned in the lathe, the rod being centered on its axis. (In Fig. 90 the dotted arcs, m and n, complete the circle traced out by the tool in cutting these surfaces.)

The radius F is equal to $D+E=7\frac{1}{8}''+3''=10\frac{1}{8}''$. The longitudinal contour of the fork is continued to the right of the parallel portion (a) by running into the circular arc (b) of the same radius F. The center of this arc should be from $\frac{1}{2}$ inch to $2\frac{1}{2}$ inches to the right of the center of D. This strengthens the fork at the curved part, and it is still further strengthened where it joins the middle length of the rod by rounding in with a large radius, H, about equal to F. In this case $G=2''$ and $H=10''$.

The distance J is measured on the accurate drawing of the rod. It is the true distance of the neck from the center of the cross-head pin. If it is less than the assumed 30 inches the error is on the safe side and no recalculation is needed. In this case $J=28\frac{3}{4}''$.

71. *Cross-head Pin.*—The bearing length of the pin, 14 inches, has been specified to suit the design of the cross-head brasses. The diameter K_1 has been found by allowing a pressure of 1200 pounds per square inch of *projected area.* Thus $14\times 9\frac{3}{4}\times 1200=163,800$ pounds. The bore hole K_2 lightens the pin, and enables it to be oil-tempered. The pin is of the same steel as the rod and is shrunk or forced into the eyes of the fork. A hole is then drilled and tapped half in the pin and half in the eye, as shown at P. A "screw plug" or "stud" is screwed tight in the hole and the head cut off flush with the surface. This plug prevents the pin from working loose and turning in the fork. For a small rod it may be as small as $\frac{1}{2}''$ diameter $\times\frac{3}{4}''$ long. For a large rod two plugs are used as here shown. These are $1\frac{1}{2}''$ diameter $\times 2''$ long.

The diameter L is slightly larger than K_1 to allow for a small fillet at the end of the bearing length of the pin. In large and medium rods this increase is $\frac{1}{8}$ inch and $\frac{1}{16}$ inch. Small rods do not require it. In addition M is greater than L by from $\frac{1}{32}$ inch to $\frac{1}{16}$ inch. The small end L of the pin will then slip easily through the large bore M of the fork in the process of inserting the pin. In this case $L=9\frac{7}{8}''$ and $M=9\frac{15}{16}''$.

The eye of the fork is reinforced by making the depth N greater than E, and the diameter O greater than A. The area left in the plane qq after boring the hole M is equal to from 90% to 100% of the area of the neck. This rule is empirical. The exact strains cannot be calculated owing to the unknown bursting strains due to the pin. Choosing 95% we get $(O-M)N=20.55$ square inches. Assuming $O=14\frac{1}{2}''$, $N=4\frac{1}{2}''$.

The arc j is usually given a radius equal to A.

The arc k has a radius equal to or less than N − E. In this case it is $1\frac{1}{4}$ inches.

Most of the bearing length of the pin is faced at the two sides to a width Q somewhat less than K_1. The sides are useless as bearing surfaces since the force acting is always in line with the axis of the rod. By cutting them away a pocket is formed for oil to collect in, and to act as a kind of reservoir. A small length at each end, say $\frac{1}{4}$ inch, is left circular to help retain this oil.

Fig. 91 is a sketch of the cross head with the fork end of the rod shown at the extreme angle of swing. The cross-head pin and brasses are sectioned and the near side of the fork removed. The

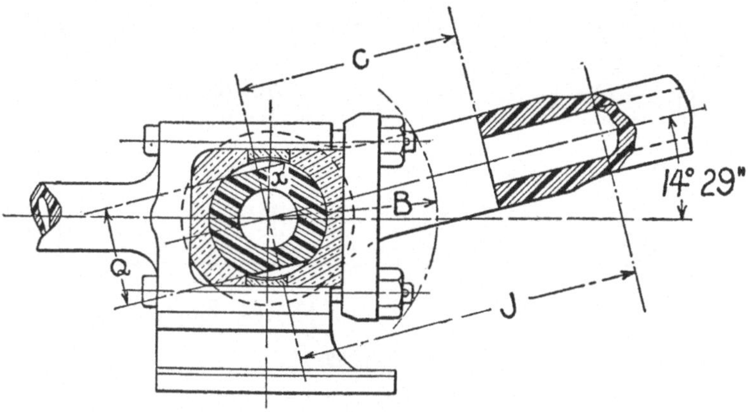

FIG. 91.

width Q is so adjusted as to allow the edge, x, of the bearing surface of the pin to slightly over-travel the bearing surface of the brass it works against. All bearing surfaces whose action is *reciprocating* should be designed with this over-travel at the point of extreme displacement. It prevents a shoulder or step being formed in the brass by the wear not extending to the edge. Without going into the design of the cross-head brasses we shall assume that we know the distance Q and that it is in this case $8\frac{7}{8}$ inches.

72. *The Size of Bolts, Crank End.*—The general shape of the crank-pin brasses is seen in Fig. 93. The cap is held by two bolts, made of the same steel as the body of the rod. We imagine that $\frac{2}{3}$ of the whole load on the rod may come on one bolt through unequal tightening of the nuts. The bolt is never in compression, but its load fluctuates with each revolution of the engine from the designed tension to zero tension. A lower factor of safety can be used than for the rod itself in which the fluctuation is from tension to compression.

For torpedo boats and destroyers (thrust in connecting rod, 0 to 100,000 pounds) the factor averages 5.8 in the designs of our naval vessels. For cruisers and battleships (thrust 130,000 to 245,000 pounds) it averages 7.8. For gunboats (thrust 0 to 40,000 pounds), 8.5.

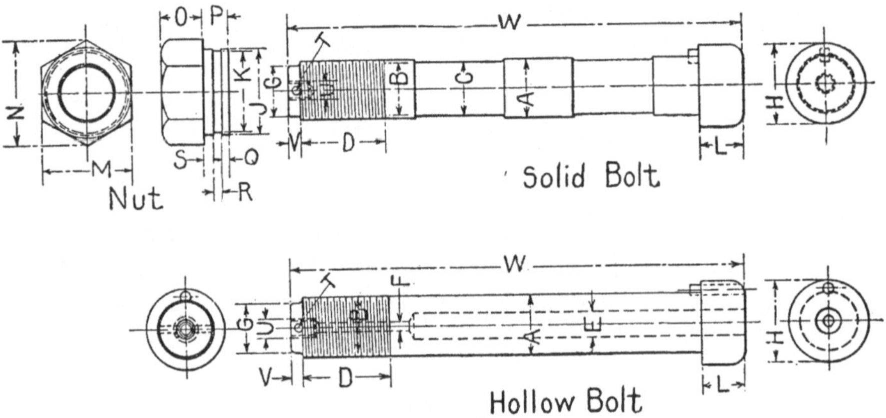

FIG. 92.

In the present case we will use a factor 7.5. Then for the area of the bolt at the weakest section we have

$$\text{Area} \times \frac{80.000}{7.5} = \tfrac{2}{3} \times 163{,}800. \quad \therefore \text{ area} = 10.22 \text{ square inches.}$$

The two styles of bolts in general use are shown in Fig. 92. In one design the solid bolt is turned down in two places so that the area remaining is about equal to the area of the threaded part of the bottom of the threads. The bolt thus reduced and made of fairly uniform strength along its length is stronger to resist sudden shocks than if left of full size. It is more yielding, more "resilient." By referring to the "Table of Standard Bolts and Nuts for the U. S. Navy," on p. I, and the "Table of Areas of Circles" on p. II. the dimensions, $A=4''$, $C=3\frac{5}{8}''$, will be found suitable for our case. The area at B is thus 10.61 square inches, and C it is 10.33 square inches.

The second design accomplishes the purpose of making the bolt of uniform strength along its length by means of the bore hole E. The 4-inch bolt has a section of 12.57 square inches or 2.34 square inches more than the 10.22 square inches required. A bore hole, E, of $1\frac{5}{8}$ inches diameter, will remove 2.07 square inches, leaving the area at A equal to 10.50 square inches.

For convenience in oil-tempering the bolt, the large bore hole E is usually continued by a small one, F, through to the end. If $F=\frac{5}{8}''$ the metal remaining at B will be $10.61-.31=10.30$ square inches.

Beyond the nut the bolt is turned to the diameter $G=C=3\frac{5}{8}''$ and is extended far enough to allow a round split pin to be passed through the hole T. Split pins vary from $\frac{1}{8}$ inch to $\frac{3}{4}$ inch in diameter, depending on the size of the bolt. In this case the diameter is $\frac{1}{2}$ inch. The pin should bear against the face of the nut when in position.

The projecting end of the bolt is $\frac{3}{4}$ inch long for the $\frac{1}{2}$-inch split pin.

The length of the bolt and the positions of the turned down places cannot be determined until the brasses and cap have been designed. The threaded length of the bolt should exceed the depth of the nut by from $\frac{1}{2}$ inch to 1 inch. Thus

$$D=P+O+1''=5\tfrac{11}{16}''.$$

73. *Eye-bolt Holes and Set Screws.*—At U a hole is drilled and tapped for a 1-inch eye bolt. The depth of the hole is $1\frac{1}{2}$ inches. Medium-sized bolts are tapped for $\frac{3}{4}$-inch eye bolts, $1\frac{1}{4}$ inches deep. Set screws are fitted in the cap or in the brasses to hold the bolts after the nuts have been slackened and removed, until the eye bolts have been screwed into place and slings fitted. No groove is required for these set screws as they always bear at one place which can be flattened with a file. Small bolts require no set screws or eye-bolt holes.

Set screws are standard square-headed steel screws (see table on p. I). The head of the $\frac{3}{4}$-inch set screw here used has a short diameter of $1\frac{1}{4}$ inches and a long diameter of $1\frac{3}{4}$ inches. If necessary, it can be made much smaller than the standard. The point is turned down to $\frac{19}{32}$ inch diameter. It projects $\frac{1}{4}$ inch and is flat across the end.

(Fig. 93), is $16\frac{3}{4}$ inches, to allow the same side clearance, $\frac{1}{4}$ inch, as is allowed for the cross-head pin. Taking 80%, we find $D = .80 \times 16''.75 = 13''.4 = 13\frac{1}{2}''$.

The distance of the butt from the center of the crank pin must now be determined. The thickness of the brass at Q is about $\frac{1}{3}$

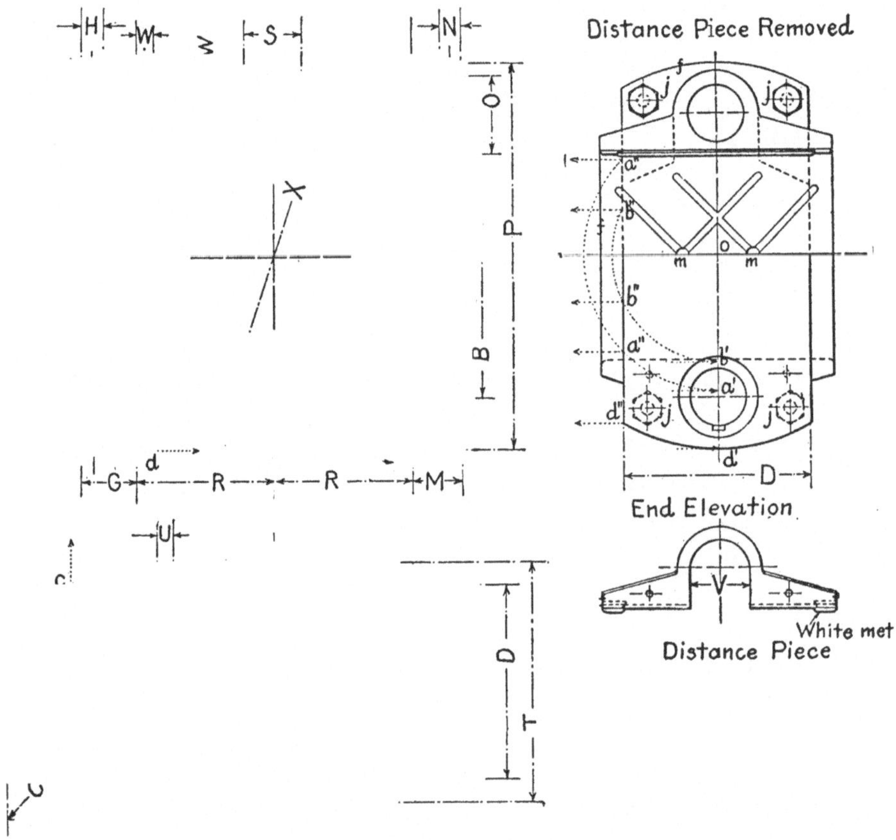

FIG. 93.

the radius of the crank pin. Thus $Q = \frac{1}{6} \times 14\frac{3}{4}'' = 2\frac{1}{2}''$ nearly. The distance R is equal to $2\frac{1}{2}'' + 7\frac{3}{8}'' = 9\frac{7}{8}''$.

The width, E, of the stub end, is equal to the distance apart of the bolts + the long diameter of the nut + a small amount for clearance. In this case let $E = 20'' + 7\frac{1}{16}'' + \frac{3}{16}'' = 27\frac{1}{4}''$. This allows $\frac{3}{32}''$ at each end for the corners of the nuts to keep within the edges of the stub end.

The thickness G must allow sufficient metal for strength after recessing the seat for the nut. If G is made equal to the diameter of the bolt (4 inches), the proportion will be a good one. The distance H is $\frac{1}{16}''$ inch less than P in Fig. 91. Thus $H=1\frac{5}{8}''$.

The radius C is determined when the laying down of the design on the drawing board has reached this point. The arc must be tangent to the tapered section of the rod and also to the surface of the stub end at a point, a, at or within the center line of the bolt. It is usual to so adjust this radius as to make F, the diameter of the rod at the point of tangency, some integral number of inches and eighths, for convenience in the shop. If necessary, the diameter at the middle section, bb, in Fig. 88, is slightly increased to allow this. In this case, C may be made $5\frac{1}{2}$ inches and F, $8\frac{1}{8}$ inches. The same radius, C, appears twice on the side elevation, the centers being on the same vertical line as the centers in the plan.

In addition to the recess for the circular part of the nut there is a recess to allow room for the corners of the hexagon as the nut turns. This recess is nearly a semicircle. Its diameter is slightly larger than the long diameter of the nut. In this case it is $7\frac{1}{8}$ inches, or $J=3\frac{9}{16}''$.

The end surfaces of the stub end are cylindrical. In the end elevation they appear as arcs of a circle whose center is O.

77. *The Cap.*—The cap is of oil-tempered nickel steel and is regarded as a beam supported at the ends and loaded with a uniformly distributed load. In other words, the bolts are not counted on to keep the ends of the cap from assuming a slight angle with their unstrained position. Undoubtedly they do resist such a tendency, but the strengthening effect gained may be offset by the fact that the load is in reality distributed over the middle three-quarters of the length of the beam only. (The length of the beam is the distance between the centers of the bolts.)

The greatest bending moment is at the middle, therefore, and is equal to $\frac{Wl}{8}=\frac{163{,}800\times 20}{8}=409{,}500$ inch-pounds.

The moment of resistance of the rectangular section at the middle is: $f_t\times\frac{bh^2}{6}$. (b=the width, D, of the cap, and h=the thickness L.) Using the same factor of safety as for the rod, 8.5,

$$f_t, \text{ the safe working stress,} = \frac{80,000}{8.5} = 9410 \text{ pounds.}$$

The equation for h, or L, becomes, therefore:

$$409,500 = 9410 \times \frac{13\frac{1}{2} \times h^2}{6}, \text{ or } h^2 = \frac{409,500}{9410 \times 13.5} = 19.34,$$

whence h=4″.4. We shall increase this to 4″.5.

The thickness of the cap at the ends, M, is about ¾ L. In this case 3½ inches. For small caps, M=L.

Large caps have the bolt heads sunk in a distance N, a little less than H. For our cap, let N=1½ inches. The diameter of the recess is ⅛ inch larger than the diameter of the bolt head. Thus O=5⅝ inches. A recess receives the snug on the bolt.

P is equal to E. When the bolt heads are not recessed. P= distance apart of the bolts+diameter of bolt head+⅛″.

The set screw which holds the bolt when the nut is slackened may set to press against the bolt head or the bolt itself, as is most convenient.

78. *The Crank-pin Brasses.*—The dimensions Q and R have been determined.

S, the distance taken up by the "distance piece" and tin liners, varies greatly with the individual designer. A good value for a 14¾-inch pin is 4 inches. Of this amount ¼ inch is taken up by the tin liners.

The two large brasses are alike, save for the boss Y on one, which fits into the bore hole of the rod. The thickness U of the brasses at the overhang is about 10% of the length of the crank pin. In this case, we take it as 1½ inches as the unsupported "lip" of the brasses is small. The outer surface of the lip is conical, the taper being 15° on each side.

Most of the other visible surfaces of the brasses are obtained by keeping the thickness as constant as possible. In this case a thickness, W, of 1 inch, is taken, at the places shown in Fig. 93.

Frequently tap bolts are used to fasten the "cap brass" to the cap and the "rod brass" to the stub end. Four are used for each, and are so placed as to utilize the waste material in the corners of the cap and rod. They are here shown at jj on the end view and j′j′ on the plan, and are 1 inch in diameter. The holes in the cap and forging are drilled and tapped 1½ inches deep.

Dowel pins projecting from the cap brass serve to keep the distance pieces and liners in place. Two pins are used for each distance piece. The pins may be $\frac{1}{2}''$ diameter $\times 2''$ long. The holes in the brass are $\frac{1}{2}'' \times 1''$ deep and are a "driving fit." The holes in the distance pieces are $\frac{1}{2}'' + 1\frac{1}{2}''$ deep and are a "working fit."

The distance pieces are of horse-shoe shape, for convenience in removing them. The cap bolts need only be slackened back enough to free the distance pieces from the dowel pins. The diameter, V, of the concave semicircle is equal to the bolt diameter $+\frac{1}{16}''$.

79. *White Metal Lining.*—The brasses and the ends of the distance pieces are lined with white, or anti-friction, metal. The finished thickness ranges from $\frac{3}{16}$ inch to $\frac{7}{16}$ inch with the size of the rod. Taking it as $\frac{3}{8}$ inch, the inside diameter, x, of the unlined brasses becomes $14\frac{3}{4}'' + \frac{3}{4}'' = 15\frac{1}{2}''$. In addition, many grooves, $\frac{1}{4}''$ deep $\times \frac{1}{2}''$ wide, with undercut edges, are spaced at equal angles around the circle. Two at least must be put in each distance piece. These grooves hold the white metal firmly in place.

The middle of the distance pieces and adjacent parts of the brasses are not lined, thus leaving a shallow cavity ($\frac{3}{8}$ inch deep) in which oil collects and from which it is redistributed over the crank pin.

The white metal, when cast over the concave surfaces of the brasses, is made considerably thicker than the finished work. It is hammered down to a dense condition with a riveting hammer. The parts having been assembled and bolted together, the metal is bored out and scraped to an exact fit with the crank pin.

80. *Oil Holes and Ducts.*—The provision for oiling the crank pin is usually given in detail on a separate drawing. Two copper tubes of $\frac{3}{4}$ inch outside diameter, one running down each side of the rod, lead the oil from the cross-head end, where it is received drop by drop in a narrow funnel-shaped opening, to holes bored in the main forging along the lines ll. The holes extend nearly to the brass with the diameter of $\frac{3}{4}$ inch and are continued through the brass and white metal with a smaller diameter.

From the openings on the surface of the white metal (see points mm on the end elevation, Fig. 93), oil ducts are cut by a grooving chisel leading in some pattern intended to spread the oil over the surface of the crank pin.

Sometimes the oil is introduced down the center line of the rod by a tube inside the bore hole. In such a case the hole through the brass and white metal is in the center of the boss Y and the oil ducts radiate from the point o in the end elevation.

ALTERNATIVE DESIGNS.

81. *Solid Circular Middle Section.*—This section is now little used in naval vessels, as it is not a favorable form for oil-tempering. It may be used for gunboats.

Since k, the radius of gyration of a disc about a diameter, is one-fourth the diameter, the formula can be written

$$W=\frac{\frac{C}{N}\times\frac{\pi}{4}d^2}{1+\frac{16C}{\pi^2 E}\times\frac{l^2}{d^2}}, \tag{106}$$

and solved by trial and error, but the application of the general formula, making $d_2=0$, is just as simple.

82. *Flattened Circular Sections.*—These sections (c and d of Fig. 89) are modifications of the sections a and b. The regular

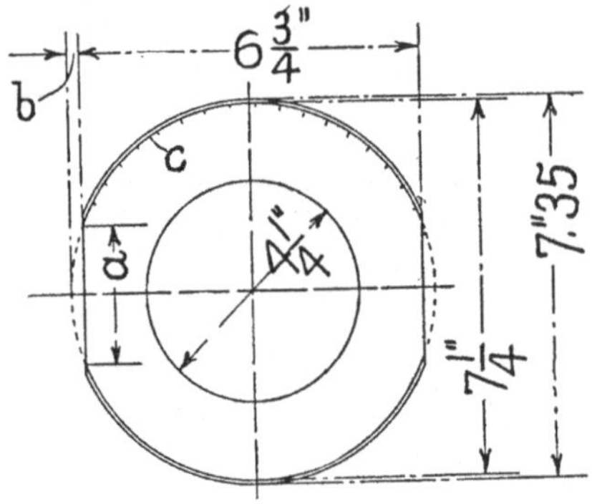

FIG. 94.

solution for the section a or b is first made including the size at the neck. Usually the section at the neck is not flattened at all and the middle section is flattened at the sides to a width equal to the outside diameter at the neck. (Of course the whole middle length is then faced to this width.) The flattening diminishes the area of the section, and to compensate, the outside diameter is increased. Graphical tables of the radii of gyration of flattened circular sections, solid and hollow, have been constructed and by their use the exact increase can be determined.

A fair solution, of which the error is on the safe side, is obtained by making the added area *equal* to that removed. Fig. 94 shows by this method the determination of the increase in the diameter of the rod of the Alabama at the middle, if it is desired to face the sides to the diameter at the neck. The area removed is approximately $\frac{2}{3} \times$ the chord a (which is by measurement $2\frac{5}{8}''$) $\times$ the altitude b, or $\frac{2}{3} \times 2\frac{5}{8}'' \times \frac{1}{4}'' = .437$ square inch. By stepping along the arc c with the bow spacer set to $\frac{1}{2}$ inch we find the arc equal to $8\frac{3}{4}$ inches. The increase required is, therefore, $\frac{.437}{8.75} = ''.05$. The new outside diameter is, therefore, 7''.35.

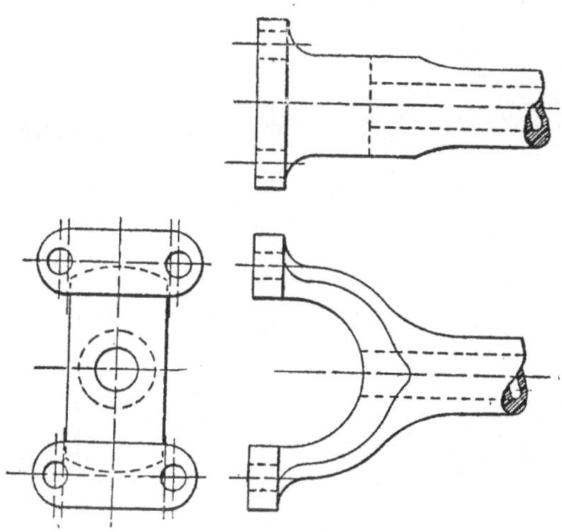

FIG. 95.

83. *Rectangular Section.*—Values for the width and thickness of the section are assumed and the formula used to test them in the regular trial and error method. The values of t and w must be such that $w \gtreqless 2t$. Since $A = w \times t$, and $K^2 = \frac{w^2}{12}$, the formula becomes:

$$W = \frac{\frac{C}{N} \times t \times w}{1 + \frac{12C}{\pi^2 E} + \frac{l^2}{w^2}}. \tag{107}$$

The use of it is illustrated in the following example:

For a torpedo boat, given $W = 52{,}000$ pounds, $l = 40''$, $C = 80{,}000$ pounds, $N = 8.5$, assume first that $w = 3\frac{1}{2}''$, $t = 2''$. Then $W =$

46,500 pounds. Assume next, $w=3\frac{3}{4}''$, $t=2$. Then $W=51,700$ pounds, a good solution.

The thickness t is generally kept constant and the width w made to taper uniformly from the neck to the crank end. Assuming the neck to be at 13 inches from the cross-head pin, we find that, if $w=3\frac{3}{8}''$, and $t=2''$, W will equal 53,400 pounds. This results in a taper of $\frac{3}{8}$ inch in 7 inches, an unusual amount, changed with advantage to $\frac{3}{4}$ inch in one foot. This change will increase the width at the middle by $\frac{1}{16}$ inch, *i. e.*, to $3\frac{13}{16}$ inches.

84. *Open Fork, Cross-head End.*—Large rods are sometimes designed with the jaws of the forked end fitted with small "marine

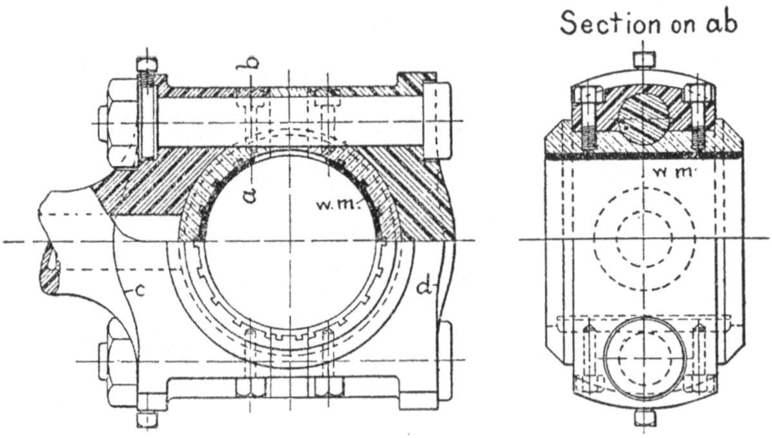

FIG. 96.

ends," reduced copies of the crank end of the rod. In such cases the cross-head pin is shrunk in the cross head at the middle of its length, the two projecting ends forming end-journals for the brasses on the jaws. It is supposed that $\frac{2}{3}$ of the load may come on one jaw from irregularity in fitting, and the ends are designed for that load. The end of each jaw is forged out to the shape shown in Fig. 95, to give room for the bolts. White metal is often omitted from these brasses. The *nuts* of the cap bolts bear on the *caps,* not the fork.

"Solid" Crank End.—The crank end and cap are sometimes designed in the manner known as a "solid end." The brasses are reduced to thin semicircular pieces held in the steel by "nipping bolts." The end of the main forging and the inner surface of the cap are concave semi-cylindrical surfaces instead of flat surfaces.

To reduce the distance apart of the cap bolts, they are made to pass partly through the brasses. Fig. 96 shows this design. The section on ab is that through the nipping bolts.

85. *Cap with Rounded Ends.*—The cap and the similar portion of the main forging are often finished off with semicircular ends, instead of ends formed by turning them on the center line of the rod. Thus in Fig. 93 we should have instead of the arc d″d′, whose center is o, a semicircle of diameter D, passing through the point d′. The two ends of the fork in Fig. 95 have been rounded in this manner.

86. *Brass Cap.*—Small rods very frequently have a brass cap in lieu of a steel one. The cap in such cases takes the form of the steel cap in Fig. 96 combined in one piece with the cap brass. The rest of the design may follow either Fig. 93 or Fig. 96. In calculating the thickness of the brass cap, a low-working fiber stress is taken, its value being about half that for the steel. "Brass" caps are generally made of some form of bronze.

Original Calculations and Sketches for a Connecting Rod.

87. *Note Book, Sketches, etc.*—Sets of specifications like those in Art. 67 will be given out in the drawing room. From them each one will make his own design of a connecting rod for a naval vessel, following in a general way the methods given in the pages just preceding, but depending on his own judgment for a selection from among the alternatives mentioned for the different details.

These specifications are the first entry on the interleaved pages and the calculations, as they are made, are neatly entered in conjunction with free-hand sketches of the details of the rod to which they refer. The sketches are of detached parts of the whole, like Figs. 90, 92 and 93, or of even smaller portions. Dimension letters are not used, but the values as determined by the calculations are recorded on the sketches which then become the basis from which the drawing of the rod is ultimately made.

88. *Order of Recording Calculations.*—The following table gives the order in which the required dimensions are classified and recorded. It is, as far as possible, that in which they are calculated or determined:

In sketch books place dimensions on the left.

I. *Specifications* (see Art. 67).

II. *Middle Length.*

1. Thrust on piston rod.
2. $\text{Sec}\left\{\sin^{-1}\frac{\text{crank}}{\text{connecting rod}}\right\}$.
3. Thrust on connecting rod.
4. Factor of safety of connecting rod.
5. Type of cross section, middle.
6. Dimensions of middle section.
7. A, K^2, and W for middle section.
8. Assumed position of neck.
9. Dimensions of section at the neck.
10. A, K^2, and W for section at the neck.
11. Strength of the rod as a tie rod.
12. Diameter of rod at cross-head end (to be filled later).
13. Corrected size at middle section (to be filled later).
14. True distance of neck from center of cross-head pin (tc be filled later).

III. *Fork End.*

15. Width of fork in plane of motion of rod.
16. Width between jaws of fork.
17. Area of fork required.
18. Thickness of fork at center.
19. Distance between centers for inside and outside circular contours of fork.
20. Radius for rounding in at the neck.
21. Diameter of cross-head pin within top eye of fork.
22. Diameter of cross-head pin within bottom eye of fork.
23. No. and size of screw plugs for fastening cross-head pin in eyes of fork.
24. Outside diameter of eye of fork.
25. Depth or thickness of eye of fork.
26. Radius for rounding in side lines of fork and outside diameter of eye of fork.

IV. *Cap Bolts, etc.*

27. Factor of safety of cap bolts.
28. Area of cap bolts.

29. Diameter and area of turned-down part of bolt.
or
Diameter and area of bore hole in bolt.
30. Diameter and area of small continuation of bore hole.
31. Diameter of bolt beyond nut.
32. Length of bolt beyond nut.
33. Diameter of hole for split pin.
34. Threaded length of bolt (filled later on).
35. Total length of bolt under head (filled later on).
36. Position of turned-down places (filled later on).
or
Depth of large bore hole in bolt (filled later on).
37. Diameter and depth of eye-bolt holes.
38. Diameter of bolt head.
39. Depth or thickness of bolt head.
40. Dimensions of snug.
41. Short and long diameters of nut.
42. Depth of hexagonal part of nut.
43. Depth of cylindrical part of nut.
44. Width and depth of groove on nut.
45. Width of collars on each side of groove.
46. Size of set screws.
47. Size of heads of set screws, if not standard.
48. Diameter of point of set screws.
49. Length of point of set screws.
50. Length of set screws under head (to be filled later).
51. Distance apart of bolts.

V. *Stub End and Cap.*

52. Width of stub end parallel to crank pin.
53. Width of stub end perpendicular to crank pin.
54. Thickness of stub end parallel to center line of rod.
55. Depth of recess for cylindrical part of nut.
56. Diameter of recess for hexagonal part of nut.
57. Radius of arc of enlargement by which middle length joins stub end.
58. Thickness of cap at center.
59. Thickness of cap at ends.
60. Depth of recess for bolt heads.
61. Diameter of recess for bolt heads.
62. Length of cap over all, perpendicular to crank pin.

VI. *Brasses, etc.*

63. Length of crank pin covered by brasses.
64. Length of overhang, or lip of brasses.
65. Thickness of brasses at overhang.
66. Thickness of brasses between crank pin and stub end.
67. Distance from center of crank pin to stub end.
68. Thickness of distance pieces.
69. Thickness of tin liners.
70. Thickness of brasses over bolt.
71. Thickness of flanges.
72. Size and position of tap bolts fastening brasses to forgings.
73. Size and position of dowel pins for aligning distance pieces.
74. Width of horse-shoe opening in distance pieces.
75. Thickness of white metal lining.
76. Depth and width of undercut grooves for holding white metal.
77. Number and angular position of grooves.
78. Dimensions of cavity where white metal is not placed.
79. Diameter and direction of oil holes through main forging.
80. Diameter of oil holes through brass and white metal.
81. Pattern and size of oil ducts or grooves.

The different groups into which these dimensions have been collected should have their own sketch or sketches to illustrate them and to show the exact details.

Some of the dimensions are here entered in advance of their natural order of determination. Such are left blank for the moment. A few are determined only on the accurate drawing of the connecting rod. With the notes otherwise complete proceed to lay down the design on the drawing board, but finally return to the notes and fill in those dimensions here marked "to be filled later."

The Working Drawing of the Connecting Rod.

89. *Arrangement of Views and Sections.*—The figures in the preceding pages illustrate special points and are not models to be followed in the drawing.

The connecting rod should be shown with *all parts assembled.*

For a *vertical* engine the *plan,* in an architectural sense, would be an end view of the rod from the fork end. Practically the rod is considered as lying on one side, and that view in which the cross-head and crank pins show as circles, is taken as the "plan." The (other) side elevation is placed under the "plan." Contrary to the general rule it is advisable to put the end view or elevation (looking on the *crank* end) on a line with the "plan," not with the "side elevation." This arrangement is shown on a small scale in Fig. 88. It makes a compact drawing, as the end view is narrowest in the direction parallel to the crank pin and it leaves a space convenient for the legend.

When necessary to show the interior structure, parts of the plan and elevations are made "in section," as in Figs. 93 and 96. (In Fig. 93 the bolts are absent, as is often the case in *sketches.* In the *drawing* they should be in place, as in Fig. 96.

Often small additional sections are drawn, without showing any part beyond the plane of the section itself, like the sections of the cross-head pin and the fork in Fig. 90. These are intended to show clearly to the eye the strength of the metal at those places.

90. *The Scale of the Drawing.*—The scale depends on the size of drawing space to be devoted to the rod. The cutting, border and working lines are first laid down to the sizes specified in the drawing room. The *actual* distances, c and d, on the connecting rod (see Fig. 88) are computed and added together and the result compared with the horizontal clear space, e, between the two side-working lines. Such a scale is chosen as will make the *apparent* size of $c+d$ a little less than e. The distance left over must provide first for a good separation between plan and end elevation, and then, if still sufficient, for additional space at the ends like f and g in Fig. 88.

91. *Center Lines, Circles, etc.*—The vertical center line through the cross-head pin is first laid out by measuring from the left-hand working line. From it is measured the center line of the crank pin. The vertical center line of the end view is measured from the right-hand working line. The horizontal center line of the plan is measured down from the top working line. The center line of the side elevation can be put in by eye alone unless the vertical height available is scanty. The center lines of the bolts are then drawn. The circles representing the cross-head and crank pins on the plan, the dotted circles for the diameters of the rod on the end view, the arcs for the ends of the cap and the full and dotted circles for the diameters of bolt heads and bolts are then drawn. The circles for the contour of the fork inside and out follow. As a general rule, circles are drawn first. Very often by projecting tangent lines from these circles from one view to another double measurements of the same distances are saved.

The points aa and bb are now established. Their horizontal distances from the cross-head pin are measured off, and their vertical positions projected from the end view. The rod as finally designed may not follow these points exactly, but the contour must include them.

All the fundamental dimensions so far drawn should be verified and altered until absolutely correct.

The details of the ends of the rod are now taken up in much the same order as that of performing the calculation.

92. Some of the lines on the drawing are obtained, not by measured distances, but by a *geometrical construction*. Thus, in the plan in Fig. 93, the line marked d‴ is derived by projection from the point d″ in the end view.

Many of these geometrical lines are of one class and represent the appearance of the edge or intersection of a surface of revolution and a plane parallel to its axis. Such curves are a‴, b‴, c′ in Fig. 93, c and d in Fig. 96, and fgh in Fig. 90.

The method by which points are found on such a curve is illustrated in Fig. 93, by finding the points b‴. The "surface of revolution" is in this case the "flared" or "bell-shaped" surface by which the middle length of the rod is enlarged to join the stub end. The plane is the flat side of the stub end. The contour of the flared part is formed, both on the plan and on the side elevation, by arcs of the *same* radius C, whose centers all lie on the *same* vertical line.

One point on the edge, c′, can be taken at once by projecting up from c on the side elevation. If any plane is passed perpendicular to the axis of the flared surface at a point between c′ and h, it will cut out a circle. bb is such a plane and it appears as a line on the plan, while on the end elevation the circle it cuts out is seen in its true shape. A part of the circle is the dotted arc b′b″b″, whose center is o, of which the point b′ is taken by projection from b on the plan. The two points, b″, lie both on this circle and on the flat edge of the stub end. Project from b″ to the plane bb on the plan. The points b‴ so marked are the same two points and arc, therefore, on the required edge between the two given surfaces.

The plane ha is the limiting plane to the right, the plane to which the bell-shaped surface becomes tangent. In the same way, therefore, a, the point of tangency of the arc of radius C, gives rise to the arc a′a″a″, and the points a″ give the points a‴ on the line ha. Outside of the points a‴ the curve is straight.

If the dimensions F and D are more nearly equal than in Fig. 93, the curve becomes a more marked one, and it may be well to find more points than those here found, a‴, b‴ and c′. When enough points have been found, by passing new perpendicular planes, the curve can be drawn in with irregular curves, or arcs of circles found to approximate to it.

When the curve is as flat as in this case, since nothing depends on the accurate drawing of it, determine only the points a‴ and c′. Pass through them three arcs of circles tangent to each other, finding the centers by trial. The center of the middle one will be on the center line of the figure to the right of c′, the centers of the others to the left on horizontal lines through the points a.

The curve c in Fig. 96 is like that just described.

The curve d (Fig. 96) arises when the entire surface of the cap visible on the end view is turned on the lathe with the rod centered on end centers. Both c and d can be represented by three arcs of circles.

The curve fgh in Fig. 90 is of the same kind. The sectional view in Fig. 90 is not needed on the completed drawing, but some portions of it must be drawn in pencil for use in finding the curve fgh. A convenient place for it is at x in Fig. 88.

In Fig. 90 the outer surface of the fork between the planes rr and ss is cylindrical, and in consequence the edge f is straight and is given by projecting from the corner f′ in the sectional view. Between the planes ss and tt the surface is truly spherical. The edge g is, therefore, a circular arc whose center is O. Between tt and uu the surface is bell-shaped. The point h″ is projected down from h′ on the plan. Intermediate points can be found by passing planes between tt and h″, but the edge h is usually represented by three circular arcs tangent to each other and to the arc g, the middle arc passing through h″.

The line f (Fig. 90) is continued to the left by the line l, which meets the upper surface of the fork at a point projected from the plan, as shown. The exact drawing of the curve is immaterial It is usually drawn a circular arc and need not continue quite to the upper surface of the fork (see Fig. 88). The line represents in reality the edge produced by the hand filing necessary to finish neatly the cutting out of the corners at j.

In Fig. 93 the recess cut in the stub end for the corners of the nut to turn in, is almost a semicircle. The line of the edge appears on the side elevation approximately as the half ellipse ge′g, in which the major axis gg is equal to $2 \times J$ (on the plan) and the point e′, the extremity of the minor axis, is projected from e on the plan. Any approximate method of drawing the ellipse by circular arcs will be sufficiently accurate.

93. *Marking Finished Surfaces.*—A working drawing should show definitely which surfaces are machined and which are left in a comparatively rough condition ("rough forged" or "cast" surfaces). The general rule is to mark every planed, turned or slotted surface with an "f" (meaning "finished") on that part of the drawing where it is seen *on edge* and is represented by a *line.* The f is put right on the line it refers to. The letter "b" or the word "bore" is used to distinguish holes bored or drilled from those made by a core in the process of casting. It is put after the dimension figures giving the diameter of the bore hole thus |← $\frac{3}{4}''$ b →|.

In most connecting rods, the brasses, which are castings, alone have unfinished surfaces, and it is customary, therefore, to put on the drawing in a prominent place a note to this effect, "All surfaces finished except on brasses, which are finished where marked." The brasses, in Fig. 93, have been marked in conformity with this practice to show where to put the "f's."

CHAPTER XIII.

SHAFTS, JOURNALS.

TORSION, STRENGTH OF SHAFTS. TWIST OF SHAFTS. SIZE OF SHAFT PER GIVEN HORSE-POWER. STRENGTH OF JOURNALS AND PINS. COMBINED TURNING AND BENDING MOMENTS. COUPLINGS AND BOLTS.

94. *The General Theory of Torsion.*—Fig. 97 represents two shafts, B and C, fitted with couplings and connected together by the pin A, the two couplings being very close together, the shaft B driving the shaft C, the power being the pull W acting at the distance l from the axis. It is evident that the pin is in shear, and

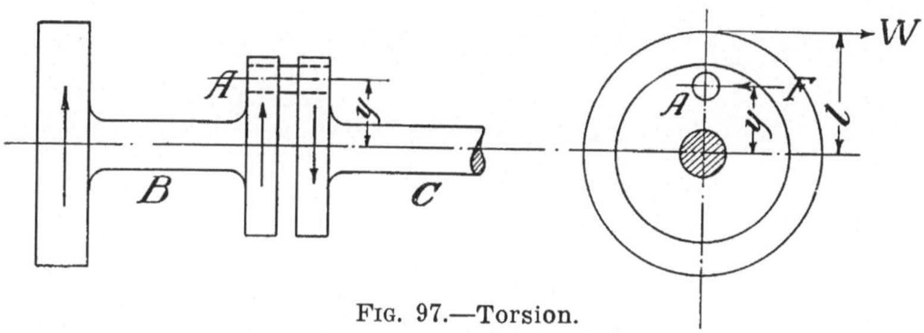

FIG. 97.—Torsion.

if we call F = the shearing resistance of the pin, we have, by taking moments about the axis,

$$Wl = Fy. \tag{108}$$

Let a = area of the pin.

f_s = shearing stress of the pin. Then

$$Wl = f_s ay. \tag{109}$$

Now let us have two driving pins between the couplings; then

$$Wl = Fy + F_1y_1 = f_s ay + f_{s_1}a_1y_1. \tag{110}$$

The dotted circles in the figure represent the pin holes in the driving coupling when the rotating force W is acting. Before W is applied the holes are exactly in line, but when the force is applied the yielding of the pins allows a slight movement of the couplings relative to each other.

This yielding, or the strain, varies directly as the distance of the pin from the axis, and since the material is supposed to be elastic the stress varies directly as the strain, and we have

$$\frac{f_{s1}}{f_s} = \frac{y_1}{y} \text{ or } f_{s_1} = \frac{f_s y_1}{y}; \tag{111}$$

substituting this value of f_{s1} in equation (110) we have

$$Wl = f_s ay + \frac{f_s a_1 y_1^2}{y} = \frac{f_s}{y}(ay^2 + a_1 y_1^2). \tag{112}$$

Suppose now instead of only two pins we connect the couplings with a very great number of very small pins, so great a number as to form a solid section between the couplings. Let the areas of each

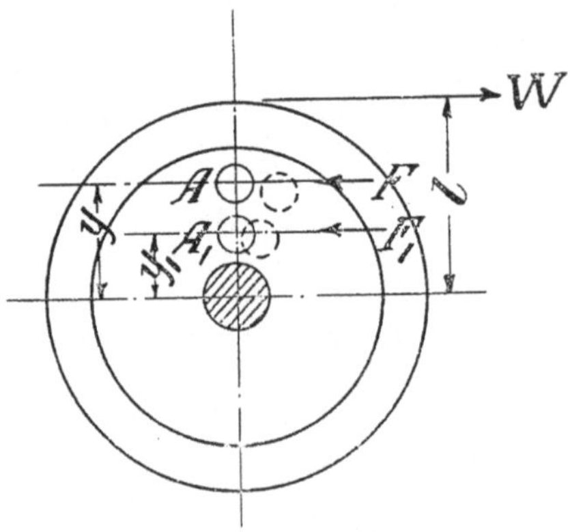

FIG. 98.—Torsion.

of these small pins be a, a_1, a_2, etc., and their distances from the axis be y, y_1, y_2, etc., respectively. Then we have

$$Wl = \frac{f_s}{y}(ay^2 + a_1 y_1^2 + a_2 y_2^2 + \text{etc.}). \tag{113}$$

The quantity in the parenthesis is the product of each very small area by the square of its distance from the axis; in other words, it is the polar moment of inertia of the section. This is usually denoted by the symbol I_p. Then

$$Wl = \frac{f_s I_p}{y}, \tag{114}$$

but $\frac{I_p}{y}$ may be called the *polar* modulus of the section which is denoted by Z_p. Hence

$$Wl = f_s Z_p. \tag{115}$$

The moment Wl is called the *twisting* or *turning* moment, and is denoted by T, so we have

$$T=f_sZ_p, \tag{116}$$

which is similar to equation (18) for bending moment.

In this case, however, f_s is the shear stress of the material furthest from the axis and Z_p is the *polar* modulus of the section in distinction from the modulus for bending.

95. *The Strength of Circular Shafts to Resist Torsion.*—From equation (114) above it is seen that the strength of a shaft depends upon the value of $\frac{I_p}{y}=Z_p$, but $I_p=\frac{\pi d^4}{32}$, where d is the diameter of a solid circular section (see Art. 44), and $y=\frac{d}{2}$; hence we have

$$Z_p=\frac{\frac{\pi d^4}{32}}{\frac{d}{2}}=\frac{\pi d^3}{16}=\frac{d^3}{5.1}=0.196d^3.$$

This is just twice the value of Z for bending. It is easy to recollect which value to apply by considering the fact that in a circular *shaft* the metal is disposed in the very best manner to resist torsion, but is in a bad form for bending; hence the torsion modulus will be greater than that for bending.

For a hollow shaft d external, and d_1 internal diameter

$$I_p=\frac{\pi(d^4-d_1^4)}{32},$$

$$Z_p=\frac{\frac{\pi(d^4-d_1^4)}{32}}{\frac{d}{2}}=\frac{\pi(d^4-d_1^4)}{16d}=0.196\left(\frac{d^4-d_1^4}{d}\right).$$

96. *The Twisting of Shafts.*—Referring back to Chapter III, Art. 23, it was shown that the amount of slide x when a piece of material was under the action of a shearing force was determined by the relation

$$\frac{x}{l}=\frac{f_s}{G} \tag{117}$$

f_s = the shearing stress of the material.
G = the coefficient of rigidity.

To apply this to the case of a shaft under a twisting action, the slide x is measured around the circumference, as indicated in Fig. 99, and it is usually expressed in terms of the twist angle, thus:

The circumference of the shaft $=\pi d$.

The arc which subtends $1°$ at the axis $=\frac{\pi d}{360}$.

The arc which subtends $\theta°$ at the axis $=\frac{\pi d\theta}{360}=x$.

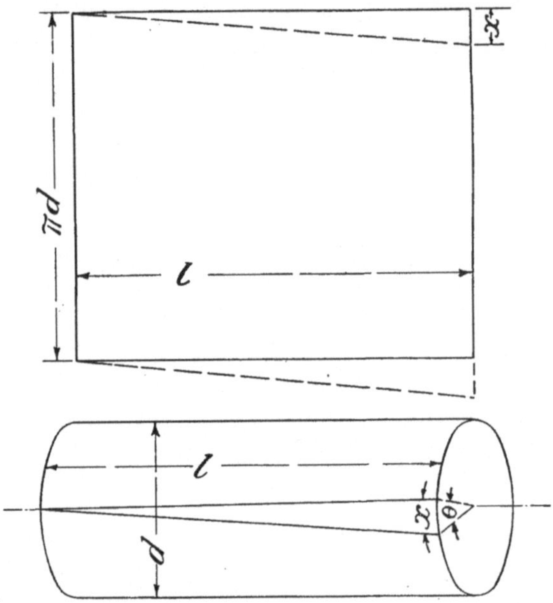

FIG. 99.—Twist.

Substituting this value of x in equation (117) gives

$$\frac{\pi d\theta}{360l}=\frac{f_s}{G} \text{ or } \theta=\frac{360f_s l}{\pi dG}. \tag{118}$$

But from equation (116) we have

$$T=f_s Z_p=f_s\frac{\pi d^3}{16},$$

whence

$$f_s=\frac{16T}{\pi d^3}. \tag{119}$$

Substituting this value of f_s in equation (118) gives

$$\theta=\frac{360\times16\times Tl}{\pi^2 Gd^4}=\frac{584Tl}{Gd^4} \tag{120}$$

for a solid circular shaft. Similarly, for a hollow circular shaft,

$$\theta=\frac{584Tl}{G(d^4-d_1^4)}. \tag{121}$$

The above *twisting* formulæ are used when it is desired to keep the spring of a shaft within a given amount, the usual limiting value of the spring or twist being 1° in a length equal to 20 diameters. Thus from equation (118)

$$\theta=\frac{360f_sl}{\pi dG},$$

when

$$\theta=1°,\ l=20d;$$

then

$$f_s=\frac{\pi dG}{360\times 20d}=\frac{G}{2292}. \tag{122}$$

From equation (122) we get the following:

For steel,	$G=13,000,000$;	$f_s=5670$ lbs. per sq. in.
For wrought iron,	$G=11,000,000$;	$f_s=4800$ lbs. per sq. in.
For cast iron,	$G=\ 6,000,000$;	$f_s=2620$ lbs. per sq. in.

These values of f_s are those that must be used to keep the twist not greater than 1° in 20 diameters, and show that for long shafts the diameter will be considerably greater than is necessary for mere strength.

When the shaft is short, so that the consideration of spring or twist does not have to be considered, the following values of the working stress may be used:

Nickel steel,	$f_s=12,000$ lbs. per sq. in.
Steel,	$f_s=10,000$ lbs. per sq. in.
Wrought iron,	$f_s=\ 8,000$ lbs. per sq. in.
Cast iron,	$f_s=\ 3,000$ lbs. per sq. in.

97. *The Size of Shaft to Transmit a Given Horse-Power.*—The problem that is most frequently presented is to find the diameter of shaft of a given material necessary to transmit a given horse-power. The following method is probably the clearest and most direct for the solution of this problem.

Suppose the shaft is being revolved against a resistance by a force of F * pounds acting at a distance l inches from its axis; then

* In a steam engine the force F and the turning moment are not uniform throughout the revolution, but the method gives the mean turning moment. The maximum turning moment, in naval engines is usually from 1.25 to 1.5 of the mean, according to the type of engine.

F(pounds) $\times l$(inches) = twisting moment on the shaft; in inch-pounds = T.

In each revolution the force F acts through a distance of $2\pi l$ (inches) or $\frac{2\pi l}{12}$ (feet); hence

$$\frac{F\times l\times 2\pi}{12} = \text{work done per revolution in foot-pounds.}$$

If N = number of revolutions per minute

$$\frac{F\times l\times 2\pi\times N}{12} = \text{work done per minute in foot-pounds,}$$

but 33,000 foot-pounds per minute = 1 H. P.; hence

$$\text{H. P. transmitted} = \frac{Fl\times 2\pi N}{12\times 33{,}000},$$

or

$$Fl = T = \frac{\text{H. P.}\times 33{,}000\times 12}{2\pi N}. \qquad (123)$$

From equation (116) we have

$$T = f_s Z_p = \frac{f_s\pi d^3}{16}; \qquad (124)$$

hence

$$\frac{f_s\pi d^3}{16} = \frac{\text{H. P.}\times 33{,}000\times 12}{2\pi N}$$

or

$$d^3 = \frac{\text{H. P.}\times 33{,}000\times 12\times 16}{2\pi^2 N f_s}. \qquad (125)$$

In all reference books this formula is found given in the form

$$d = C\sqrt[3]{\frac{\text{H.P.}}{N}}$$

where C is a constant depending on the value of f_s. This is really the same as formula (125), the constant

$$C = \sqrt[3]{\frac{33{,}000\times 12\times 16}{2\pi^2 f_s}}.$$

98. *Journals and Pins.*—The *journal* is that part of a machine which rotates in the *bearing,* the latter being the stationary part. In some cases the motion of the journal in its bearing is occasional only, and then the size of the journal must be calculated principally for *strength.* In other cases the motion of rotation is continuous, and in these cases, in addition to strength, considerations of heating are equally important. Again, in some cases the load acts on the journal in such a manner as to produce shearing or bending only; and in other cases the journal is subject to torsional stresses at the same time as the bending stresses, when the size must be calculated by the method of combined stresses. Fig. 100 represents a journal under a load P uniformly distributed over its length l. Let d=diameter of journal and f=allowable working stress of the material. Then the bending moment at the fixed end of the journal is $\frac{1}{2}Pl=M$.

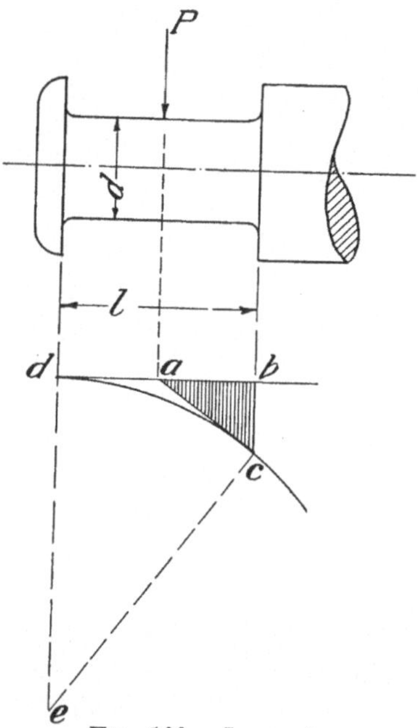

FIG. 100.—Journal.

From equation (18)

$$M=fZ=f\frac{\pi d^3}{32};$$

hence

$$\tfrac{1}{2}Pl=\frac{\pi d^3}{32}f,$$

or

$$Pl=\frac{\pi d^3}{16}f. \qquad (126)$$

Suitable values of f are:

Steel,	f=9000 to 12,000.
Wrought iron,	f=6000 to 9,000.
Cast iron,	f=3000 to 4,500.

Unwin gives the following graphic method of finding the bending action:

"Suppose, first of all, that the load on a journal (Fig. 100) is P acting at the center of its length. Then the bending moment diagram is the triangle abc, in which bc=$\frac{1}{2}$Pl, is the bending moment at the root of the journal. If the load is uniformly distributed the bending moment curve is the parabola dc, which may be approximated to by drawing ce perpendicular to ac and describing a circular arc dc with center e. The bending moment at the root of the journal is the same as before, and on this the diameter necessary for strength depends."

In most cases the value of l, the length of the journal, is not known, and, therefore, there are two unknown quantities, l and d, in equation (126). However, it is generally possible to fix on a ratio of $\frac{l}{d}$, the following being customary values:

For journals working intermittently, or on which the pressure acts for a part of a revolution only, $\frac{l}{d}=1$; for journals working at speeds below about 150 revolutions per minute, $\frac{l}{d}=1.5$ to 1.25 for composition "brasses," or $\frac{l}{d}=1.75$ for cast-iron bearings.

With a fixed ratio of l to d, equation (126) reduces as follows:

$$Pl=\frac{\pi d^3}{16}f;$$

dividing through by d gives

$$P\frac{l}{d}=\frac{\pi d^2}{16}f,$$

or

$$d=\sqrt{\frac{16P}{\pi f}\times\left(\frac{l}{d}\right)}. \tag{127}$$

This formula is used for finding the dimensions of an over-hung crank pin such as is found on the ice machine on board ship for *strength*.

Many cross-head pins are fixed solidly into the forks of the connecting rod, and are then in the condition of uniformly loaded beams, fixed at the ends. If, then, as before, P=the total load, l=length of the pin and d=its diameter, we have

$$M=\frac{Pl}{12},$$

also

$$M=fZ=f\frac{\pi d^3}{32};$$

whence

$$\frac{Pl}{12}=f\frac{\pi d^3}{32}$$

$$d^3=\frac{8}{3}\times\frac{Pl}{\pi f}, \tag{128}$$

or when the ratio $\frac{l}{d}$ is fixed,

$$d^2=\frac{8P}{3\pi f}\times\frac{l}{d} \tag{129}$$

99. In calculating the dimensions of journals of all kinds it is not sufficient to consider only the torsional and bending stresses to which they are subject due to the load, but in addition the intensity of the surface pressure under which they act must be taken into consideration, and also the rubbing speed of the bearing surfaces. This is usually done in practice by fixing a maximum pressure per square inch of projected surface which must not be exceeded. The dimensions are first calculated for *strength* and are then checked to make sure that they come within the allowed pressure per square inch of projected area.

The projected area is merely the area obtained by the product $l \times d$ of the bearing.

The following limiting pressures per square inch of projected area, taken mostly from Unwin, are very generally accepted:

TABLE 17.

	Pressure per sq. in. of bearing surface in lbs.
Bearings carrying intermittent loads at low speeds, such as pins of shearing machine	3000
Cross-head neck journals; motion oscillation; not complete rotation (1000 commonly used)	800 to 1400
Crank pins (slow engines)	800 to 900
Crank pins (fast engines)	500 to 800
Crank pins (marine engines)	400 to 500
Crank pins (small land engines)	150 to 200
Crank shaft bearings (slow)	300 to 450
Crank shaft bearings (fast)	200 to 300
Eccentric sheaves	80 to 100
Thrust collars (slow)	70 to 80
Thrust collars (fast)	50 to 70
Cross-head slides (Babbitt on cast iron)	200 to 300
Steel or iron shaft lignum vitæ (under water)	350

100. *Combined Turning and Bending Moments. Equivalent Turning and Bending Moment.*—In many shafts, particularly the crank shaft of a marine engine, there is not only the torsion due to the horse-power transmitted, but there are, in addition, severe bending stresses due to the thrust of the connecting rod on the crank pins.

In such cases there is a shearing stress due to torsion and a tensile stress due to the bending action. In Art. 50 it is shown that the simultaneous action of shearing and tension produces a maximum stress, $p_{max}=\frac{1}{2}f_t+\sqrt{\frac{1}{4}f_t^2+f_s^2}$, see equation (61). To apply this formula to the case of a shaft of circular cross section,

Let T = twisting moment at any given section.

M = bending moment at the same section.

We have now to find the *equivalent twisting* moment T_e or the equivalent bending moment M_e, that is to say, the single simple twisting or bending moment which would produce an effect on the shaft equivalent as regards strength, to that which is actually produced by the twisting and bending acting together.

But

$$T=f_sZ_p, \text{ where } Z_p \text{ is the torsion modulus;}$$

$$f_s=\frac{T}{Z_p}=\frac{16T}{\pi d^3}.$$

$$M=f_tZ, \text{ where } Z \text{ is the bending modulus;}$$

$$f_t=\frac{M}{Z}=\frac{32M}{\pi d^3}.$$

Also

$$p_{max}=\tfrac{1}{2}f_t+\sqrt{f_s^2+\tfrac{1}{4}f_t^2}.$$

Then the equivalent twisting moment is

$$T_e=P_{max}\times Z_p,$$

$$=Z_p[\tfrac{1}{2}f_t+\sqrt{f_s^2+\tfrac{1}{4}f_t^2}]=\frac{\pi d^3}{16}[\tfrac{1}{2}f_t+\sqrt{f_s^2+\tfrac{1}{4}f_t^2}],$$

$$=\frac{\pi d^3}{16}\left[\frac{1}{2}\cdot\frac{32M}{\pi d^3}+\sqrt{\left(\frac{16T}{\pi d^3}\right)^2+\frac{1}{4}\left(\frac{32M}{\pi d^3}\right)^2}\right],$$

$$=\frac{\pi d^3}{16}\cdot\frac{16}{\pi d^3}(M+\sqrt{T^2+M^2}),$$

or

$$T_e=M+\sqrt{T^2+M^2}. \qquad (130)$$

Similarly

$$M_e=p_{max}\times Z,$$

$$=Z(\tfrac{1}{2}f_t+\sqrt{f_s^2+\tfrac{1}{4}f_t^2}),$$

which reduces to

$$M_e=\tfrac{1}{2}M+\tfrac{1}{2}\sqrt{T^2+M^2}. \qquad (131)$$

The formula $M_e=\frac{3}{8}M+\frac{5}{8}\sqrt{T^2+M^2}$ is sometimes given in text and reference books. This formula is arrived at by considering the greatest *strain* produced in distinction to the greatest *stress*, and is based on a value of Poisson's ratio of 0.25. See footnote, Art. 50.

The common American and British practice is to use the form given in equation (131). In designing shafts the equivalent twisting moment, T_e, is used rather than the equivalent bending moment, M_e, but the use of either will give the same result.

101. *Shaft Coupling.*—The several lengths of shafting are connected together by means of couplings and coupling bolts. When an engine has two or four cranks the complete crank shaft is usually in two sections, and the number of bolts in the coupling is then an even number. When three cranks are used the complete crank shaft is usually in three interchangeable sections, and the number of bolts in the couplings is then a multiple of three, six being the number commonly used.

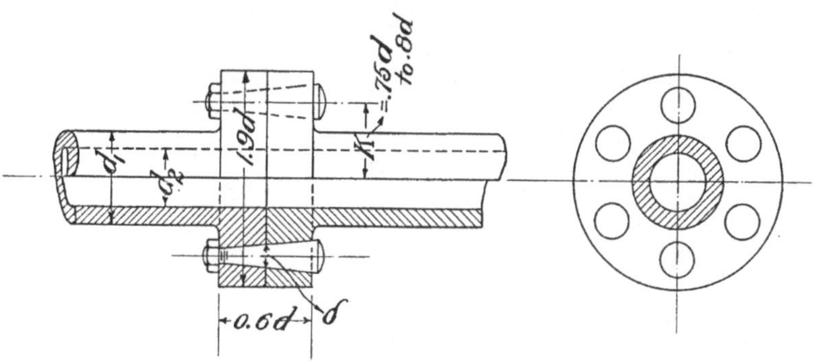

FIG. 101.—Shaft Coupling.

Fig. 101 shows the usual proportions of a shaft coupling the unit d being the diameter of the solid equivalent shaft, such that

$$d = \sqrt[3]{\frac{d_1^4 - d_2^4}{d_1}}.$$

To find the diameter δ of the coupling bolts proceed as follows:

Let H. P. be the horse-power transmitted.
- n be the number of bolts.
- K be the *radius* of the pitch circle.
- N be the number of revolutions per minute.
- A be the area of section of each bolt.
- f_s be the shearing strength of the material of the bolt.

The turning moment on the shaft is, then, $T = \frac{H.P. \times 33{,}000 \times 12}{2\pi N}$, see Art. 97, equation (123). This turning moment produces a force F acting at the arm K, giving a moment$=F\times K$; this latter moment is equal to the turning moment.

$$\therefore T=F\times K. \qquad (132)$$

But this force F is a shearing force acting on the bolts, whose combined area is $n\times A$.

$$\therefore n\times A\times f_s=F. \qquad (133)$$

Whence

$$T=n\times A\times f_s\times K, \qquad (134)$$

or

$$A=\frac{T}{n\times f_s\times K};$$

but

$$A=\frac{\pi\delta^2}{4},$$

$$\therefore \delta=\sqrt[2]{\frac{4T}{\pi n f_s K}},$$

$$=\sqrt{\frac{4\times H.P.\times 33,000\times 12}{2\pi^2 N n f_s K}}. \qquad (135)$$

A simplified formula for finding the diameter of the bolts in terms of the diameter of the shaft will be given later.

The outside diameter of the coupling is large enough so that the corners of the nuts of the bolts do not project beyond the flange. Usually 1.9 to 2.0d will be found a suitable value.

After finding the diameter, as shown above with an assumed value of K of from .75d to .8d, it may be found that the corner of the nut will not turn clear of the surface of the shaft. This is not likely to occur, however, as it is the usual custom to turn down the threaded end of the bolt to less diameter than its general body. This can safely be done, as the bolt is in shear only, and the only function of the nut is to draw the sections of the shaft closely together. In the case of the line and propeller shafts, however, in backing the thrust of the propeller produces tension in the bolts, and consequently there is a force tending to strip the threads of the nuts, which must be taken into consideration.

The diameter δ is the bolt diameter at the joint between the two coupling disks. The bolts are given a taper of $\frac{1}{4}$ inch on the diameter per foot length of bolt.

Questions and Problems.

Explain the general theory of torsion. Deduce an expression for turning moment in terms of Z_p and f_s.

Deduce an expression for the twist angle of a shaft in terms of T, l, G and d.

Show how to obtain the limiting values of f_s in order to keep the twist of the shaft within given limits, say 1° in a length of 20 diameters.

Show how to deduce an expression for turning moment given I. H. P. and N, the revolutions per minute; thence deduce an expression for the diameter of shaft necessary to transmit a given H. P. in terms of H. P., revolutions per minute, and f_s.

What is the difference between a journal and a bearing. Deduce an expression for the diameter of an end, or over-hung journal in terms of load, length of journal and f. What are suitable values of f, for steel, wrought iron, cast iron? Show how the bending moment may be shown graphically.

Deduce an expression for the diameter of an end, or over-hung journal, in terms of load, length of journal, and f, when the ratio of l to d is fixed.

Deduce an expression for the diameter of a "neck" journal, such as the cross-head pin when fixed in the forked end of a connecting rod.

A shaft is subject to torsion and bending at the same time. What stresses are produced in the material? Given the expression, $p_{max}=\frac{1}{2}f_t+\sqrt{\frac{1}{4}f_t^2+f_s^2}$, deduce an expression for the equivalent turning moment T_e in terms of T (turning moment) and M (bending moment).

Make a sketch of a shaft coupling and show how to find the diameter of the coupling bolts, to transmit a given horse-power.

Problems.

1. Find the diameter and length of the over-hung crank pin of an engine of the following principal dimensions: Diameter of cylinder = 45 inches; stroke = 42 inches; initial pressure = 50 pounds per gauge; length of connecting rod = 12 feet; R. P. M. about 50; wrought-iron pin working in composition brasses.

NOTE.—Make two calculations, first for strength, then check for maximum allowable pressure from table (slow engine). Select ratio of l to d and value of f from text.

2. Diameter of steam piston=20 inches; steam pressure=160 pounds per gauge; find the diameter of the cross-head pin, which is fixed in the forked end of the connecting rod. Select proper pressure per square inch of projected area of pin, ratio of l to d, etc., from text.

3. Diameter of steam piston=20 inches; steam pressure=160 pounds per gauge; cross-head journal consists of two over-hung pins (usual naval type). Select proper data from text, and assume that $\frac{2}{3}$ of the total load may come on either pin. Find the dimensions of the pins and mark them on a sketch of the cross head.

4. Calculate a table, the first column to show values of $\frac{M}{T}$ advancing by tenths from 0.1 to 1.0; the second column to show the *equivalent* simple turning moment in terms of the turning moment T; the third column to show the *equivalent* simple bending moment in terms of the bending moment M. (Suggestion: Let $\frac{M}{T}=x$ or $M=Tx$, and let $\frac{M}{T}=\frac{1}{y}$ or $T=My$.)

5. A solid steel shaft transmits 10,000 H. P. at 125 R. P. M. Find the mean turning moment and the diameter of the shaft.

6. A solid shaft 12 feet long transmitting 300 H. P. at 200 R. P. M., is loosely supported at each end. Weight of shaft is 520 pounds (consider this weight as a distributed load producing bending). Find the greatest equivalent turning moment in inch-pounds.

7. The I. H. P. of an engine=5000; R. P. M.=120; stroke=48 inches; distance between main bearings=55 inches. Find (1) mean turning moment in inch-pounds; (2) maximum turning moment=1.4 mean; (3) maximum turning *force* on after crank pin $\left(\text{suggestion: force}=\frac{\text{moment}}{\text{arm}}\right)$; (4) diameter of crank pin for bending strength, $f_t=10{,}000$; (5) dimensions of crank pin, ratio l to d=1.14 and allowable pressure per square inch of projected area=650 pounds; (6) from equivalent turning moment and $f_s=8000$, find diameter of after shaft journal; ratio of axial hole$=\frac{1}{1.7}$.

8. The pitch circle of the bolts of a flanged coupling has a diameter $1\frac{1}{2}$ times that of the shaft. I. H. P.=7000; R. P. M.=130; f_s=8000; number of bolts=8. Find the diameter of each bolt.

9. The diameter of a shaft transmitting 5000 I. H. P. at 130 R. P. M. is 16 inches, with an axial hole diameter=8 inches. There are 8 bolts in the coupling, f_s for bolts=8000 pounds per square inch. Make a neat, fully dimensioned sketch of the coupling.

10. A shaft, 4 inches diameter and 30 feet long, is found to twist 6.2° when transmitting power at 130 R. P. M. G=12,000,000. What horse-power is being transmitted?

CHAPTER XIV.

Friction and Lubrication of Bearings.

GENERAL EFFECT. COMPARISON BETWEEN DRY AND LUBRICATED SURFACES. NOMINAL AREA OF BEARING. WORK OF FRICTION. VALUE OF μ. PRESSURE IN BEARING. FORCED LUBRICATION. ANTI-FRICTION METAL. THRUST BEARING.

102. *The General Effect of Friction and Lubrication.*—The essential condition for a properly working bearing is that it should run cool and not seize. It is impossible to have a frictionless journal in practice, and friction always produces heat, and, consequently, a rise of temperature. This rise of temperature, in well designed and carefully lubricated bearings, reaches a certain point at which the rate of generation of the heat by friction is equal to the rate at which the heat is conducted away, when the temperature remains constant. This temperature is the so-called "working heat," and so long as it is moderate the journal will work properly. The amount of friction is reduced to a minimum by supplying the bearing with a lubricant, which forms a film between the journal and its bearing. This not only greatly reduces the heat generated, but it also greatly reduces the amount of wear.

If, from any cause, a rise of temperature above the normal occurs, the first effect is generally a reduction of friction by making the lubricant more fluid, but this reduction of the viscosity of the oil allows it to be squeezed out of the bearing by the pressure between the surfaces, and the friction then takes place between metal and metal, and seizing occurs. The heat now becomes so great as almost to weld the surfaces together, and pieces are cut out of both the journal and the brasses, causing deep scores in each. The load at which seizing occurs depends upon the smoothness of the surfaces, and the nature of the materials, but mostly on the viscosity of the oil. If the viscosity can be kept up by artificial means, such as water circulation, very high pressures may be safely carried. A case is recorded by Dr. Goodman of a journal running continuously at a surface velocity of 230 feet per minute, with a load of *two tons* per square inch, the temperature being kept at 110° F.

The brasses of the main bearings of naval engines are made hollow, and are fitted for the circulation of water through them, which greatly assists in conducting away the heat of the friction, and thus prevents the oil from becoming too fluid.

103. *Comparison of Laws of Friction between Dry Surfaces and Well-Lubricated Surfaces.* (From Goodman, "Mechanics Applied to Engineering.")—The laws which express the behavior of well-lubricated surfaces are practically the reverse of those for dry surfaces, as is shown by the following parallel columns:

DRY SURFACES.	LUBRICATED SURFACES.
1. The frictional resistance is nearly proportional to the normal pressure between the surfaces.	1. The frictional resistance is almost independent of the pressure with bath lubrication, and approaches the behavior of dry surfaces as the lubrication becomes more meager.
2. The frictional resistance is nearly independent of the speed for low pressures. For high pressures it tends to decrease as the speed increases.	2. The frictional resistance varies directly as the speed for low pressures. But for high pressures the friction is very great at low velocities, becoming a minimum at about 100 ft. per minute and afterwards increases approximately as the square *root* of the speed.
3. The frictional resistance is not greatly affected by the temperature.	3. The frictional resistance depends more upon the temperature than on any other condition—partly due to the variation in the viscosity of the oil, and partly to the fact that the diameter of the bearing increases with the rise of temperature more rapidly than the diameter of the shaft, and thereby relieves the bearing of side pressure.
4. The frictional resistance depends largely upon the nature of the material of which the rubbing surfaces are composed.	4. The frictional resistance with a flooded bearing depends but slightly upon the nature of the material of which the surfaces are composed, but as the lubrication becomes meager, the friction follows much the same laws as in the case of dry surfaces.

Dry Surfaces.	Lubricated Surfaces.
5. The friction of rest is slightly greater than the friction of motion.	5. The friction of rest is enormously greater than the friction of motion, especially if thin lubricants are used, probably due to their being squeezed out when standing.
6. When the pressure between the surfaces becomes excessive, seizing occurs.	6. When the pressure between the surfaces becomes excessive, which is at a much higher pressure than with dry surfaces, the lubricant is squeezed out and seizing occurs. The pressure at which this occurs depends upon the viscosity of the lubricant.
7. The frictional resistance is greatest at first and rapidly decreases with the time after the two surfaces are brought together, probably due to the polishing of the surfaces.	7. The frictional resistance is least at first, and rapidly increases with the time after the two surfaces are brought together, probably due to the partial squeezing out of the lubricant.
8. The frictional resistance is always greater immediately after reversal of direction of sliding.	8. Same as in the case of dry surfaces.

The well-known expression for the friction of sliding is

$$F=\mu P. \tag{136}$$

Where F is the resistance to sliding, or the friction, P is the total pressure between the surfaces, and μ is a constant called the *coefficient of friction.* This same expression will hold for the frictional resistance of lubricated bearings, but in this case μ is not the coefficient of friction of dry sliding, but is a constant depending upon the velocity of rubbing and the intensity of pressure between the surfaces.

104. *The Nominal Area of Bearing.*—It will be shown further on how the actual pressure on a journal varies from point to point, being a maximum at the crown and least at the two sides. For practical calculations, however, the pressure is assumed to be evenly distributed over the projected area of the bearing. Thus, in Fig. 102, d=diameter of bearing; l=length; and the projected area is $d\times l$. If P=the total load on the bearing, then the pressure per square inch of projected area is

$$p=\frac{P}{dl}. \tag{137}$$

105. *The Work Expended in Friction.*—If a shaft or journal makes N revolutions per minute, its surface velocity is

$$\frac{\pi dN}{12\times 60} \text{ feet per second.} \tag{138}$$

(12 inches = 1 foot; 60 seconds = 1 minute; d in inches.)

From equation (136) we have resistance, or force, $F=\mu P$, but $F\times$dist. = work; hence work of friction $= F\times$dist. $= F\times\frac{\pi dN}{12\times 60}$ foot-pounds per second, or

$$\text{Work}=\mu P\frac{\pi dN}{12\times 60} \text{ foot-pounds per second,} \tag{139}$$

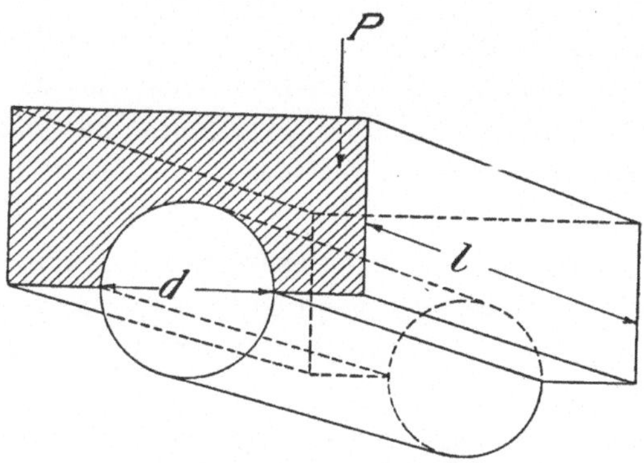

FIG. 102.—Area of Bearing.

from which the work lost in friction can be found, provided the value of μ is known.

106. *The Value of μ for Journals.*—This value has been determined by numerous elaborate and carefully conducted experiments, the most generally accepted being those by Mr. Beauchamp Tower, with oil-bath lubrication (that is to say, with practically perfect lubrication) from which it is found that

$$\mu=c\frac{\sqrt{v}}{p}, \tag{140}$$

where c = a constant; v = velocity of journal surface in feet per *second;* p = pressure per square inch of projected area of journal in pounds.

In a general way the experiments consisted of a very great number of trials at varying speeds and pressure, during which the friction was measured, and the value of μ calculated from the relation given by equation (136); μ, v and p being thus known, the value of c is calculated from equation (140).

The mean values of c thus obtained for different lubricants are given as follows:

TABLE 18.

Lubricant	Olive oil.	Lard oil.	Sperm oil.	Mineral oil.	Mineral grease.
Mean value of c	.289	.281	.194	.276	.431

During these experiments the pressures at which seizing occurs with the different lubricants were also determined, and were found to vary from about 300 to 600 pounds per square inch, according to the viscosity of the lubricant. It is from such experiments as these, modified by the results of practical experience, that tables of limiting bearing pressures, such as that given in Art. 99, Table 17, are obtained.

It must be understood that the above values of c are for practically perfect lubrication, such as oil bath, or forced systems. When the lubrication is not so perfect the friction is much higher and is generally very erratic.

From the above, then, we have the frictional resistance with perfect lubrication,

$$F=\mu P=cP\frac{\sqrt{v}}{p},$$

but

$$\frac{P}{d\times l}=p,$$

or

$$\frac{P}{p}=d\times l;$$

hence

$$F=c\times d\times l\times \sqrt{v}, \qquad (141)$$

where d = diameter of journal; l = length of journal, both in inches; v = velocity of journal surface in feet per *second;* and c = a constant depending on the lubricant, values of which are given in Table 18.

For a more detailed discussion of this subject see Unwin, "Elements of Machine Design."

107. *The Distribution of Pressure in a Bearing.*—In Beauchamp Tower's experiments, oil-bath lubrication was used in order to insure a regular and uniform supply of oil. It was found that the condition of the journal and brass was, as shown, greatly exaggerated by Fig. 103. The oil is carried in at the "on" side between the journal and the brass and forms a continuous film, being carried out again at the "off" side. The brass is completely oil-borne by this film, which supports the whole load on the bearing.

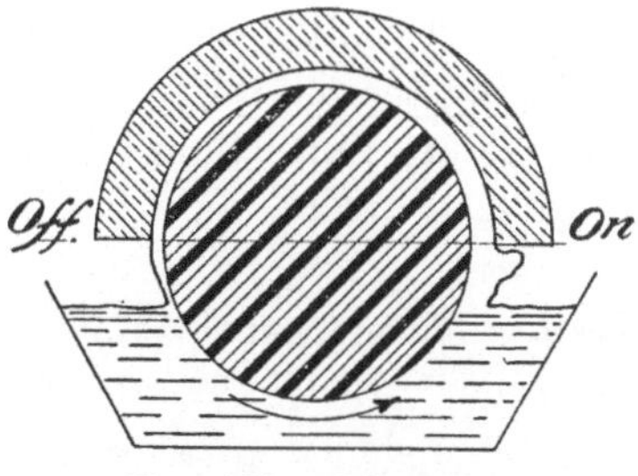

FIG. 103.—Oil Film.

The film is thinnest, and the pressure greatest, along a line parallel to the axis of the journal, but on the "off" side of the center line at the top.

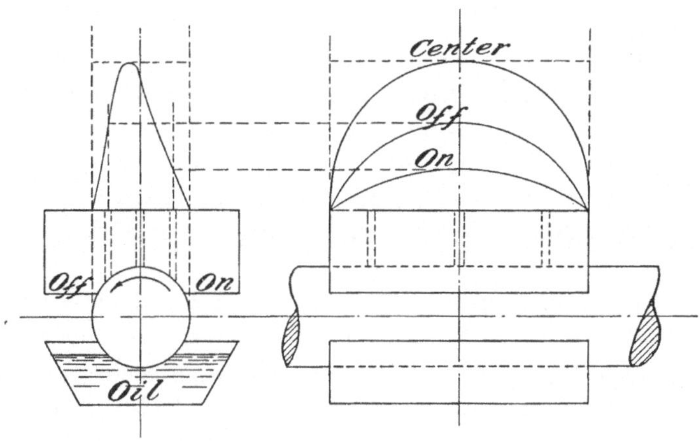

FIG. 104.

Fig. 104 shows the distribution of the film pressure, the pressure being measured at the several points indicated. No attempt is made to draw Fig. 104 to scale, as it is intended only to show in a general way the relative pressures at the different points.

It will be seen that it is impossible to feed oil into a bearing carrying a load constant in direction by means of a hole in the top of the cap brass, as this will lead the oil close to the point of greatest pressure. In fact, in such a case the oil instead of entering, will be forced out. The oil must be led to points near the sides of the brass. If the load changes its direction in each revolution, as is the case with the bearings of marine engines, the lubrication is very good even with imperfect methods of introducing the oil. The journal or pin pressure then acts alternately on the top and bottom brasses, and the oil readily distributes itself to each during the time the pressure acts on the other.

108. *Forced Lubrication.*—The engines of our naval ships are now being fitted with so-called "forced lubrication," the oil being pumped into the main bearings under a moderate pressure; the oil then going through holes in the main journals to the hollow of the shaft, thence through channels drilled in the crank webs, to the hollow of the crank pins and through holes in the pins to the surface; and then, further, in some cases, through pipes connected to the wrist pins. The surplus oil escapes at the ends of the bearings and falls into the crank pits, whence it is pumped through strainers, coolers, etc., and used over again. In this system the film pressure in the bearings is greatly in excess of the pressure under which the oil is supplied, but the oil gets into the bearings on the relief of pressure on the return stroke, and it is not driven out in the short period required by the forward stroke.

This system is commonly called "forced lubrication," but more properly it is "flooded lubrication." In forced lubrication, properly speaking, the oil is supplied under a pressure greater than the "film" pressure.

109. *Anti-Friction Metals.*—Closely connected with the question of lubrication, as it affects the practical engineer, is the question of the lining metal for bearings. If the lubrication could always be counted on to be perfect so that the load would always be fluid-borne, the nature of the metal would be of no importance. In practice, however, such is far from being the case, and the engineer must give careful consideration to the conditions under which a bearing is to work before deciding on the material to use. Thus if the bearing is to run at a slow speed with a light load almost any material can be used. If the bearing is to work at a high speed, or with a heavy load, or both, then the bearing metal must be carefully selected. If the bearing is subject to blows a hard, tough metal must be used, but if the bearing has a steady constant load a soft white metal is the best possible.

For naval engines a rather hard white metal is used, but care must be taken that the metal is not so hard as to be brittle at the same time.

The relative advantages and disadvantages of white metal linings for bearings are summarized as follows by Dr. Goodman:

SOFT WHITE METALS FOR BEARINGS.

ADVANTAGES.	DISADVANTAGES.
The friction is much lower than with hard bronzes, cast iron, etc., hence is less liable to heat.	Will not stand the hammering action that some shafts are subject to.
The wear is very small indeed after the bearing has once got well bedded (see disadvantages).	The wear is very rapid at first if the shaft is at all rough; the action resembles that of a new file on lead. At first the file cuts rapidly, but it soon clogs, and then ceases to act as a file.
It rarely scores the shaft, even if the bearing heats.	It is liable to melt out if the bearing runs hot.
It absorbs any grit that may get into the bearing, instead of allowing it to churn round and round, and so cause damage.	If made of unsuitable material it is liable to corrode.

110. *Thrust Bearing.*—Consider first a flat pivot as illustrated by Fig. 105.

The total thrust P is assumed to be evenly distributed over the whole surface of the end of the shaft; then the pressure in pounds per square inch is

$$p = \frac{P}{\pi R^2}. \tag{142}$$

The pressure on an elementary ring whose radius is r and width dr, is

$$2\pi r dr p; \tag{143}$$

the friction of this ring is

$$2\pi r dr p \times \mu, \tag{144}$$

and the work of friction, if N = revolutions per minute is

$$2\pi r dr p \mu \times 2\pi r N \text{ inch-pounds.} \tag{145}$$

The work of friction over the whole surface is

$$4\pi^2 p\mu N\int_0^R r^2 dr = \frac{4\pi^2 p\mu NR^3}{3}. \qquad (146)$$

Substituting the value of p from equation (142) gives

$$\frac{4\pi^2 P\mu NR^3}{3\pi R^2} = \tfrac{4}{3}\pi\mu NPR = \tfrac{2}{3}\pi\mu NPD,$$

or the work in foot-pounds per minute is

$$\frac{2\pi\mu NPD}{3\times 12} = \frac{\mu PDN}{5.73} \qquad (147)$$

or the horse-power absorbed by the pivot friction is

$$\text{H.P.} = \frac{\mu PDN}{189,000}. \qquad (148)$$

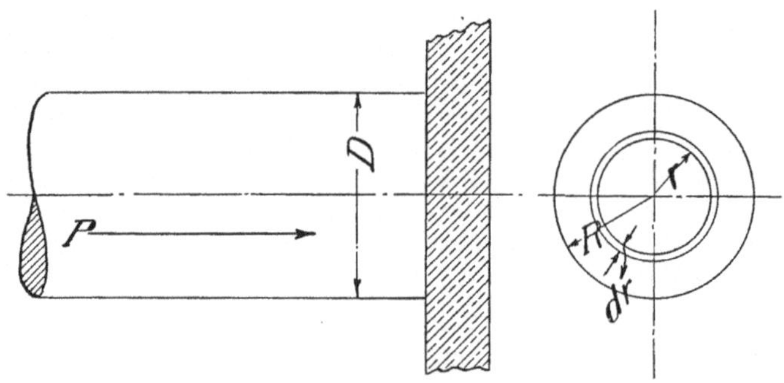

Fig. 105.

When the thrust cannot be taken on the end of the shaft, as is the case with the propeller thrust in marine engines, a number of thrust collars are provided as in the familiar thrust bearing. In some cases, where the thrust is not great, a single collar is used, but where the load is large, the single collar necessitates a large diameter in order to keep the thrust pressure within allowable limits. This increases the loss due to friction on account of the increased velocity of rubbing, and the difference of velocity at the outer diameter of the collar and that at the surface of the shaft causes unequal wear. In practice, therefore, the outer diameter of the collars is made not more than one and a half times the diameter of the shaft. The necessary area of thrust surface is obtained by using a number of collars.

In naval engines the thrust bearings are made of the horse-shoe type (see Barton's "Naval Machinery"), and each ring can be independently adjusted. Each ring has its own water circulation and oil feed, and the lower parts of the thrust collars on the shaft dip into a bath of oil, which is also cooled by circulating water. The horse shoes are faced with white metal.

The work of friction in the case of the collar-thrust bearing is obtained as follows:

Starting with equation (146) above, the limits of integration are between R_1 and R_2, see (Fig. 106).

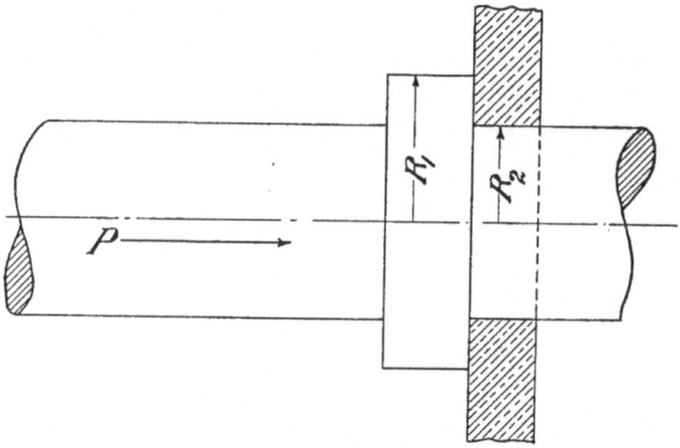

FIG. 106.—Thrust Collar.

The expression for the work of friction thus becomes

$$4\pi^2 p\mu N\int_{R_2}^{R_1} r^2 dr = \tfrac{4}{3}\pi^2 p\mu N(R_1^3 - R_2^3); \tag{149}$$

also equation (142) takes the form

$$p = \frac{P}{\pi(R_1^2 - R_2^2)}; \tag{150}$$

so for this case we have the horse-power absorbed

$$H.P. = \frac{\mu PN(R_1^3 - R_2^3)}{94538(R_1^2 - R_2^2)}. \tag{151}$$

The lubrication of such bearings is always very good, being bath lubrication in character, so the coefficient of friction is rather low, the value of μ being from .007 to .01,

$$\mu = .01$$

being a value commonly used in practice.

111. *Design of Thrust Collars.*—Example: Design the thrust journal for a steamer from the following data, and find the power lost in friction of the thrust bearing.

Speed knots	22
I. H. P. (each engine)	11500
Diameter of shaft	17¾″
Outside diameter of collars	27″
R. P. M.	120
Combined efficiency of engine, shaft and propellers	65%, say ⅔

It is evident that the thrust multiplied by the distance the ship advances in a minute must equal the work delivered to the thrust block per minute. This work, in foot-pounds per minute, $=\frac{2}{3}\times$ I. H. P. $\times 33{,}000$.

Let P = total forward thrust (each engine).

S = speed in knots per hour.

then

$$\frac{S\times 6080}{60}=\text{speed in feet per minute.}$$

$$\therefore \frac{2}{3}\times \text{I. H. P.}\times 33{,}000=\frac{P\times S\times 6080}{60},$$

or

$$P=\frac{\text{I. H. P.}\times 33{,}000\times 2\times 60}{3\times 6080\times S}=\frac{11{,}500\times 33{,}000\times 2\times 60}{3\times 6080\times 22}$$

$$=113{,}500 \text{ (by omnimeter).}$$

The area of each collar is

$$\frac{\pi}{4}(D_1^2-D_2^2)=\frac{\pi}{4}(\overline{27^2}-\overline{17.75^2})=\frac{\pi}{4}(729-315)=\frac{\pi\times 414}{4}=325.8.$$

The allowable pressure per square inch of thrust collar is 60 pounds (see Table 17) and the total pressure on each collar is, therefore, $325.8\times 60=19{,}548$ pounds. The number of collars is then

$$\frac{113{,}500}{19{,}548}=5.8, \text{ or say } 6.$$

But in the horse-shoe type of thrust bearing the horse shoes encircle only about one-half of the shaft, and, therefore, we must double the above number, giving 12.

Use 12 collars.

In naval practice the thickness of each collar is

$$t=\frac{D_1-D_2}{4}.$$

This is much greater than is required to resist the shearing force, but is used to allow for the possibility of the whole thrust coming on a few of the collars due to improper adjustment, and also to provide a sufficient space between adjacent horse shoes for the independent adjusting nuts.

From equation (151) the horse-power lost in friction on one thrust block is

$$\text{H. P.} = \frac{\mu PN}{189{,}000}\left(\frac{R_1{}^3 - R_2{}^3}{R_1{}^2 - R_2{}^2}\right) = \frac{.01 \times 113{,}500 \times 120}{189{,}000}\left(\frac{\overline{13.5}^3 - \overline{8.87}^3}{\overline{13.5}^2 - \overline{8.87}^2}\right)$$

$$= \frac{.01 \times 113{,}500 \times 120 \times 1761}{94{,}538 \times 103.8} = 24.45.$$

Questions and Problems.

Explain in general terms the effect of friction in bearings and how it is reduced and controlled. What is meant by "working heat" of a bearing? Explain the "seizing" of a bearing and how it may be prevented.

Explain the difference in the coefficient of friction for dry sliding and the corresponding constant for lubricated surfaces. What is the nominal area of bearing? Deduce an expression for the work of friction of a bearing, in terms of μ, load, diameter of journal and R. P. M. State the relation between μ, the velocity of rubbing and the intensity of pressure of a bearing.

Make a sketch showing the action of the oil between a journal and its brasses when lubrication is ample. Show also by a sketch the distribution of the oil pressure in a loaded bearing, both transversely and longitudinally. Where should the oil enter a bearing carrying a steady load? How and why does this differ from a crank-pin bearing?

Describe the naval system of forced lubrication.

Discuss the use of anti-friction metal.

Problems.

1. The diameter of the cylinder of an engine is 30.7 inches; mean effective pressure = 40 pounds per square inch; R. P. M. = 140; diameter of crank pin = 10″; length of crank pin = 10″. Perfect lubrication, using olive oil, find

(a) Rubbing speed per second.

(b) Pressure per square inch of bearing surface on pin.

(c) Value of μ.

2. An engine has a cylinder whose diameter=30.″7; mean effective pressure=40 pounds per square inch; R. P. M=140; stroke =24″; diameter of crank pin=10″; length of crank pin=10″. Lard oil is used with a system giving perfect lubrication. Find μ. Find the percentage of the total work lost in crank-pin friction. (Suggestion: Total energy in foot-pounds per minute=2pLAN.)

3. An engine, whose cylinder diameter=$38\frac{1}{2}$″; stroke=48″; diameter of crank pin=19″; length of crank pin=$22\frac{1}{2}$″; M. E. P. =118 pounds per square inch; R. P. M.=128, uses mineral oil with flooded lubrication. Find the value of μ; and the percentage of the total energy lost in crank-pin friction.

4. Find the number and thickness of the thrust collars and the horse-power lost in friction of the thrust bearing of a naval engine from the following data:

Speed in knots	18½
I. H. P. (each engine)	8250
R. P. M.	125
Diameter of thrust shaft—inches	15½

CHAPTER XV.

Notes on the Design of Crank Shaft, Cross-head Pins, etc.*

PRACTICAL PROBLEM V.

112. *Crank Shaft.*—In calculating the dimensions of a crank shaft for strength, the theoretical aggregate mean effective pressure is found and the actual pressure expected is determined by comparison with the results given by the trials of engines of the same type, pressure and rate of expansion. In other words, a value for the efficiency of the valve gear, or the percentage of the actual to the theoretical mean pressure must be obtained (Seaton).

This efficiency is obtained as follows: Let r=total rate of expansion$=\dfrac{\text{area of L. P. cylinder}}{\text{area of H. P. cylinder}\times\text{cut-off in H. P. cylinder}}$.

Then the total mean pressure P_m is found from

$$P_m = p_1 \times \frac{1+\log_e r}{r} \qquad (152)$$

or, by using table, Barton, p. 522.

$$P_e = P_m - p_o. \qquad (153)$$

This calculated value of P_e is compared with the actual value found from the trial trip, and the efficiency is found from

$$x = \frac{\text{actual}}{\text{calculated}} P_e.$$

The above method may be used when the I. H. P. is not previously known, but in all practical cases the I. H. P. is the first thing determined, and all detail calculations are based on it.

113. *Bearings.—Pressure per Square Inch of Projected Area with Regard to Heating: Seaton.*—The bearing surface of *crank pins* must be such that the pressure per square inch does not exceed 500 pounds, and, in the case of merchant ships where room will permit of larger pins, 400 pounds should not be exceeded. When the brass is recessed so that it bears only on parts of the shaft, the actual bearing surface should not be exposed to more than 600 pounds per square inch. *Main bearings* of screw engines, when room admits,

* From "Notes on Machine Design," revised and enlarged.

should be of such a size that the pressure does not exceed 200 pounds, measuring the whole of the bearing, or 300 pounds on the actual bearing surface.

It is found that the channel ways take up about 20% of the bearing surface, leaving about 80% as the actual bearing surface.

The length of crank pins is from 1 to $1\frac{1}{2}$ times the diameter of the pins. Vertical engines usually have space for a length of bearing equal to $1\frac{1}{2}$ diameters. The length of each main journal is from 1 to $1\frac{1}{2}$ times the diameter of the shaft. Vertical engines have usually a length of bearing equal to $1\frac{1}{2}$ diameters.

Foley's Engineers' Reference Book.—"No hard and fast rule can be laid down. The engineer must be guided by circumstances and experience. The velocity with which the surfaces move on each other must be considered, and also whether the direction of the pressure alternates. In the case of a cross-head pin, the motion is not only slow, but the pressure alternates as well; hence the high pressure permissible. The following may be a rough guide:

"Cross-head pins, 750 pounds; not to exceed 1000 pounds. Crank pins, 300 pounds; not to exceed 500 pounds. Main bearings, 250 pounds; not to exceed 500 pounds."

The Bureau of Steam Engineering uses the following formulæ as guides:

For crank-shaft journals: $p\sqrt{v}<8000$ (1). Where p=pressure per square inch and must not exceed 500 pounds; v=velocity of surface in feet per minute.

For crank pins: $p\sqrt{v}<15{,}000$ (2) and p must not exceed 700 pounds.

114. *Hollow Shafts and Pins.*—The diameter of the equivalent hollow shaft may be calculated by the equation $d^3=\dfrac{d_1^4-d_2^4}{d_1}$. (See Art. 95.)

Foley's practice is:

When dia. of hole=.4 of d add 1% to dia. of solid shaft.

"	=.5	"	2	"
"	=.6	"	5	"
	=.7	"	10	

This brings very nearly the same result as solving the equation above, which must be done by assuming some ratio between the diameters of shaft and bore hole or by assuming one diameter.

For convenience of interpolation the above table from Foley has been plotted as a chart, which is used as follows: Having first found the diameter of the *solid* shaft necessary and having fixed on the diameter of the hole through the axis of the shaft, find the decimal fraction from the ratio diameter of hole to diameter of *solid* shaft, this decimal is shown at the left hand edge of the chart; run across horizontally to the curve; drop down vertically to the bottom edge of the chart, where will be found the percentage to be added to the diameter of the solid shaft to give the diameter of the equivalent hollow shaft.

Graphic approximate solution of $d = \sqrt[3]{\dfrac{d_1^4 - d_2^4}{d_1}}$ to be used instead of Foley's practice:

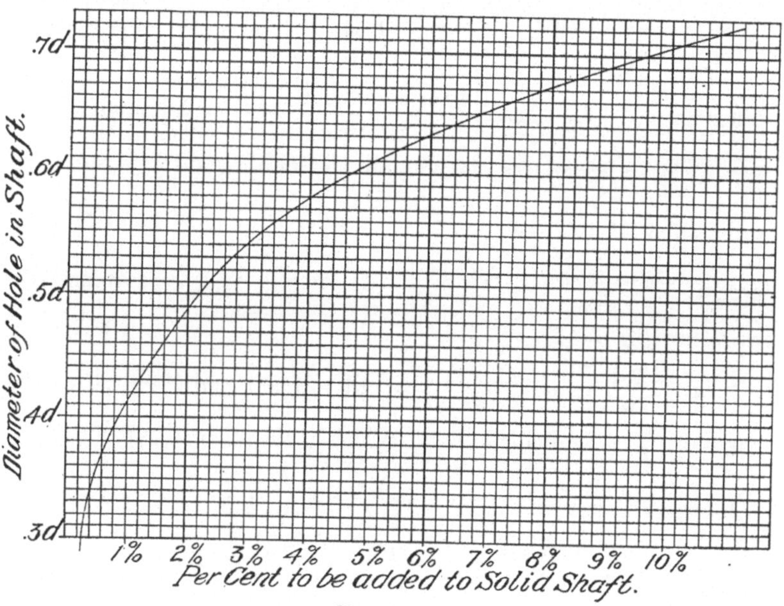

FIG. 107.

The Bureau of Steam Engineering makes the diameter of the crank pins one inch larger than that of the shaft journals. The length of the pins is calculated so that a certain pressure, about 400 pounds per square inch, is not exceeded; this length being generally $1\frac{1}{4}$ times the diameters of the pins.

115. *Calculations.*—The engine for which the following calculations are made is that of the U. S. S. "Raleigh."

Data: Vertical, inverted cylinder, direct-acting, triple-expansion, twin-screw engines. Cylinder diameters, 36 inches, 53 inches, 80⅝ inches. Stroke, 33 inches. Revolutions per minute, 164. Boiler pressure 160 pounds, by gauge. First receiver pressure, absolute, 60 pounds (assumed). Vacuum, 24 inches. Cut-off in H. P. cylinder, .7 stroke, I. H. P., 9863.2.

$$\frac{\text{Max. T. M.}}{\text{Mean}}=1.5; \text{ dist. between bearings, } 53''. \quad f=9000.$$

116. *Crank Shaft.*—For one set of engines

$$\text{I. H. P.}=4931.6.$$

$$\text{Mean T. M. of after engine}=\frac{\text{I. H. P}\times 33{,}000\times 12}{2\pi\times \text{Rev.}}$$

$$=\frac{\text{I. H. P.}}{\text{Rev.}}\times 63{,}000.$$

$$\text{Maximum T. M.}=\frac{4931.6\times 63{,}000\times 1.5}{164}=2{,}840{,}000 \text{ and}$$

$\overline{\text{Max. T. M.}^2}=T^2=8{,}075{,}000{,}000{,}000$ (by omnimeter).

M, the bending moment, $=\frac{Wl}{8}$, where W = maximum turning force $=\frac{\text{Max. T. M.}}{\text{crank}}=\frac{2{,}840{,}000}{16.5}=172{,}200$ and l = distance between bearings = 53″.

$\therefore M=\frac{Wl}{8}=\frac{172{,}200\times 53}{8}=1{,}140{,}000$, and $M^2=1{,}302{,}000{,}000{,}000$. T_e = equivalent T. M. $=M+\sqrt{M^2+T^2}=4{,}202{,}000$.

$$T_e=f^2{}_p=f\,\frac{\pi d^3}{16} \quad \therefore \text{Dia. of shaft}=d=\sqrt[3]{\frac{T_e\times 16}{\pi\times f}}$$

$$=\sqrt[3]{\frac{T_e\times 5.1}{f}}=\sqrt[3]{\frac{4{,}202{,}000\times 5.1}{9000}}=13''.35.$$

This is for a solid shaft. The shaft in question is to have a 6-inch axial hole. Referring to Foley's practice, as above:

Dia. of hole $=\frac{6}{13.35}\%=45\%$ of dia., nearly. By interpolating in the table or by the use of the chart we get 1.5% increase for $d_2=.45\,d_1$. Then $13.35\times 1.016=13''.56$, or dia. of shaft, by calculation for strength, $=13\frac{9}{16}''$.

Length of Main Journals.—In order to obtain this, we must find first the thrust R on the connecting rod. This is calculated for the H. P. cylinder.

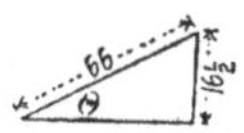

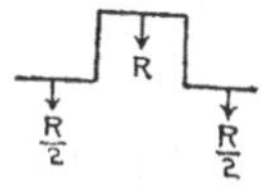

Effective load on H. P. piston$=P=$area$\times(p_1-p_r$, or initial abs. press.$-$rec. press.$)=1018\times(175-60)=117,100$. Then $R=P\sec(\sin^{-1}\frac{1}{4})=121,000$ pounds.

Then $\frac{R}{2}$, the thrust on each bearing, $=\frac{121,000}{2}=60,500$ pounds. From notes on "Bearings," allowing a pressure of 190 pounds per square inch of projected area.

The projected area required$=\frac{60,500}{190}=318.1=l\times d$.

But $d=13\frac{9}{16}$ whence $l=\frac{318.1}{13\frac{9}{16}}=23.45$.

From this, $\frac{l}{d}=\frac{23.45}{13\frac{9}{16}}=1.73$, which is too large, as there is seldom fore and aft length allowed for such a coefficient. Therefore, the dimensions obtained must be changed as follows:

$$l=1.5d \therefore l\times d=1.5d^2=\text{projected area.}$$

$$\therefore 1.5d^2=318.1 \text{ and } d=14.57, \text{ or } 14\tfrac{9}{16}''.$$

$$\therefore l=14\tfrac{9}{16}\times1.5=21.82, \text{ or } 21\tfrac{13}{16}''.$$

The projected area is, then, $14\frac{9}{16}\times21\frac{13}{16}=316.5$ square inches. The pressure per square inch in this area$=\frac{60,500}{316.5}=191$ pounds, which is within the limit of 200 pounds. The area of the actual bearing surface$=316.5\times.8=253.5$ square inches, and the pressure per square inch of this area$=\frac{60,500}{253.5}=238.4$ pounds, which is within the limit of 300 pounds.

The Bureau of Steam Engineering's check for speed: velocity $=v=\frac{\pi d\times164}{12}$ where $d=$diam. main bearing in inches, 164 is revolutions per minute$=\frac{\pi\times14\frac{9}{16}\times164}{12}=626$. $\therefore\sqrt{v}=25.02$ ft.

p, the pressure per square inch of projected area (from above), $=191$ pounds. Then $p\sqrt{v}=191\times 25.02=4780$, which is within the limit of 7500.

For the dimensions of the main bearings, we have, then:

$$\text{Diameter}=14\tfrac{9}{16}''; \ \text{length}=21\tfrac{13}{16}''.$$

117. *Crank Pins.*—All crank pins are made of the same dimensions as the after pin, which is calculated for bending moment. To determine the bending moment on the after pin, consider the turning force of the H. P. and I. P. engines to be transmitted directly back to the forward end of the L. P. pin; or, what is the same thing, consider two-thirds of the turning force of the

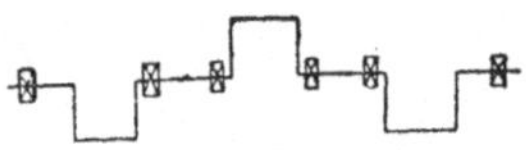

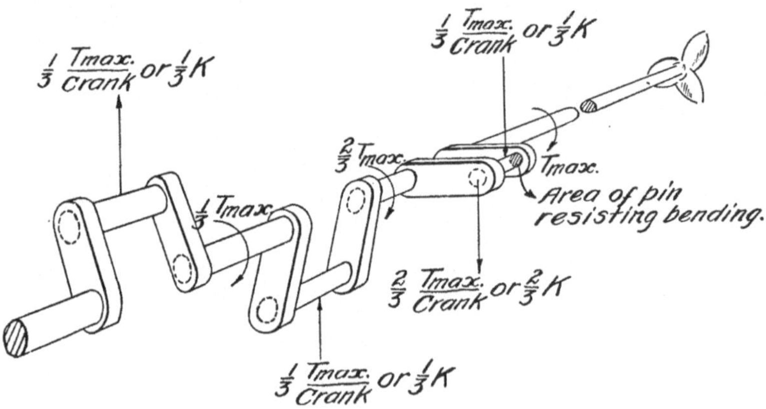

FIG. 108.

three engines transmitted to this point. Then, the pressure or load transmitted to the forward end of the L. P. pin is equal to two-thirds of the maximum twisting moment, divided by the length of the crank. Let $\frac{\text{Max T. M.}}{\text{crank}}$ or $\frac{T_{max}}{\text{crank}}=K$ and length of pin$=l$.

Then, $K=\frac{2,840,000}{16.5}=172,200$. Since the transmitted force acts on the forward end of the pin, its bending moment$=\frac{2}{3}Kl$. In addition to this load on the forward end, the pin takes the bending moment of its own engine, which is $\frac{\frac{K}{3}\times l}{2}$, since the pin is fixed at the after end and carries a load $\frac{K}{3}$ uniformly distributed.

That is to say, the turning effort of the H. P. and I. P. engines carries around the forward web of the L. P. crank, so that it offers no support to that end of the pin in resisting bending. The resistance of the propeller tends to prevent the free swing of the after web, so in effect the whole bending moment on the L. P. pin tends to break it off at its junction with the after web.

Hence the total bending moment on the pin is

$$M=2\frac{Kl}{3}+\frac{K}{3}\times\frac{l}{2}=\tfrac{5}{6}Kl.^{*} \qquad (154)$$

Substituting this bending moment in the formula for a beam subject to bending gives

$$\tfrac{5}{6}Kl=fZ=\frac{\pi d^3 f}{32}=\frac{d^3 f}{10.2},$$

$$\therefore d=\sqrt{\tfrac{5}{6}K\times\frac{l}{d}\times\frac{10.2}{9000}}. \qquad (157)$$

From notes on "Bearings" $\frac{l}{d}$ is from 1 to 1.5. Take it as 1.25, since naval engines seldom have enough fore and aft space for the ratio 1.5. Then

$$d=\sqrt{\frac{5\times 172{,}200\times 1.25\times 10.2}{6\times 9000}}=14.25=14\tfrac{1}{4}''.$$

Then, $l=1.25d=1.25\times 14.25=17.80$, or $17\tfrac{13}{16}''$.

These dimensions are for solid pins. For hollow pins apply the table given above.

Dia. of hole $=\frac{6}{14.25}\%=42\%$ of d and 14.25×1.02 (interpolating in table or using chart) $=14.55$, or $14\tfrac{9}{16}''$, = dia. of hollow pin.

Also, length of hollow pin $=14\tfrac{9}{16}''\times 1.25=18.2$, or $18\tfrac{7}{32}''$.

These are the dimensions of the pins for strength.

* In the case of the four-cylinder triple expansion engine with two L. P. cylinders, the bending moment on the after crank pin (which is always a L. P. pin) becomes—

(a) When the two L. P. engines *together* develop ⅓ of the total power of the engine

$$M=\tfrac{1}{6}Kl+\tfrac{2}{3}Kl+\frac{\tfrac{1}{6}Kl}{2}=\tfrac{11}{12}Kl \qquad (155)$$

(b) When each of the four cylinders develop ¼ of the total power

$$M=\tfrac{3}{4}Kl+\frac{\tfrac{1}{4}Kl}{2}=\tfrac{7}{8}Kl \qquad (156)$$

Check for Heating.—Pressure per square inch of projected area must not exceed 500 pounds. Assume a pressure of 450 pounds per square inch, and we have:

Projected area of pin $= \frac{\text{load or thrust R on pin}}{450} = \frac{121,000}{450} = 268.8$ square inches.

Referring to the calculated dimensions for the crank pins, $14\frac{9}{16}'' \times 18\frac{7}{32}''$, the projected area $= 265$ square inches. This area, therefore, though ample for strength, is too small to prevent heating, for which an area of 268.8 square inches is required. Therefore, the dimensions of the pin must be increased:

Then $(d \times 1.25)\,d = 268.8$. $\therefore d = 14''.67$, or $14\frac{21}{32}''$, and $l = 1.25d = 18''.3$, or $18\frac{5}{16}''$; and the projected area $= 14\frac{21}{32}'' \times 18\frac{5}{16}'' = 268.5$ square inches. The pressure per square inch on this area $= \frac{121,000}{268.5} = 450$ pounds.

The actual projected area of bearing surface is, then, $268.5 \times .8 = 214.8$ square inches; and the pressure per square inch on this area $= \frac{121,000}{214.8} = 563$ pounds, which is inside the limit of 600 pounds.

Bureau of Steam Engineering's check for speed: From notes on "Bearings," equation (2):

$p\sqrt{v} < 15,000$, p not to exceed 700 pounds.

$$\sqrt{v} = \sqrt{\frac{\pi \times d \times 164}{12}} = \sqrt{\frac{3.14 \times 14\frac{21}{32} \times 164}{12}} = 25.1 \text{ feet},$$

the slight difference in v, owing to the angularity of the connecting rod, being left out of the calculation.

Also, from above, $p = 563$ pounds; then, $p\sqrt{v} = 563 \times 25.1 = 14,300$, which is within the limit of 15,000.

It is seen, then, that the dimensions found fulfill all conditions, the changes in diameter keeping the hollow pin amply large for strength.

The dimensions of the crank pins are, then: Diameter $= 14\frac{21}{32}''$; length $= 18\frac{5}{16}''$.

Bureau of Steam Engineering's practice:

Dia. of shaft $= 13\frac{9}{16}''$. Then dia. of pins $= 14\frac{9}{16}''$. Length of pins $= 1.25 \times 14\frac{9}{16}'' = 18.2$, or $18\frac{3}{16}''$. Bearing pressure per square in.

of projected area $= \frac{121{,}000}{l \times d} = \frac{121{,}000}{14\frac{9}{16} \times 18\frac{3}{16}} = 456.2$ pounds, which is within the limit of 500 pounds per square inch. These dimensions are slightly smaller than those found by the above calculations, so the dimensions found in "check for heating" will be taken, $d = 14\frac{21}{32}''$ and $l = 18\frac{5}{16}''$.

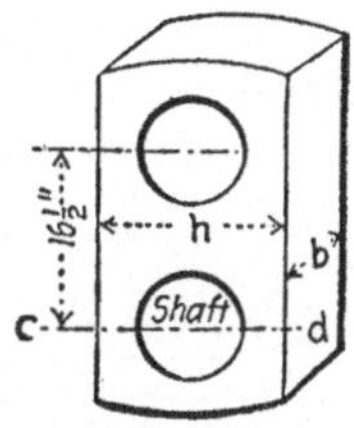

FIG. 109.

117a. *After Arm of After Crank.*—Since the leverage is at the pin, the greatest thickness is required at cd.

The bending moment at cd is the entire twisting moment on the crank shaft at the after bearing, and $\frac{bh^2}{6}$ = modulus of section (rectangular). Therefore

$$\frac{bh^2}{6} f = \text{Max. T. M.} \qquad (158)$$

$$\therefore h = \sqrt{\frac{\text{Max. T. M.} \times 6}{b \times f}} \text{ and } b = \frac{\text{Max. T. M.} \times 6}{h^2 \times f}.$$

Bureau of Steam Engineering's Method: The check used by the Bureau is to make the width of the crank arm across the face, h, about two inches more than the diameter of the shaft, and then compute the value of the thickness b, by the formula above for b, or by using a formula given by Unwin,

$$bh^2 = cd^3, \qquad (159)$$

giving c a value of .9 for steel.

1st Method: Dia. of shaft $= 13\frac{9}{16}''$. $\therefore h = 15\frac{9}{16}''$, and

$$b = \frac{2{,}840{,}000 \times 6}{(15\frac{9}{16})^2 \times 9000} = 7.81.$$

2d Method: $bh^2 = cd^3$. $\therefore b = \frac{.9 \times (13\frac{9}{16})^3}{(15\frac{9}{16})^2} = 9''.27$.

The average of these two methods gives 8.54 inches. The dimensions of the after arm of the after crank are, then, $b = 8\frac{17}{32}''$; $h = 15\frac{9}{16}''$.

If, in working out the design of the crank shaft as a whole, the sum obtained by adding the length of crank pin plus twice the thickness of the web is greater than the distance assumed between main bearings, then either this distance must be increased, or the length of the crank pin decreased. In either case the calculations must be made a second time with the new conditions.

118. *Cross-head Journals.*—Assume that two-thirds of the thrust R of the connecting rod may come on each journal. This allows for inaccurate adjustment and irregular wear.

$$\tfrac{2}{3}R = P = \tfrac{2}{3} \times 121{,}000 = 80{,}666 \text{ pounds.}$$

$$\frac{Pl}{2} = f\,\frac{\pi}{32}\,\frac{(d_1^4 - d_2^4)}{d_1}.$$

In this case, let $\frac{l}{d} = 1.1$, and the axial hole $d_2 = 4''$; $f = 9000$ for steel. The ratio of $\frac{l}{d}$ is taken small to prevent too great a spread of the jaws of the connecting rod, thus making a lighter rod. Substituting,

$$\frac{1.1 \times 80{,}666 d_1}{2} = 9000 \times \tfrac{22}{7} \times \tfrac{1}{32} \times \frac{(d_1^4 - (4)^4)}{d_1}.$$

Solving for d_1 we have $d_1 = 7''.42$, or $7\tfrac{13}{32}''$.
Then, $l = 1.1d = 1.1 \times 7\tfrac{13}{32}'' = 8.15$, or $8\tfrac{5}{32}''$.

This gives dimensions for strength. The projected area of the bearing surface $= 7\tfrac{13}{32}'' \times 8\tfrac{5}{32}'' = 60.4$ square inches and the pressure per square inch of projected area $= \frac{80{,}666}{60.4} = 1335$ pounds.

Check for Heating.—From notes on "Bearings" we find the pressure per square inch of projected area allowed for cross-head pins to be 750 pounds and not to exceed 1000 pounds. This shows that the pins are too small to prevent heating.

Assume a pressure of 750 pounds per square inch. Then the projected area required $= \frac{P}{750} = \frac{80{,}666}{750} = 107.6$ square inches.

Since $l = 1.1d$, $1.1d^2 = 107.6$ and $d = 9''.89$, or $9\tfrac{7}{8}''$. $\therefore l = 1.1d = 10''.85$, or $10\tfrac{27}{32}''$.

The projected area is, then, $9\tfrac{7}{8}'' \times 10\tfrac{27}{32}'' = 107$ square inches, and the pressure per square inch $= \frac{80{,}666}{107} = 753$ pounds, which is within the limit of 1000 pounds.

The dimensions of the cross-head pins are, then: Dia. $= 9\tfrac{7}{8}''$; length $= 10\tfrac{27}{32}''$.

119. *Shaft Couplings.*—The couplings are forged solid with the shafts, and are subjected to a twisting strain tending to shear the coupling off leaving the shaft cylindrical to the ends, so that the coupling can rotate on it like a washer on a bolt. Let t = the thickness of coupling at a radius R from the center of the shaft. Then $2\pi Rt$ is the area of the section and $2\pi Rtf$ its resistance to twisting. Its moment around the center of the shaft is $2\pi R^2tf$, and this should equal the twisting moment on the shaft, which is $\pi fd^3 \div 16$. Equating these, we have: $t = \frac{d^3}{32R^2}$. From this it will be seen that, as R increases, t decreases. When $R = \frac{d}{2}$, or where the coupling joins the shaft, we have, $t = \frac{d^3}{8d^2} = \frac{d}{8}$. This gives the thickness at this point for strength only, but is found to be too small for stiffness, and allows nothing for the loss of material due to the coupling-bolt holes. Seaton states that, from practical considerations, the thickness of the couplings should not be less than the diameter of the coupling bolts. He also states that, to allow for the decrease in strength due to cutting holes for the bolts, the thickness should not be less than $.3 \times$ dia. of a shaft subjected to twisting only. This is the same as given in Art. 101, Fig. 101, where the usual dimensions of couplings are given and are the ones to be used in the design.

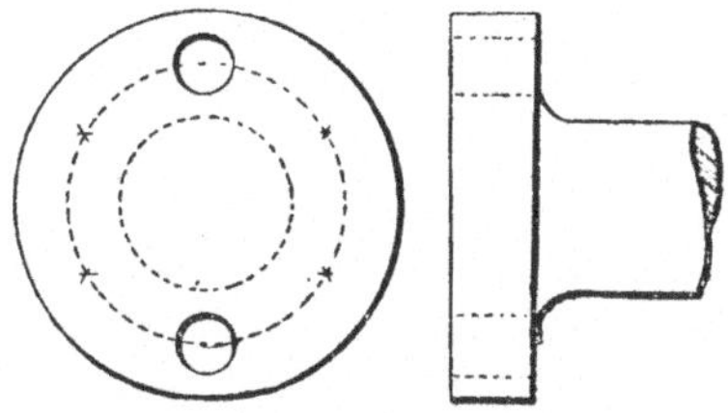

FIG. 110.

120. *Coupling Bolts.*—The number of bolts depends, to a certain extent, on circumstances. If it be desirable to keep the outside diameter of the coupling as small as possible, the number of holes should be increased; this decreases their diameter, and brings their centers nearer to the center of the shaft, thus reducing the outside diameter of the coupling. In general, when engines have two cranks with the shafts in duplicate pieces, there should be an even number of bolts. When there are three cranks and three interchangeable parts of the crank shaft, the number of bolts should be a multiple of three. In this case six is the number usually taken, and this applies to the couplings of the line shaft also.

Diameters of Coupling Bolts.—Let K=radius of bolt circle, which is about .80 of the diameter of the shaft. The coupling bolts, if well fitted, will be subjected to shearing only.

Let f_s=shearing strength of material$=\frac{3}{4}$ to $\frac{4}{5}$ of f_t.

$P\times K=T.\ M.=\frac{\pi}{16}f_t d^3$, and, at bolt circles, the force P is $\frac{\pi f_t d^3}{16K}$.

Let δ=dia. of bolts, and n=number of bolts, then:

$$n\frac{\pi}{4}\delta^2 f_s=\frac{\pi f_t d^3}{16K}. \quad \text{If } f_s=\tfrac{3}{4}f_t,\ \delta=\sqrt{\frac{d^3}{3nK}}. \qquad (160)$$

In the case under consideration, n=6, K=10.85 and $d=13\frac{9}{16}$. $\therefore\ \delta=3.57$, or $3\frac{9}{6}''$.

Questions and Problems.

Describe clearly the method of designing a crank shaft (including crank pins and webs), showing to what stress the various parts are subjected. After finding the sizes of journals and pins for strength, what further check is applied?

Problems.

1. Find the dimensions of the main journals of a vertical triple-expansion three-cylinder engine; twin screws; I. H. P.=10,000; cylinder diameters, $33\frac{1}{2}$ inches, 51 inches, 78 inches; stroke, 48 inches; initial pressure, 180 pounds per gauge; vacuum 26 inches; cut-off in H. P. cylinder, $\frac{7}{10}$ stroke; 1st receiver pressure 70 pounds (absolute); $\frac{\text{Max. T. M.}}{\text{Mean T. M.}}=1.5$; distance between bearings, 4 feet $5\frac{1}{2}$ inches; $f_t=9500$; shaft to have $7\frac{1}{2}$-inch axial hole; R. P. M.=116.

2. With the same data as problem 1, find the dimensions of the crank pins; axial hole through pins=6″; ratio l to d=1.25.

3. Find the dimensions of the after crank area of a three-cylinder triple-expansion engine from the following data: diameter of crank shaft=$14\frac{5}{8}''$; I. H. P. (one engine)=5000; R. P. M=116; f_t= 9500.

4. Find the dimensions of the crank shaft couplings and bolts for a three-cylinder triple-expansion engine; I. H. P. (one engine) =5000; R. P. M.=116; $f_t=9500$; diameter of shaft=$14\frac{5}{8}''$, with $7\frac{1}{2}$-inch axial hole; shaft in three interchangeable sections; for bolts take $f_s=\frac{3}{4}f_t$.

121. *Sequence of Calculations of the Crank-shaft Problem:*

1. Find the mean T. M. on shaft abaft the after engine.

2. Find the maximum T. M.

3. Find the bending moment, M.

4. Find the equivalent T. M., $T_e = M + \sqrt{M^2 + T^2_{max}}$.

5. Find the diameter of the *solid* shaft.

6. Find the diameter of the equivalent *hollow* shaft either from the known size of the hole, if given, otherwise by assuming $D_2 = \frac{1}{2}D_1$.

7. Find the length of the journals, pressure per square inch of projected area less than 200 pounds.

8. Check for reduction of area due to oil grooves, less than 300 pounds per square inch.

9. Check for speed of rubbing and pressure $p \times \sqrt{v} < 7500$; giving the final dimensions.

10. Find K.

11. Find bending moment at after end of after crank pin.

12. Assume a value of the ratio $\frac{l}{d}$ and solve for the diameter of crank pin.

13. Reduce diameter to diameter of hollow pin, and find length of pin.

14. Check for pressure, not to exceed 500 pounds per square inch.

15. Check again for reduction of area due to oil grooves, not to exceed 600 pounds per square inch.

16. Check for speed of rubbing and pressure, $p \times \sqrt{v} < 15{,}000$, giving final dimensions.

17. Note whether or not the dimensions of crank pin, as found above, are smaller than the Bureau practice of making diameter of crank pin 1 inch larger than diameter of shaft. If it has worked out smaller, then make diameter of crank pin = diameter of shaft + 1″; if it has worked out larger take largest dimensions.

18. Find dimensions of crank webs, making h = diameter of shaft + 2″. Using both the bending method and Unwin's method, and taking the mean of results for b.

19. Find thickness of shaft couplings.

20. Select the proper number of coupling bolts, and find their diameter.

21. Check diameter of coupling to see that the nuts of coupling bolts do not project beyond the circle of the coupling.

122. CRANK SHAFT PROBLEM.—DATA FOR PRACTICAL WORK.

Problem No.	1.	2.	3.	4.	5.
Desk No. ending in	1 or 6	2 or 7	3 or 8	4 or 9	5 or 0
Total I. H. P. (both engines)	10,000	21,000	16,500	23,000	25,000
No. of cylinders (each engine)	3	4	4	4	4
Cylinder diameters	33½, 51, 78	36, 59½, 2 of 69	32½, 53, 2 of 61	38½, 63½, 2 of 74	36, 57, 2 of 76
Stroke	48 in.	45 in.	48 in.	48 in.	48 in.
R. P. M.	120	133	120	120	125
Boiler pressure, lbs. per gauge	180	250	250	250	265
1st receiver pressure (abs.)	70	90	90	95	150
Vacuum	26 in.	25 in.	24 in.	26 in.	26 in.
Cut-off in H. P. cylinder	$\frac{7}{10}$	$\frac{7}{10}$	$\frac{7}{10}$	$\frac{7}{10}$	$\frac{78}{100}$
$\frac{\text{Max. T. M.}}{\text{Mean T. M.}}$	1.5	1.25	1.25	1.25	1.25
Distance between bearings	4 ft. 5½ in.	5 ft. 8 in.	4 ft. 11½ in.	4 ft. 4½ in.	4 ft. 3 in.
f	9,500	11,000	12,000	10,000	10,000
Diameter hole in shaft	7½ in.	11½ in.	9½ in.	10 in.	11 in.
Diameter hole in pins	7½ in.	12 in.	10 in.	12 in.	12 in.
Breadth of crank arm	10 in.	11¾ in.	10¼ in.	12 in.	11¾ in.
Distance center F. L. P. to H. P. cylinder	....	5 ft. 6 in.	4 ft. 11½ in.	5 ft. 10 in.	5 ft. 9½ in.
Distance center H. P. to I. P. cylinder	8 ft. 10 in.	10 ft. 11½ in.	10 ft. 0½ in.	12 ft. 6 in.	11 ft. 2½ in.
Distance center I. P. to A. L. P.	8 ft. 10 in.	6 ft. 4 in.	5 ft. 9 in.	6 ft. 11 in.	6 ft. 6½ in.

No. of sections of shaft	3 interchangeable.	2	2	2	
Sequence of cranks	H. P., I. P., L. P.	H. P., I. P., F. L. P., A. L. P.	H. P., I. P., F. L. P., A. L. P.	F. L. P., I. P., H. P., A. L. P.	H. P., I. P., F.L.P., A.L.P.
Total length of forward section of crank shaft	8 ft. 10 in.	15 ft. 11¼ in.	14 ft. 10 in.	17 ft. 5½ in.	16 ft. 11 in.
Total length of after section of crank shaft	....	17 ft. 8½ in.	16 ft. 3 in.	19 ft. 9½ in.	18 ft. 3 in.
Distance between centers of—					
Main bearings No. 1 to No. 2	4 ft. 5½ in.	5 ft. 7 in.	4 ft. 10½ in.	5 ft. 10 in.	5 ft. 9 in.
Main bearings No. 2 to No. 3	4 ft. 4½ in.	5 ft. 6 in.	4 ft. 10½ in.	5 ft. 10 in.	5 ft. 9 in.
Main bearings No. 3 to No. 4	4 ft. 5½ in.	5 ft. 4 in.	5 ft. 0 in.	6 ft. 8 in.	5 ft. 4½ in.
Main bearings No. 4 to No. 5	4 ft. 4½ in.	6 ft. 0 in.	5 ft. 3 in.	6 ft. 4 in.	5 ft. 4½ in.
Main bearings No. 5 to No. 6	4 ft. 5½ in.	6 ft. 0 in.	5 ft. 4 in.	6 ft. 4 in.	1 ft. 6½ in.
Main bearings No. 6 to No. 7	....	....	....	....	5 ft. 4½ in.
Length of seatings for eccentric	10 in.	12 in.	10 in.	15 in.	14 in.
Diameter eccentric seatings greater than diameter of shaft, by	1 in.	1 in.	1 in.	1 in.	1 in.
Distance C. L. of eccentric seating from corresponding cylinder axis (+) Forward (—) Aft.					
F. L. P.	....	+ 4 ft. 8 in.	+ 4 ft. 4 in.	+ 5 ft. 0½ in.	+ 4 ft. 10½ in.
H. P.	+ and — 3 ft. 5 in.	— 4 ft. 2 in.	— 3 ft. 8½ in.	— 4 ft. 6¼ in.	— 4 ft. 4 in.
I. P.	+ and — 3 ft. 5 in.	+ 4 ft. 4 in.	+ 4 ft. 0 in.	+ 4 ft. 9½ in.	+ 4 ft. 5½ in.
A. L. P.	+ and — 3 ft. 5 in.	— 4 ft. 5 in.	— 4 ft. 4 in.	— 5 ft. 0 in.	— 4 ft. 10½ in.

NOTE.—Problem No. 5 design for equal work in each of the 4 cylinders. Make all necessary calculations and enter the work with results in text book. Then lay down the shaft to as large a scale as the paper will permit, showing longitudinal and end elevations.

CHAPTER XVI.

I. H. P. Required for Given Vessel. Cylinders. Valves and Valve Chests. Pistons and Rods. Steam and Exhaust Pipes.*

123. The general method is as follows:

Data.—Displacement; length; breadth (extreme); mean draught; area of immersed midship section; speed required.

Problem.—To design the propelling machinery of a naval vessel with the following dimensions and data:

Displacement	13,700 tons
Length	502 feet
Breadth (extreme)	69 feet 6 in.
Mean draught	24 feet.
Area of immersed midship sect'n	1596 sq. ft.
Speed required	22 knots

Indicated Horse-power.—This is determined by two methods: Kirk's "Analysis" and Fronde's "Law of Comparisons."

Kirk's analysis: From a comparison of the trial trips of several vessels of similar lines, it is found that, reduced to a general basis of ten knots per hour, from 5.0 to 5.5 I. H. P. are required for each 100 square feet of wetted surface.

Wetted surface: Kirk's Analysis—

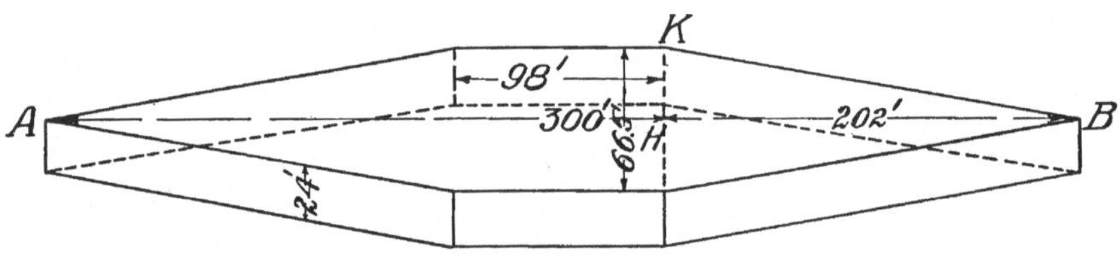

Fig. 111.

$$13{,}700 \times 35 = \text{Vol. in cubic feet},$$

$$\frac{13{,}700 \times 35}{1596} = 300 = AH_1,$$

* From "Notes on Machine Design," rewritten.

$$\frac{1596}{24}=66.5=EK,$$

$$EB=\sqrt{\overline{EH}^2+\overline{BH}^2},$$

$$BH=502-300=202.$$

Then,

$$EB=\sqrt{\overline{33.25}^2+\overline{202}^2}=204.9.$$

Area of bottom............	$300\times66.5=19{,}950$
Area of sides..............	$2\times24\times98=\ \ 4{,}704$
Area of entrance and run	$4\times24\times204.9=19{,}670$
Area of wetted surface..........	$=44{,}324$

For a speed of 10 knots, 5 I. H. P are required for 100 square feet of wetted surface, and, as the power varies as the cube of the speed, we have:

$$\text{I. H. P.}=\left(\frac{22}{10}\right)^3\times5.0\times\frac{44{,}324}{100}=23{,}600.$$

124. *Indicated H. P. by Froude's Law of Comparisons.*—Fronde's law may be stated as follows: If two ships have the same coefficient of fineness, same water lines, etc., but their dimensions have a ratio n, then for speeds s_1, s_2, s_3, etc., of one with resistances r_1, r_2, r_3, etc., the corresponding speeds of the other will be $s_1\sqrt{n}$, $s_2\sqrt{n}$, $s_3\sqrt{n}$, etc., with resistances n^3r_1, n^3r_2, n^3r_3, etc.

In similar ships, the cube of the ratio of dimensions equals the ratio of displacements, hence n^3=ratio of displacements and $\sqrt{n}$=(ratio of displacements)$^{\frac{1}{6}}$.

Let d, r, s and p represent the displacement, resistance, speed and I. H. P respectively of a given vessel whose data are known, preferably from a progressive speed trial accurately conducted; or, the same quantities may be obtained experimentally from a model of the proposed ship.

Let D, R, S and P represent the corresponding quantities for the vessel whose power is to be determined.

Then $S=s\sqrt[6]{\frac{D}{d}}$, and $R=r\frac{D}{d}$, also $p=rs$.

Now, $P=RS=rs\times\frac{D}{d}\times\sqrt[6]{\frac{D}{d}}=p\times\frac{D}{d}\times\sqrt[6]{\frac{D}{d}}$.

This law applies only to exactly *similar* ships. It is based on two fundamental assumptions:

(a) The resistances of two similar ships vary as their displacements, that is:

$$\frac{R}{r}=\frac{D}{d}, \text{ but } \frac{D}{d}=n^3, \text{ and } n=\sqrt[3]{\frac{D}{d}}; \qquad (161)$$

also

$$R=r\frac{D}{d}. \qquad (162)$$

(b) The *corresponding* speeds of the two ships are proportional to the square roots of their lineal dimensions, that is:

$$S=s\sqrt{n}=s\sqrt[6]{\frac{D}{d}}. \qquad (163)$$

Power is rate of doing work: $\therefore P=RS$, and $p=rs$.

$$P=RS=r\frac{D}{d}\times s\sqrt[6]{\frac{D}{d}}=p\frac{D}{d}\sqrt[6]{\frac{D}{d}}=p\left(\frac{D}{d}\right)^{\frac{7}{6}}. \qquad (164)$$

Applying this method to the engine under design and assigning as a model a ship of similar lines and fineness and whose trial was accurately conducted, we obtain the following:

Model Ship.	Proposed Ship.
d=6000.	D=13,700.
s= 16.5.	S= 22.0.
p=5820.	P= ?.

Substituting and solving for S and P we have

$$S=s\sqrt[6]{\frac{D}{d}}=16.5\times\sqrt[6]{\frac{13,700}{6000}}$$

$$=16.5\times 1.148=18.95 \text{ corresponding speed.}$$

$$P=p\frac{D}{d}\sqrt[6]{\frac{D}{d}}=5820\times 2.283\times 1.148=15,250.$$

This is the power necessary for the proposed ship *at the speed corresponding to the speed of the model.*

But our new ship is to have a speed of 22 knots. The power must therefore be increased in proportion to the cube of the speed.

Indicated horse-power by Fronde's law $=15,250\times\left(\frac{22}{18.95}\right)^3=23,850.$

Indicated horse-power by Kirk's analysis=23,600.

Mean of the two methods=23.725.

In designing for large powers use mean to nearest hundred, or 23,700; in this case, indicated horse-power of one engine=11,850.

125. *Engines.*—The type of engine, number of engines and other data are found from experience and best practice, and for this vessel the following are decided upon:

Twin screws. Triple-expansion, vertical, direct engines, with the low-pressure stage in *two* cylinders. Independent pumps. Boiler pressure, 265 pounds (p. g.). Cut-off in H. P. cylinder .75 stroke. Ratio L. P. to H. P. cylinders about 8.5 (including clearances). Clearance H. P., 15%; I. P., 14%; L. P., 12%. Stroke, 48 inches. Revolutions per minute about 125. Vacuum, 26 inches. I. H. P. (each engine), 11,850.

126. *Diameters of the Cylinders.*—From the above data, the actual ratio of expansion in the H. P. cylinder, taking clearance into account, is: $\frac{.15+1}{.15+.75}=1.278.$

The total ratio of expansion of the engine is: Ratio L. P. to H. P. × ratio of expansion in H. P. $=8.5\times1.278=10.86$.

The total theoretical mean pressure is: $P_m=p_1\times\frac{1+\log_e r}{r}$ $=280\times\frac{1+2.3\log 10.86}{10.86}=280\times.312=87.36$. Or the table in "Barton," p. 522, may be used to obtain the value of $\frac{1+\log_e r}{r}$.

A vacuum of 26 inches corresponds to an absolute pressure of 2 pounds. Assume a back pressure of 2 pounds. Then the theoretical mean effective pressure is

$$P_e=87.36-2=85.36.$$

From the data of many trial trips of various ships, the actual mean effective pressure is found to be from 50% to 60% of the theoretical. This percentage is commonly called the "card efficiency." If we assume a card efficiency of 55%, the expected mean effective pressure is $85.36\times.55=46.95$.

Using the well-known horse-power formula,

$$\frac{P\times L\times A\times 2N}{33,000}=11,850=\frac{45.95\times4\times2\times125\times A}{33,000},$$

we get

$$A_{L.P.}=\frac{11,850\times33,000}{46.95\times1000}=8322 \text{ square inches}$$

(for *both* cylinders) or 4161 square inches for each L. P. cylinder.

The area of the low pressure is thus calculated as if the whole expansion of the engine took place in this cylinder. The reason for this is fully explained in Barton, "Naval Engines and Machinery."

Area H. P. Cylinder.—Area H. P. cylinder (1+H. P. clearance)

$$= \frac{\text{area L. P. cylinder } (1+\text{L. P. clearance})}{\text{cylinder ratio}},$$

or

$$A_{H.P.}\,(1+.15) = \frac{A_{low}(\text{both})\,(1+.12)}{8.5}.$$

$$\therefore A_{H.P.} = \frac{8322\times 1.12}{8.5\times 1.15} = 954 \text{ square inches.}$$

Area I. P. Cylinder.—The size of this cylinder has no effect on the total power of the engine and good practice shows considerable variation in its relative size, which is determined principally with the object of obtaining a proper distribution of work and considerations of balancing the engine as a whole. The following formula is frequently used and gives good results:

$$\text{Area I. P.} = \frac{\text{Area L. P.}}{1.0 \text{ to } 1.1\sqrt{\text{ratio of L. P. to H. P. areas}}},$$

or, using 1.0,

$$= \frac{8322}{1.0\times\sqrt{8.5}} = 2850 \text{ square inches.}$$

In determining the above areas the area of the piston rod has not been taken into consideration, therefore, the sizes of the cylinders must be modified.

Assume the size of the piston rod to be $8\frac{1}{2}$ inches, which is about the size necessary for an engine of this size and power.

One-half area corresponding to a diameter of $8\frac{1}{2}'' = 28.37$ square inches. Then,

Total area H. P. ...... = 954+28.37 = 982.37.
Total area I. P. ...... = 2850+28.37 = 2878.37.
Total area L. P. (each) = 4161+28.37 = 4189.37.

The corresponding calculated diameters are:

H. P. ...... = 35.4.
I. P. ...... = 60.6.
L. P. (each) = 73.1.

Then the diameters of cylinders used are:

$$\left.\begin{array}{r} \text{H. P. } \ldots\ldots = 35\tfrac{1}{2}'' \\ \text{I. P. } \ldots\ldots = 61'' \\ 2\text{ L. P. (each)} = 73'' \end{array}\right\} \text{by } 48'' \text{ stroke.}$$

Cylinder ratios $\dfrac{\text{L. P.}}{\text{H. P.}} = 8.46$; $\dfrac{\text{I. P.}}{\text{H. P.}} = 2.96$; $\dfrac{\text{L. P.}}{\text{I. P.}} = 1.43$.

NOTE.—In case superheated steam is to be used the area of the H. P. cylinder must be increased in proportion to the increased specific volume of the superheated steam. Thus if 100° superheat is to be used, the H. P. area must be increased about 11%.

127. *Thickness of Cylinders and Various Parts of the Cylinder Castings.*—The cylinders of modern engines are steam-jacketed, the jacket being formed by the space between the cylinder barrel and the liner. The Bureau of Steam Engineering uses the following formula:

$$\text{For cylinder liners } t = \frac{Pd}{2f} + C. \qquad (165)$$

P = boiler pressure; d = diameter of cylinder (inside diameter of liner) inches; $f = 4000$ (f is large as liners are made of special iron, and usually have steam pressure outside); C = constant, to allow for reboring $= .5$ for large cylinders, $= .25$ for less than 20″ diameter.

For barrels:

$$\text{When liner is used, } D = d + 2t + 2c. \qquad (166)$$

d = diameter of liner (inside) inches.

t = thickness of liner.

c = width of steam space around liner = about $\frac{3}{4}''$.

$$T = \text{thickness of barrel} = \frac{PD}{2f_1}. \qquad (167)$$

P = boiler pressure.

D = diameter of barrel inches.

$f_1 = 2250$ for vertical cylinders.

The thickness of the remaining parts of the cylinders are calculated by taking certain percentages of a constant K as follows:

$$K = \frac{PD}{2f_1} + .01D. \qquad (168)$$

Thickness of various parts of cylinders using above constant:

Part.	H. P.	I. P.	L. P.
Steam Ports	$K_{H.P.}$ x .65	K_I x .65	$K_{L.P.}$ x .65
Cylinder bottoms, double	" x .7	" x .7	" x .7
Cylinder bottoms, single	" x .85	" x .85	" x .85
Cylinder flanges	" x 1.2	" x 1.2	" x 1.2
Cylinder covers, double	" x .65	" x .65	" x .65
Valve chests	" x .65	" x .65	" x .65

The table above represents the practice of the Bureau of Steam Engineering for a general guide, but it gives rather heavy results, particularly for the thickness of valve chests.

For valve chests use K×.58; this will give more reasonable thickness.

In formula (168) D is found by formula (166). The following table gives values of the pressure P used in formulæ (165), (166), (167) and (168), in the latest Bureau design.

Steam pressures to use in preparing design. At H. P. valve chest 265 pounds p. g. = (280 pounds abs.). 100° superheat:

Pressures.	H. P.	I. P.	L. P.	Condenser.
Test pressure	425	200	100	30
Cylinder relief valves set at	275	130	40	..
For cylinder liners and barrels	265	125	100	..
For load on piston	265	100	35	..

When the covers are steel use a value of f_1 of 9000.

The thickness of small cylinders, having no liners, is calculated in the same way, and a small allowance added for wear, reboring, etc.

From practical considerations of stiffness, and to insure reliable castings, the I. P. details are usually made about the same thickness as the H. P. The L. P. details are somewhat less, in the case of large modern engines about ½ inch less. If the solution of the formula gives much smaller values for L. P. cylinder than about ½ inch less than I. P., then the thickness should be arbitrarily increased.

Very frequently the inner wall, that is, the wall on which the cylinder pressure acts, of double bottoms and covers, is made slightly thicker than the outer wall. When this is done it is usual to make the outer wall $\frac{1}{8}$ inch less than found by the formula, and to add $\frac{1}{8}$ inch to the inner wall.

When double covers and bottoms are used they are strengthened and stiffened by being connected by radial webs. The height of radial webs in single covers, or bottoms, and the height of the clear space between walls in double covers and bottoms is given by

$$h=7t \text{ to } 9t, \qquad (169)$$

t being the thickness of wall.

The thickness of webs may be made three-quarters of the thickness of cylinder liner, or thickness of web $=.75\times$ thickness of liner.

Determine the number of webs by the following:

$$\text{Number of webs}=\begin{cases} \text{H. P. } \dfrac{D+20}{9} \\ \text{I. P. } \dfrac{D+20}{10} & \text{where } d=\text{dia. of corresponding cylinder.} \\ \text{L. P. } \dfrac{D+20}{11} \end{cases}$$

When single covers are used, the material is generally cast steel, giving a light cover, but when the cover is double it is necessary to use cast iron on account of the rather complicated form. The double cover also affords a convenient method of steam jacketing the cylinder ends. Double covers and bottoms are used in all large modern naval engines.

128. *Cylinder-head Bolts or Studs.*—For the method of determining these see Art. 45.

129. *The Pistons.*—The H. P. piston of modern engines is made of cast iron, the I. P. and L. P. pistons of cast steel.

Cast iron is used for the H. P. because it is desired that the weight of the reciprocating parts for each cylinder should be as nearly as practicable constant, in order to obtain a well-balanced engine. Therefore, if steel were used the piston would still be made much heavier than necessary from considerations of strength only, and there would be a waste of money in using the more expensive material. For a good example of a modern piston see Barton's Plate IX.

The pistons are of conical form, to give increased stiffness and to allow the stuffing box to be fitted partially within the lower head of the cylinder, thus reducing the total height of the engine for a given length of connecting rod. The height of the cone, L, in Fig. 112, is approximately, but not necessarily exactly, equal for all pistons of a given engine; thus the slope of the H. P. is much steeper than that of the I. P. and L. P.

High Pressure Piston. Cast Iron.—To determine the thickness of the H. P. piston (cast iron) near the rim, just inside the follower recess, where the metal is thinnest, the following formula may be used:

$$t = .004 \text{ to } .005 D \times \sqrt{p}. \qquad (170)$$

t= thickness in inches. D=diameter of H. P. cylinder in inches. p=boiler pressure.

In practice the low pressure piston is designed first, then the I. P., and these designs checked by the method given later. The H. P. piston is not calculated for strength, but is made as heavy and as thick as practicable in order to give weight, on account of considerations of engine balancing. For the first tentative design, however, the formula for H. P. piston given above should be used.

Cast Steel Pistons.—The accompanying chart shows a method of designing cast steel pistons. The dimensions T, h and d_1 are the same for the H. P., I. P. and L. P. L, for the H. P., may be made from .2 to .25×diameter of H. P. cylinder, the constant being the result of practical experience. Then

$$\left.\begin{aligned} L_{H.P.} &= .2 \text{ to } .25 \text{ diameter H. P. cylinder.} \\ L_{I.P.} &= .2 \text{ to } .25 \text{ diameter H. P. cylinder} + 1\tfrac{1}{2}''. \\ L_{L.P.} &= .2 \text{ to } .25 \text{ diameter H. P. cylinder} + 2\tfrac{1}{2}''. \end{aligned}\right\} \qquad (171)$$

As an illustration of the above, assume the following data: Triple-expansion engine, cylinders 36 inches, 60 inches and 2 of 70 inches; boiler pressure 265 pounds per gauge; 1st receiver, 140 pounds absolute; 2d receiver, 55 pounds absolute.

Then for H. P. (cast iron) thickness near rim, $t = .004D \times \sqrt{p} = .004 \times 36 \times \sqrt{265} = 2.35$, would be made about $2\frac{1}{2}''$.

For I. P., $i = K \times c$; to find K enter the chart with pressure 140 pounds (absolute), run up to curve (3) and squaring across to scale, values of K, $4\frac{1}{8}''$ is found.

L for I. P. $= .2$ diameter H. P. $+ 1\frac{1}{2}'' = .2 \times 36 + 1\frac{1}{2} = 8''.7$, say $8\frac{3}{4}''$.

The slope of I. P. is then $\tan^{-1}\frac{8.75}{30}$ = about 16°, then c = .725.

i = 4.125 × .725 = 2.99, or 3″.
a = .45 × i = .45 × 3 = 1.35, or about $1\frac{3}{8}$″.

The method of design given above is based on the general formula for a circular plate, supported at the center and uniformly loaded.

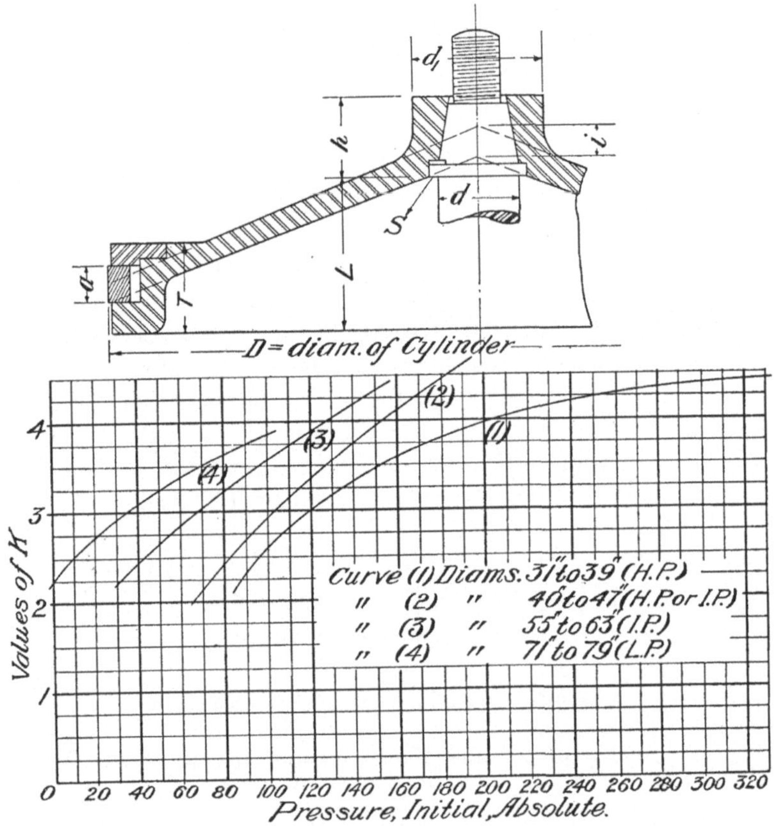

FIG. 112.

This formula in its general form is $t = C \times D\sqrt{p}$. The modifications are in the values of the constant based on the coned shape given the piston, considerations of inertia and practical allowances. Proportions of cast-steel pistons (Bauer): *

* Curves plotted from Bauer's tables.

h=.25×diameter H. P. (not less than 1.1K).

d_1=1.6 to 1.7 diameter rod (not less than 1.7K for a small piston).

i=K×C (for values of c see below).

a=.45i to .55i (use .45 for large modern engines).

Values of C:

Coned pistons, inclination from 6° to 18° C= .85 to .59. Use .725 for I. P.

Coned pistons, inclination from 18° to 28° C= .75 to .85.

Coned pistons, inclination from 28° to 35° C= .65 to .75.

Flat pistons, inclination from 0° to 6° C=1.0 to 1.1.

Use this value for L. P., even if inclination exceeds 6°, for stiffness.

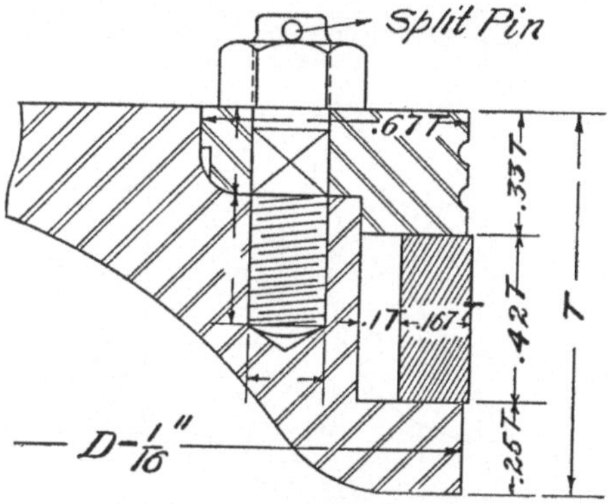

FIG. 113.—Piston Details.

The following method may be used for designing the details of the rim, follower and packing rings:

The depth of the rim will, in ordinary cases, give satisfactory results if made=D×.155 to .145, where D is the diameter of the H. P. cylinder, or T=.155 to .145D (Figs. 112 and 113).

The sketch shows the proportion of the other details, based on T as the unit, which conforms to good modern practice. The follower bolts are commonly made 1¼ inches with a square neck. The H. P. and I. P. packing rings are plain solid rings with tongue and groove slot to allow for expansion. For the L. P. two separate rings are used, one on top of the other, each ring being made of a number of sections, usually about eight pieces to each complete ring. The packing rings in all pistons are set out by steel springs, to give an equal bearing pressure between ring and cylinder wall, and also to hold the piston centrally in the cylinder.

130. *Piston Rods.*—The formula used by the Bureau of Steam Engineering is:

$$P = \frac{f \times S}{1 + \frac{16}{9} a \frac{l^2}{d^2}}. \qquad (172)^*$$

P = total load on piston rod in pounds.

S = sectional area of rod sq. in.

$a = 1/2250$.

$\frac{l}{d} = r$ = usually 10 to 12.

$f = 4500$ (steel) to 6500 (high-grade steel).

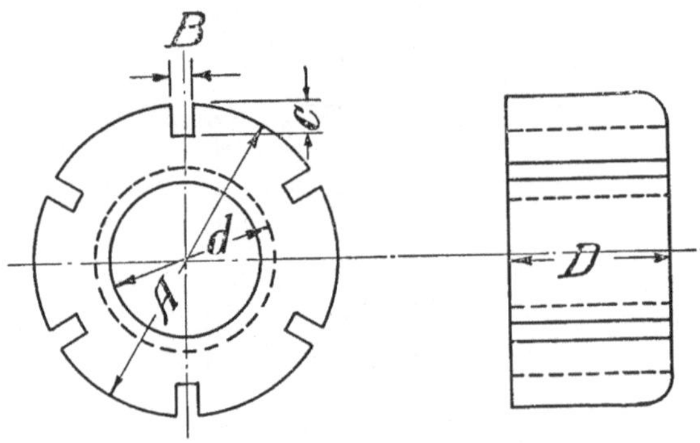

FIG. 114.

The rods are made of "high grade" steel (nickel steel) and are usually hollow to facilitate tempering. When the rods are to be hollow, solve for a solid rod and find the diameter of the equivalent hollow rod, by the formula

$$d^3 \text{(solid)} = \frac{d_1^4 - d_2^4}{d_1},$$

the size of the hole being known or assumed. The usual practice is to make the diameter of the hole $= \frac{1}{2}$ the outside diameter of the rod. Applying this formula to the I. P. and L. P. rods, generally gives smaller sizes than for the H. P., but the I. P. and L. P. rods are actually made of the same dimensions as the H. P., except that the

* This will be recognized as a modified form of Gordon's Strut Formula.

diameter of the hole of the L. P. rod is sometimes made larger than the H. P. and I. P.

Collars.—The rods are forged with a collar, sketch in a previous note, and is flush with the inside, or lower face of the piston when set up. The diameter of this collar at the piston end of the rod is made 1.2×diameter of rod, and the thickness of the collar is about .15×diameter of the rod.

Taper of Rods.—Where the rods pass through the hub of the piston they are given the conventional taper of 1½ inches to 1 foot. The end of the rod projecting above the piston is turned down and threaded for the Bureau standard slotted nut. The diameter of the end is such that area at bottom of thread $= \frac{\text{load on piston}}{11{,}500}$. A shearing stress, tending to strip the threads of the nut, of 11,500 pounds per square inch is allowed; this makes the height of the nut=the nominal diameter of the threaded part of the rod.

The standard proportions of the nut are tabulated by the Bureau, but, in case the table is not available, the following rough approximation may be used: A=1.7d; B=.2d; C=.1d; D=d.

131. *Details of H. P. Cylinders, Valves, etc.*—For purposes of illustration the details of an actual modern engine of about 12,000 I. H. P. is used. This engine has the following principal dimensions, viz: Cylinder diameters—H. P., 38½ inches; I. P., 63½ inches; two L. P., 74 inches (each) by 48-inch stroke; R. P. M., 120.

Shape of Bottom Heads and Covers.—This is determined by the shape of the piston, and is so arranged that the clearance is kept uniform at ½ inch at the top and ¾ inch at the bottom, these clearances being called for by the specifications and based on experieuce. The opening in the bottom head for the piston-rod stuffing box is determined by the kind of packing to be used, and is made to conform to the sizes given by the makers catalogue: in this case 12½ inches for the upper (or inner wall) and 13½ inches for the lower (or outer wall), so that the stuffing box can be entered into place from below. The diameter of the flange or boss, for securing the stuffing box is 18 inches. Fig. 2, Plate VII, Barton's "Naval Engines and Machinery," shows a high pressure cylinder similar to the one under discussion.

Counterbore.—The cylinder liner is counterbored at each end to a diameter of 39 inches, which is ½ inch greater diameter than the bore of the liner, and the length of the counterbore at each end is such that the piston ring overruns $\frac{1}{4}$ inch at the top and $\frac{1}{8}$ inch at the bottom. The object of this counterbore is to prevent wearing shoulders at the ends of the stroke. The exact position of the counterbore is found by drawing the piston in position with the proper clearance, when the length of the counterbore can be laid off.

Note on Velocity of Steam.—The sizes of the cross-sectional areas of steam and exhaust passages, receiver pipes, etc., are determined by fixing a maximum rate of flow of steam, which must not be exceeded. In ordinary engines these limiting velocities are:

1. Main steam pipe, v=6000 to 7500 feet per minute. If the steam pipe is very long, the velocity should be taken a little less.

2. In steam passages of—

H. P. cylinder, v=5000 to 6000 feet per minute.
I. P. cylinder, v=6000 to 7000 feet per minute.
L. P. cylinder, v=7000 to 8000 feet per minute.

3. In exhaust passages and receiver pipes of—

H. P. cylinder, v=4000 to 5000 feet per minute.
I. P. cylinder, v=5000 to 5500 feet per minute.
L. P. cylinder, v=5500 to 6500 feet per minute.

4. The velocities given above are increased from 10 to 20% in very light high speed engines, where the saving of weight and space is of more importance than economy at the highest speeds.

5. In naval engines these velocities are often somewhat exceeded.

Port Area for H. P. Cylinder.—Since the steam has to exhaust through the same port by which it enters, the size of the ports must be governed by the velocity of the exhaust steam, rather than by the velocity of steam at admission. For the H. P. this area then must be such that the velocity of exhaust does not exceed 5000 feet per minute (see note on velocity of steam). Design the valve to open wide for exhaust.

$$\text{Area of port} = \frac{\text{Area of H. P. cylinder} \times \text{piston speed}}{5000}.$$

NOTE.—The numerator of this faction represents the volume swept by the H. P. piston per minute.

$$\text{In this case area of port} = \left\{\frac{\pi \times (38\frac{1}{2})^2}{4} \times \frac{120 \times 48 \times 2}{12}\right\} \div 5000 = 223.5.$$

The width of the port, parallel to the plane of the cylinder bottom is, in this case, nearly equal to the diameter of the cylinder. (See Fig. 2, Plate VII, Barton.) It is here about 91% of the diameter of cylinder, or 35 inches. Then the dimension of the port in the other direction is $\frac{223.5}{35}=6''.4$. As these ports must be strengthened and stiffened by ribs, which will reduce the effective area, it will be well to increase this to $6\frac{1}{2}$ inches, which was actually done in the case selected. The dimensions of the ports and passages connecting the cylinder to the valve chest will then be 35 inches by $6\frac{1}{2}$ inches.

The total length of cylinder can now be laid off. See Barton's Plate VII. Sketch in the piston at the lowest point of the stroke; allow the required clearance below it. This will locate the inner surface of the bottom head, also the flange of the cylinder liner, and the edge of the counterbore. Now sketch the piston again at the top center, and since it is specified that the ring is to overrun at top $\frac{1}{4}$ inch, the upper counterbore can be located. To this add about $\frac{5}{8}$ inch for the depth of counterbore $+\frac{1}{2}''$ for thickness of liner packing gland. This brings us up to the point marked E on the sketch of the valve-chest liner, which follows (Fig. 116). To this must be added the length of the steam port, which has just been found to be $6\frac{1}{2}$ inches, bringing us to the inner face of the upper cylinder flange; adding the thickness of this flange, which in this case is $2\frac{3}{4}$ inches, we reach the extreme upper surface of the cylinder casing. From this it is seen that the outer cylinder wall extends $9\frac{1}{4}$ inches above the top of the liner (the liner gland being considered as part of the liner). In the same way at the bottom, the thickness of the liner flange (which is equal to the thickness of the liner itself), plus the thickness of the inner wall of the bottom head, plus the length of the port, plus the thickness of the outer wall of the bottom head.

Valve Chest and Piston Valve Details.—The dimensions of the valve-chest liners and the piston valve are the next steps to be considered.

The specifications require a mean cut-off for each end of each cylinder of .75 of the stroke, and the eccentricity is specified as $5\frac{1}{4}$ inches, given a maximum travel of $10\frac{1}{2}$ inches. A good illustration of a piston valve, with a development of the valve-chest liner, showing the bridges, is found on p. 85, Barton, Fig. 36. The H. P. valve takes steam on the inside and exhausts at the ends, the valve being hollow for free passage of the exhaust steam through it. See Barton, Fig. 37. The Bureau of Steam Engineering specifies that the bridges must take up not more than from 20% to 25% of the opening, leaving from 80% to 75% of clear area for the passage of the steam.

Diameter of H. P. Valve.—This diameter must be such as to give an area equal to that of the ports. The maximum port opening for steam is assumed in this calculation, and for an engine of this size experience shows that about 3 inches is right.

A Zeuner valve diagram is now constructed. Given length of connecting rod, 8 feet. Stroke, 48 inches. Travel of valve (by specification) 10½ inches; assume, lead (angular), 10°; top piston

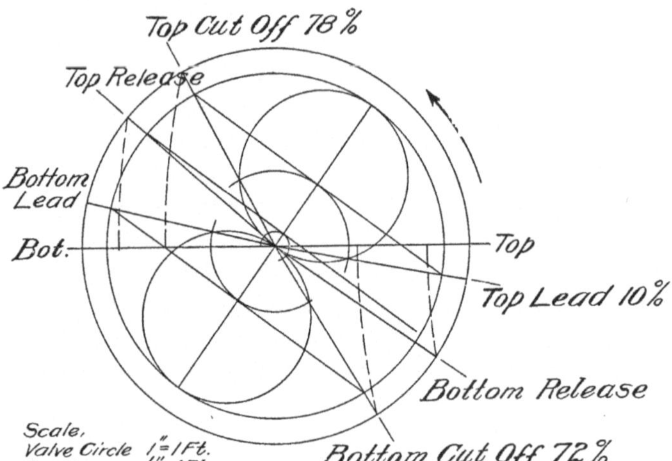

FIG. 115.—Zeuner Valve Diagram.

position at cut-off, .78 of stroke. To allow for the angularity of the connecting rod take the top cut-off at 78% of the stroke, and the bottom cut-off at 72%, giving a mean value for both strokes of 75%. Release at $\frac{1}{10}$ of the stroke from each end. These assumptions are based on experience gained from previous engines, and from a general knowledge of practical requirements. Then the following points and dimensions are obtained from the diagram.

Steam lap, top$=2\frac{5}{16}''$, bottom$=2''$, exhaust lap top (−) ½″ bottom 0. Maximum opening to steam top, 2″.9, bottom 3¼″, exhaust opening, top$=5\frac{3}{4}''$, bottom 5¼″.

In order that the total length of the valve chest may be kept within practicable limits, the length of the opening in the liners will be taken as 4 inches. The circumference of piston valve × length of opening × per cent of clear opening = area of steam port. Or

$$\pi D \times 4 \times .775 = 223.5,$$

$$D = \frac{223.5}{\pi \times 4 \times .775} = 22.95, \text{ or } 23''.$$

Working Liner.—There will be at each end of the valve chest a working liner made of hard close-grained cast iron, as hard as can be worked. These are forced into place and held by screws tapped

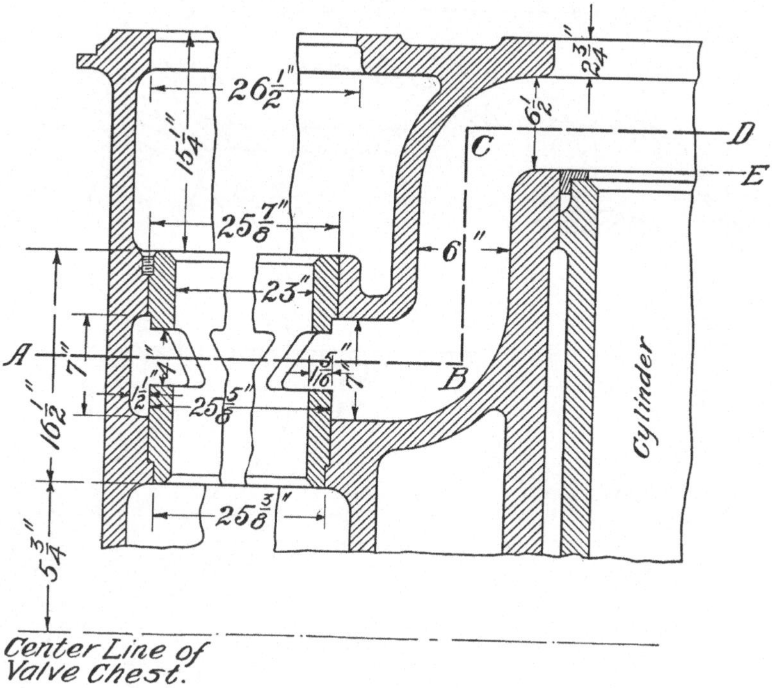

FIG. 116.—Top Valve-Chest Liner and Steam Port.

half into the liners at the ends and half into the metal of the castings. See Barton, p. 86. The upper liner is driven into place from the top, the lower liner from the bottom. The outside diameters of these liners are turned down in steps, as shown in the sketch so as to enter and seat accurately (see Fig. 116).

The thickness of the liners is found from the formula $t = \frac{D}{30} + .5$ where D = internal diameter of liner and t = thickness, both in inches. Thus

$$t = \tfrac{23}{30} + .5 = 1''.27 \text{ say } 1\tfrac{9}{32}.$$

The thickness of the valve-chest casting has been determined by the formula given previously.

The narrowest part of the pocket surrounding the liner and which forms part of the steam post is $1\frac{1}{2}$ inches in the direction of the diameter of the chest, and 7 inches vertically.

Fig. 116 shows the method of fitting the liner in the casting,

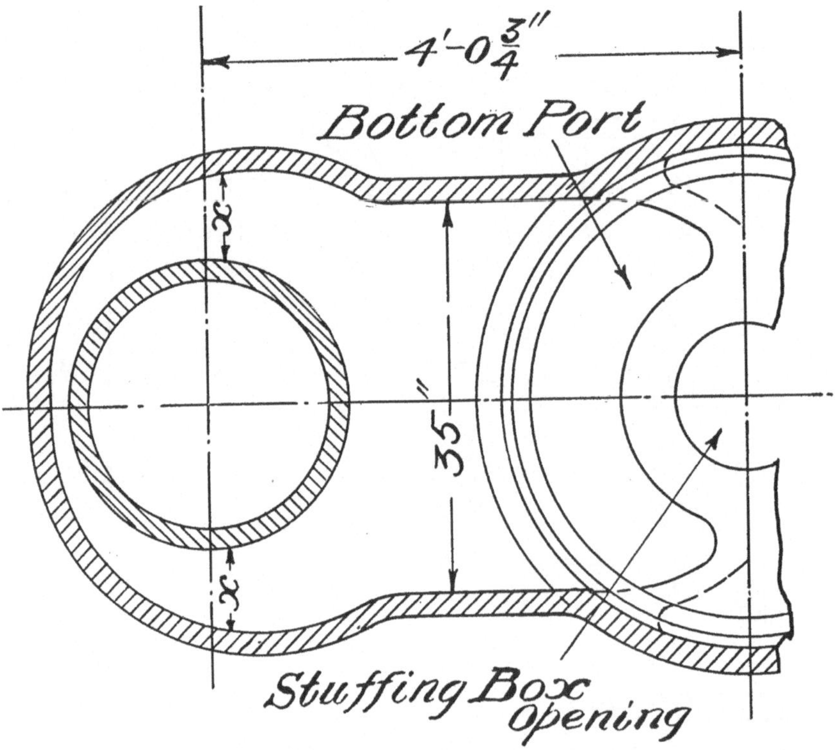

FIG. 117.—Section on ABCD (Fig. 116).

together with the dimensions used. It will be noticed that ample clearance for entering the liner through the end of the casting is provided, the diameter of the opening being $26\frac{1}{2}$ inches. Note that in the sketch all webs have been omitted for the sake of clearness. The lengthwise dimension of the port, where it runs vertically, has been reduced to 6 inches. This may be done without materially decreasing the area, as the width at this point has a circular form, surrounding the cylinder wall.

The distance between the center lines of the H. P. cylinder and the H. P. valve chest depends upon the length of the crank pin and the main bearings, and the thickness of the crank web. These have been already calculated, and the items are added as follows: $\frac{1}{2}$ length of crank pin+width of crank web+clearance+length of main bearing+distance from edge of main bearing to vertical center line of go-ahead eccentric. This works out, for the engine used as an illustration, as $4'-\frac{3}{4}''$.

It is desirable to keep this distance as small as possible, in order to reduce the clearance volume, which is always a large percentage of the piston displacement, especially in the H. P. cylinder.

Port in Valve Chest.—To determine the shape and size of the port immediately surrounding the valve-chest liner, proceed as follows: The dimensions of the port at the cylinder walls, as previously found, are $6\frac{1}{2}$ inches by 35 inches, giving an area of 223.5 square inches. Referring to the sketches, Figs. 116 and 117, it is seen that each quadrant of the liner must furnish one-fourth of the steam; hence the sectional area at xx (each side) must be equal to one-fourth of 223.5 square inches, or 55.9 square inches. The depth of the port around the liner has been made 7 inches. x then $=\frac{55.9}{7}=8$ inches (nearly). The outline of the passage can now be put in, using arcs of circles and straight lines and rounding off the corners.

NOTE.—The value of x of 8 inches as found above, was considerably reduced in this case, in order to keep down the clearance volume, even at the expense of a local increase in the velocity of the steam, between the bridges of the liner. This is a good example of one of the many compromises necessary in designing machinery.

Length of Working Faces of Valve and Liner.—The length of the valve must equal the opening of the port in the liner+steam lap +exhaust lap. In this case, from the Zeuner diagram previously constructed, length top$=4+2\frac{5}{16}+(-\frac{1}{2})=5\frac{7}{16}''$, length of bottom disk$=4+2+0=6$. The length thus found is that of the valve rings. For the length of the liner we now know the travel of the valve and the length of the valve face, so the length must equal the sum of these, or $10\frac{1}{2}+6=16\frac{1}{2}$. Each end of the liner is counterbored, or chamfered off $\frac{1}{2}$ inch to allow the valve rings to overrun in order to prevent the formation of shoulders (see Fig. 116).

To Fix the Position of the Liners in the Valve-chest Casting.—The total length of the valve chest is made the same as the total length of the cylinder. See Barton, Plate VII, Fig. 2. In order to reduce the length of the steam passages to each end of the cylinder it would be well to separate the two liners, axially, as far as possible, but here again a compromise must be made in order to allow space for the opening of the exhaust nozzle to the I. P. cylinder, the valve taking steam at the middle and exhausting at the ends.* The liners are placed equally distant from the transverse center line of the cylinder and valve chest. The lower valve-chest cover carries the valve-stem stuffing box and the upper cover carries the balance piston cylinder, and these project into the valve chest in the manner shown by Fig. 37, p. 87, Barton.

The size of the exhaust nozzle will be found to be about 14 inches internal diameter, and this, together with the necessity of allowing sufficient clearance between the valve and the lower edge of the balance cylinder, fixes the highest point at which the upper edge of the upper liner can be located. This is $15\frac{1}{4}$ inches from the upper flange of valve chest. The total length of the valve chest is 6 feet 3 inches, or $37\frac{1}{2}$ inches measured each way from the middle line. So we now have $37\frac{1}{2}-(15\frac{1}{2}+16\frac{1}{2})=5\frac{3}{4}$ inches. This $5\frac{3}{4}$ inches locates the inner edge of the liner, from the middle lines. As shown in Fig. 37, Barton, the edges of the openings in the liner are not flush with the edges of the ports in the casting, in order to allow of slight alterations should any be found desirable after the engine has been tried.

Valve.—The valve can now be laid off in its middle position, the principal dimensions being taken from the diagrams previously constructed. The steam enters at the middle, and as there are to be two exhaust nozzles, one on each side, at the upper part of the valve chest, the exhaust from the bottom of the cylinder must pass through the valve itself. Therefore, the valve will be similar to that shown by Fig. 37, Barton.

Valve Stem.—The specifications require as follows: "The high and intermediate pressure valve stems will be $3\frac{1}{4}$ inches in diameter at the stuffing boxes, reduced to $2\frac{1}{4}$ inches where they pass through the valves and to $1\frac{3}{4}$ inches above the valves. The balance pistons

* In the latest engines the length of the valve chest has been considerably increased in order to locate the valve liners nearly on a line with the top and bottom of the cylinder, thus giving an almost straight passage and considerably reducing the clearance volume.

will be secured to the valve stems by 1¼-inch nuts, as shown." (See Barton, Fig. 37.)

Valve-stem Stuffing Box.—The opening in the lower valve-chest cover will be made to accommodate a reliable make of stuffing box for a 3¼-inch rod. United States metallic packing is very generally used.

Cylinder Cover and Valve-Chest Cover Studs and Flanges.—Experience shows that the diameter of the cylinder cover studs for the pressure used in this engine should be somewhere about 1½ inches. *To find the pitch circle and the number of studs.* The thickness of the cylinder liner has already been found, which doubled and added to the inside diameter with a further addition for the seatings of liner, gives, in this case, 42¾ inches. The clear entrance at the top of the cylinder must be made somewhat larger than this maximum diameter, to provide clearance for entering the liner. Make this 43½ inches. The width of the flange should be from two and one-half to three times the diameter of the stud, or $1\frac{1}{2}'' \times 2\frac{1}{2}'' = 3\frac{3}{4}''$. This makes the outside diameter of the flange $43\frac{1}{2} + 7\frac{1}{2} = 51$ inches and the diameter of the pitch circle (the mean between inside and outside diameters) $(51 + 43\frac{1}{2}) \div 2 = 47\frac{1}{4}$. The pitch is roughly taken at between 4 and 5 inches in order to make a steam-tight joint. For steel studs of 1½ inches diameter a working stress of 7500 may be allowed. Then for a first approximation. Pitch=4 and number of studs $\frac{\pi \times 47.25}{4} = 37.1$, or say 36 bolts. To determine the exact diameter of studs:

$$\frac{\pi \times (47.25)^2}{4} \times 265 = 36 \times \frac{\pi d^2}{4} \times 7500,$$

$$d^2 = \frac{(47.25)^2 \times 265}{36 \times 7500} \quad d = 1.48 \text{ (effective) or use 36 } 1\tfrac{3}{4}'' \text{ studs.}$$

And for convenience the same size will be used for all cylinders. In the same way the studs for valve-chest covers are found. A stud not less than 1 inch should be used, so the width of the flange will be from 2½ to 3 inches. Using the larger value, and the diameter of valve chest previously found, viz, 26½ inches, the diameter of the pitch circle will be 29½ inches.

$$\frac{\pi \times 29.5}{4} = 23.2 \text{ or use 24 studs.}$$

NOTE.—The allowable stress per square inch decreases with the decrease in size of the bolt, and in this case, since the valve takes steam on the inside, the pressure on the cover for calculation is 125 pounds per square inch.

Then

$$\frac{\pi}{4}\times(29.5)^2\times125=24\times\frac{\pi d^2}{4}\times5000,$$

$$d^2=\frac{(29.5)^2\times125}{24\times5000},\ d=.952\text{ (effective) or use 24, }1\tfrac{1}{8}''\text{ studs.}$$

Valve-Chest Cover.—A cover of this kind is considered as secured around the edge in a manner between merely "supported" and "fixed," so that the formula may be of the form $t=\sqrt{\frac{3r^2p}{4f}}$, t=thickness in inches, r=radius of cover in inches (take radius of pitch circle of bolts or studs); p=maximum pressure in pounds per square inch; and f the greatest permissible stress of the material. For cast-iron covers, when a suitable value of f is inserted and a constant added for practical purposes, this becomes

$$t=\sqrt{\left\{\frac{3\times r^2\times p}{4\times45{,}000}\right\}}+.5.$$

This will give satisfactory results for dished covers, with ribs. The thickness of the ribs is made $.9\times t$, and the number found as given for cylinder covers.

This same formula may be used for the cover of the manhole in the cylinder cover.

Intermediate Pressure Cylinder.—Diameter, 63½ inches; area =3170 square inches. The thickness of liner, of barrel, of bottom head and cover are to be calculated by the rules and formula already given. Also the piston, piston rod, etc.

Valves.—In order to save length and to keep the diameter of the valve within reasonable limits two piston valves will be used, the stems being connected to a common yoke. (See Barton, Plates I, II and III.)

Dimensions of Steam Ports at I. P. Cylinder.—For this cylinder the width of the port may be taken slightly larger than the bore, which is 63½ inches, say make the width 66 inches.

$$\text{Area}=\frac{\text{Area I. P. cylinder}\times\text{piston speed}}{5500}$$

$$=\left\{\frac{\pi\times(63\tfrac{1}{2})^2}{4}\times\frac{120\times48\times2}{12}\right\}\div5500=552.$$

Then length of port$=\frac{552}{66}=8.36$ inches. Now since the corresponding dimension of the H. P. has been found to be $6\frac{1}{2}$ inches, and since the facings for the stuffing boxes must be in line for all cylinders, and the upper flanges of cylinders must also be in one horizontal plane, it will be impracticable to use so long a port as found above. In the engine under consideration this was reduced to 7 inches. Using this valve the port area becomes $66\times7=462$ square inches, and the velocity will then be $\frac{3170\times960}{462}=6580$ feet per minute. As this falls within the limits for *steam* velocities in I. P. cylinder, it may be used without fear of unduly increasing the back pressure on I. P. piston during exhaust.

Diameter of I. P. Valves.—Make the length of port in the valve liner 4 inches, as for the H. P. and assume 20% of the opening to be taken up by the bridges, leaving 80% clear opening. Then, since the port area has been decided as 462 square inches, and there are to be two valves, $2(\pi d\times4\times.8)=462$, or $d=22.9$, say 23″. Then the thickness of liners, and the inside diameter of valve chest, and outside diameter of valve-chest flange will be the same as for the H. P.

Cylinder Cover Bolts.—The diameter of these, for the H. P. was decided as $1\frac{3}{4}$ inches and for convenience the same size will be used for this cylinder. Calculating the thickness of the I. P. cylinder liner by the formula previously given, it is found to be $1\frac{1}{2}$ inches, therefore, proceeding as in the case of the H. P., the width of flange should be the same, or $3\frac{3}{4}$ inches, and the inside diameter of cylinder, at flange $67\frac{1}{4}$ inches and diameter of pitch circle$=71$ inches. Then.

$$\frac{\pi\times(71)^2}{4}\times125=n\times\frac{\pi\times(1.49)^2}{4}\times7500.$$

Number of bolts$=n=37.8$, say 38; pitch$=\frac{\pi\times71}{38}=5.87''$. This is too wide a spacing to make a reliable joint, and although 38 $1\frac{3}{4}$-inch bolts are enough for strength, the number must be increased to make a tight joint. Assume a pitch of 4 inches. Then

$$n=\frac{\pi\times71}{4}=55.8, \text{ or } 56, \text{ use } 56 \text{ bolts.}$$

From the calculation for the H. P. the diameter of valve-chest cover (outside diameter of flange), is $32\frac{1}{2}$ inches.

To Fix the Position of the Axes of Valve Chests relative to Axis of I. P. Cylinder.—Lay off the cylinder representing the outside diameter of the I. P. cylinder cover, the diameter in this case being 74¾ inches. Now the distance between the centers of the two valves, in the athwartships direction is often taken at from 1½ to 1¾ of the diameter of the valve itself. This gives good proportions for the passages connecting the two valve chests, but the exact dimension is not of great importance, so long as it is sufficient to give space for the rather complicated core work in moulding the casting, and at the same time is not so great as to unduly increase the clearance volume. Also it is desirable to keep the length of the yoke connecting the lower ends of the valve stems as short as practicable. Taking

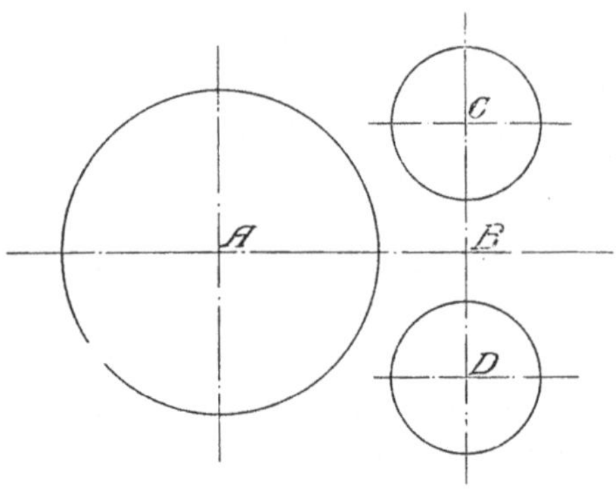

FIG. 118.

1¾ times the diameter of the valve gives $1\frac{3}{4} \times 23 = 40.25$ inches, or say 40¼ inches.

Therefore, the center of each valve is 20⅛ inches from the longitudinal center line of the engine. The minimum distance longitudinally from the center of the cylinder to the transverse line connecting the centers of the two valves, is fixed, as in the case of the H. P. by the length of shaft necessary for crank pin, width of crank web, length of main bearing, etc., up to the center of I. P. go-ahead eccentric. Lay off this minimum distance AB on the fore and aft center line of the engine, and draw a transverse line through the point B. On this transverse line, from B, in each direction lay off 20⅛ inches, giving the points C and D. With C and D as centers, and with radii equal to ½ the diameter of valve-chest covers, strike circles. If these circles clear the circle of the cylinder cover by, say 3 or 4 inches, the distance AB thus found may be used. The clearance of 3 or 4 inches is allowed for convenience in getting a wrench on the nuts of the cover studs. If the circles of the valve covers cut the circle of the cylinder cover, then the point B must be moved back from A, until the valve-chest covers clear the cylinder cover by 3 or 4 inches.

Proceeding as above in the case used; the diameter of the I. P. cylinder cover is 74¾ inches. The minimum possible length for AB is 4 feet 0¾ inches. The diameter of the valve-chest covers is 32½ inches. Using these dimensions it is found that the valve-cover circles cut the cylinder-cover circle. Therefore, B must be moved back. By actual trial it is found that AB must be given a value of 4 feet 5¼ inches, in order to give a clearance of 3½ inches at E. F.

Low-Pressure Cylinder.—The details of this cylinder are worked out in a way exactly similar to that given for the H. P. and I. P. The diameter of each L. P. cylinder in the engine used for illustration, is 74 inches. The diameter of the upper cylinder flange, and, therefore, of the cylinder cover is 85½ inches. The diameter of each L. P. valve, there being two piston valves, for each L. P. cylinder works out at 28½ inches, using a steam velocity of 7500.

NOTE.—In the engine used in this illustration, the steam velocities used were higher than those given in the note on velocity of steam. The engine was evidently designed by using the velocities allowed in *steam passages.*

The transverse distance between centers of the two valve chests is 45 inches, and the distance, fore and aft, between centers of valve chests and center of cylinder is 5 feet 2 inches.

Distance between Cylinders.—These are given by the specifications: F. L. P. to H. P., 5 feet 10 inches; H. P. to I. P., 12 feet 6 inches; I. P. to A. L. P., 6 feet 11 inches.

Steam Pipe, Receiver Pipes and Exhaust Pipes and Nozzles.—The specifications call for the main steam pipe to be connected to a nozzle at the middle of the H. P. valve chest, on the inboard side of the engine. There are nozzles, at the upper part of the H. P. valve chest on both the inboard and outboard sides for the receiver pipers. For the I. P. there are nozzles, on the upper part of the I. P. valve chests, on both the inboard and outboard sides for the receiver pipes from the H. P. There are also, from the middle of the I. P. valve chest, on the outboard side, two nozzles, one for the receiver pipe leading to the F. L. P. and the other for the receiver pipe leading to the A. L. P. As these pipes approach the L. P. valve chests they divide into two branches, thus leading the steam to both the upper and lower parts of the L. P. valve chests. The L. P. chests will then have a nozzle at the upper part and another at the lower part to receive the steam. They, also, each have a large central nozzle, on the outboard side for the main exhaust pipes. All of these steam, receiver and exhaust pipes have slip joints.

To Find the Sizes of the Pipes.—The velocity of steam through the main steam pipe may be 7500 feet per minute.

Then

$$\frac{\text{area H. P. cylinder} \times \text{piston speed}}{\text{velocity of flow}} = \text{area of pipe.}$$

The cut-off in the H. P. cylinder is not considered, and the size of the pipe is taken as if the steam followed full stroke. Then

$$\frac{\pi \times (38\tfrac{1}{2})^2 \times 960}{4 \times 7500} = 148 \text{ square inches, or dia.} = 13\tfrac{3}{4} \text{ (from table).}$$

Make main steam pipe 14 inches diameter.

Diameter of First Receiver Pipe, and of the Nozzles to which it Connects at H. P. and I. P. Valve Chests.—Considering this pipe as the exhaust pipe from the H. P. cylinder, it must have an area at least equal to the area of the exhaust port opening of the H. P. cylinder. This area was found to be 223½ square inches. Since there are two receiver pipes, each must have an area of at least $\frac{223.5}{2} = 111.75$ square inches, or the diameter must be 12 inches (nearly). These pipes also act as steam pipes for the I. P. cylinder. Treating them as such

$$\text{Area of each pipe} = \frac{\text{area I. P. cylinder} \times \text{piston speed}}{\text{steam velocity}}$$

$$= \frac{\pi \times (63\tfrac{1}{2})^2 \times 960}{4 \times 7000 \times 2} = 217 \text{ square inches.}$$

And the corresponding diameter is 16⅝ inches nearly. The size is to a certain extent governed by the opening to exhaust of the H. P. cylinder, so that two 16⅝-inch pipes are larger than necessary. Take the diameter of the pipe part way between the two solutions, about a mean between the two, disregarding fractions. Therefore, make each pipe 14 inches diameter.

Size of Second Receiver Pipes.—The diameter of these pipes is determined in a manner similar to that used above. Thus, the area of exhaust port in I. P. cylinder has been found to be 552 square inches, and since there are two exhaust pipes, one leading to the F. L. P. and the other to the A. L. P. cylinders, the area for each is $\frac{552}{2}=276$ square inches, the corresponding diameter being $18\frac{3}{4}$ inches. Considered as the steam pipe for the L. P. cylinders, and allowing a velocity of flow of 8000 a pipe about 25 inches diameter will be required. Taking a valve between, as before, the diameter to be used is 20 inches. These pipes divide, as they approach the L. P. cylinders, to lead the steam to both ends of the L. P. valve chests. In order to maintain an equal area, each branch must be between 14 inches and 15 inches diameter. Make each branch 15 inches diameter.

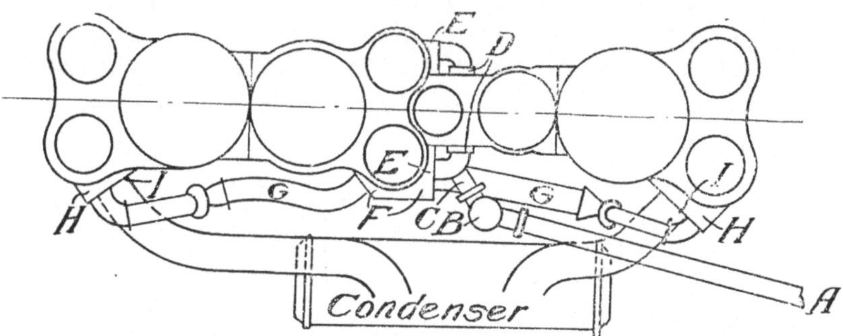

FIG. 119.—Sketch Arrangement of Steam, Receiver and Exhaust Piping.

Main Exhaust Pipe.—The exhaust from each L. P. cylinder is led to separate nozzles on the condenser. The area of the L. P. exhaust port is $7\frac{1}{2}\times 71$ inches giving an area of about 533 square inches, the corresponding diameter being slightly over 26 inches. The size of exhaust pipe actually used is 27 inches.

Laying off Pipes and Nozzles.—Having now determined the necessary sizes of the steam, receiver and exhaust piping, the sizes and angles of the nozzles on the several valve chests to which these pipes are connected can be laid off. Make a plan of the complete engine, showing the cylinders and valve chests in their correct relative positions, as in Fig. 119.

The main steam pipe, A, is led along the inboard side of the engine, as shown. B represents the main throttle valve, and C, the nozzle for steam of the H. P. valve chest. DD represent the H. P. exhaust nozzles, and EE the I. P. steam nozzles. FF are the I. P.

exhaust nozzles. GG the receiver pipes to the L. P. cylinders, and HH the steam nozzles of the L. P. valve chests. II the L. P. exhaust nozzles. Having thus obtained the general arrangement of the piping, the several nozzles are designed in detail on the drawings for the cylinders.

This practically finishes the design of the cylinders, there remaining only the flanges for bolting the various castings to each other, and the feet for attachment to the engine columns, or framing. In addition the bosses for attachment of the receiver safety valves, cylinder relief valves, drain valve connections, indicator bosses, peep holes for observing the setting of the main valves, etc., remain. These will not be described here.

Double-ported Slide Valve for L. P.—Quite frequently a double-ported flat slide valve is used for the L. P. engines instead of two piston valves. These are described in Barton's "Naval Engines and Machinery," p. 81, and the valve diagram with dimensions of the parts for such a valve is worked out in the plates (see Plate XXXVII).

132. *Check for Strength of Cast-Steel Pistons such as are Used for I. P. and L. P. of Naval Engines.*—After the piston has been designed and the thickness at different points determined, either by a method similar to that described, or by using the design of some other engine approximately similar to the new engine, the strength is checked by the following method:

The rupture of a piston is caused by the force P acting normally to the face. Rupture is resisted by the section of the piston made by the plane YY through the axis of the rod. To find the combined bending moments of the resolved forces, moments are taken from the center of gravity of the section lying in the plane YY.

The resolved force parallel to the rod (in the direction M_1) is the pressure $P \times$ the projected area of the piston $= P \times \pi R^2$.

On each side of the piston this force is $\frac{P \times \pi R^2}{2}$ and it acts at the center of gravity of a semicircle $= .4244R$ from YY, hence its moment $= M_1 = P \times \frac{\pi R^2}{2} \times .4244R = .2122\pi R^3 P$.

The resolved force in the direction M_2 is $P \times$ projected area of the section of the piston $= 2$ (triangle ABD + rectangle BCED) $\times P$ $= P \times 2\left(\frac{h}{2} \times r + h_1 \times r\right) = Pr(h + 2h_1)$.

NOTE.—The slight flat part under the hub is disregarded, and the form considered as being conical, as if the line of the inner face continued unbroken to A. This assumption does not cause an appreciable error. In the same way, as will be seen later on, the hub is disregarded and the line of the outside surface considered unbroken to the axis.

If y_c=distance of the center of gravity of the section from the base EC of the piston, then the arm of the resolved part of the ring force (projected rectangle CBDE) is $y_c-\frac{h_1}{2}$, and the moment of this force $=P\times 2h_1\times r\times\left(y_c-\frac{h_1}{2}\right)=Pr\times 2h_1\left(y_c-\frac{h_1}{2}\right)$. (A)

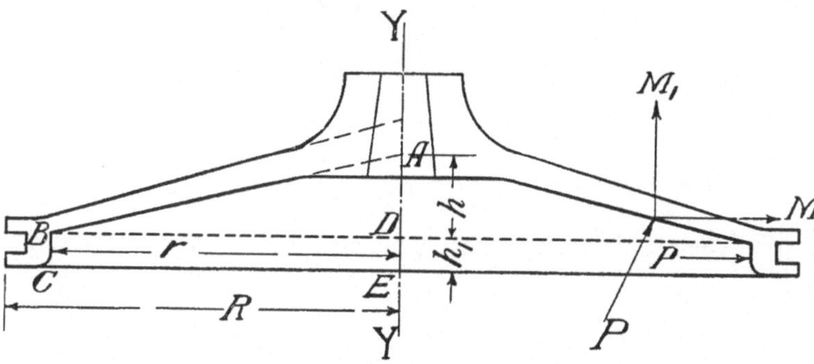

FIG. 120.—Piston (Calculations for Strength).

The arm for the conical part of the piston is at the center of gravity of the projected triangle or at $\frac{1}{3}$h above BD$=h_1+\frac{h}{3}$ above CE, or is $y_c-\left(h_1+\frac{h}{3}\right)$ distant from the center of gravity of the metal section of the piston, and the force is $2Pr\frac{h}{2}$, or Prh, and the moment $Prh\left[y_c-\left(h_1+\frac{h}{3}\right)\right]$. (B)

$$(A)+(B)=Pr\left\{2h_1\left(y_c-\frac{h}{2}\right)+h\left[y_c-\left(h_1+\frac{h}{3}\right)\right]\right\}=M_2.$$

There is also a distinct tensile pull due to the steam acting on the ring $BC=\frac{P\times 2rh}{2\times \text{area of section of piston (one side)}}=\frac{Prh}{a}$. This force is small and questionable and is probably accounted for in M_2. It is, however, added for safety.

The resistance to the sum of the moments is $f_t \times z \times 2$, where f_t = resistance of material per square inch. z is the modulus of section (one side).

$$M = M_1 + M_2, \text{ then } M = f_t \times 2z, \; f_t = \frac{M}{2z} = \frac{M_1 + M_2}{2z}.$$

To this must be added the increase for direct tension,

$$\frac{Prh}{a} = T \; \therefore \; f_t = \frac{M_1 + M_2}{2z} + T.$$

For shearing stress take any ring at radius x and thickness t, then circumference of the ring $= 2\pi x$. Force on the ring $= (\pi R^2 - \pi x^2) \times P = 2\pi x t f_s$ = strength of the ring.

$$\therefore f_s = \frac{P\pi(R^2 - x^2)}{2\pi tx} = \frac{P(R^2 - x^2)}{2tx}.$$

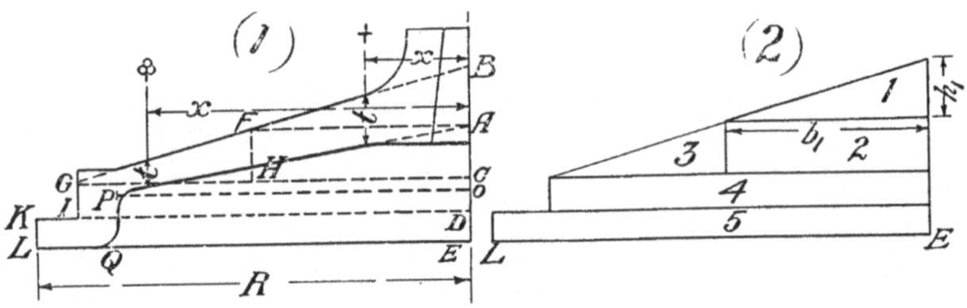

FIG. 121.—Showing Half Section of Piston Divided into Triangles and Rectangles for Finding Moment of Inertia, etc.

FIG. 122.—Showing Triangles and Rectangles Forming *Outside* Surface of Piston.

In order to find the moment of inertia and thence z, the half-section of the piston is divided into a number of simple figures (triangles and rectangles), and the moments of inertia of each figure about the base line are calculated; then by sums and differences the moment of inertia of the section of the piston is obtained.

Let k^2 = square of radius of gyration of the figures

$= \frac{1}{18} h^2$ for triangles, $\frac{1}{12} h^2$ for rectangles.

b = base of figures in (2).

h = height of figures in (2).

a = area of figures in (2).

$I = a \times k^2$ = moment of inertia of figures.

y = height of center of gravity of figures above EL.

Now, for convenience and clearness, arrange the computations in tabular form as follows:

FOR OUTSIDE FORM OF PISTON. (3)

Section.	Shape.	b	h	a	p^2	$a\times k^2=I$	y	ay	ay^2	$I+ay^2$
1	Triangle.	b_1	h_1	$\frac{b_1h_1}{2}$	$\frac{1}{18}h_1^2$	$\frac{b_1h_1^3}{36}$	y_1	a_1y_1	$a_1y_1^2$	$I_1+a_1y_1^2$
2	Rectangle.	b_2	h_2	b_2h_2	$\frac{1}{12}h_2^2$	$\frac{b_2h_2^3}{12}$	y_2	a_2y_2	$a_2y_2^2$	$I_2+a_2y_2^2$
3	Triangle.	b_3	h_3	$\frac{b_3h_3}{2}$	$\frac{1}{18}h_3^2$	$\frac{h_3h_3^3}{36}$	y_3	a_3y_3	$a_3y_3^2$	$I_3+a_3y_3^2$
4	Rectangle.	b_4	h_4	b_4h_4	$\frac{1}{12}h_4^2$	$\frac{b_4h_4^3}{12}$	y_4	a_4y_4	$a_4y_4^2$	$I_4+a_4y_4^2$
5	Rectangle.	b_5	h_5	b_5h_5	$\frac{1}{12}h_5^2$	$\frac{b_5h_5^3}{12}$	y_5	a_5y_5	$a_5y_5^2$	$I_5+a_5y_5^2$
Total.	......	....	....	Σa	....	......	....	Σay	....	$\Sigma(I+ay^2)$

FOR SPACE INCLUDED BY PISTON BEING TRIANGLE APO AND RECTANGLE OPQE. (4)

6	Triangle.	b_6	h_6	$\frac{b_6h_6}{2}$	$\frac{1}{18}h_6^2$	$\frac{b_6h_6^3}{36}$	y_6	a_6y_6	$a_6y_6^2$	$I_6+a_6y_6^2$
7	Rectangle.	b_7	h_7	b_7h_7	$\frac{1}{2}h_7^2$	$\frac{b_7h_7^3}{12}$	y_7	a_7y_7	$a_7y_7^2$	$I_7+a_7y_7^2$
Total.	......	....	....	$\Sigma a'$	....	.....	....	$\Sigma a'y'$	....	$\Sigma(I'+a'y'^2)$

To find the center of gravity:

$$\frac{\Sigma ay}{\Sigma a}=y_c \text{ (from (3)).} \qquad \text{(a)}$$

$$\Sigma a\times y_c^2=\Sigma ay_c^2. \qquad \text{(b)}$$

$$I_1=\Sigma(I+ay^2)-\Sigma ay_c^2 \text{ and } z=\frac{I_1}{y_c}. \qquad \text{(c)}$$

$M_1=.2122\pi R^3\times P$ where P= initial pressure on piston. (d)

C. of G. of internal space$=\frac{\Sigma a'y'}{\Sigma a'}=y_c'$ (from (4)). (e)

Total area of cross section of (4) $=2\times\Sigma a'$. (f)

Load on section (4) $=2\times\Sigma a'\times P$. (g)

Dist. (arm) between C. G.'s of (4) and (3) $=y_c-y_c'$. (h)

$$M_2=(2\times\Sigma a'\times P)(y_c-y_c'). \qquad \text{(i)}$$

$$T=\frac{Prh}{\Sigma a'(\text{from}(3))};$$

$$f_t=\frac{M^1+M_2}{2z}+T,$$

due to bending action;

$$f_s=\frac{P(R^2-x_1^2)}{2tx},$$

shearing force at any section, usually taken at points marked + and ⚘ in (1).

To allow for shock, due to water in the cylinder, the value of f obtained above must be less than 8000. If it works out greater than 8000 the thickness must be increased and another trial design made.

In actual practice, in determining the thickness of pistons, the design for an engine using about the same pressures, and of about the same size as the new engine and which has proved satisfactory in service will be at hand. The pistons are then laid down with small changes, according to the judgment of the designer. The method above given is then applied as a check, and the necessary corrections are made if required.

Questions and Problems.

Describe the method of determining the I. H. P. for a new vessel by Kirk's "Analysis." Illustrate by assuming the necessary data.

Explain the method of determining the I. H. P. required for a new vessel by Froude's "Law of Comparison." Illustrate by assuming data.

Make a sketch of the H. P. cylinder and valve chest of an engine of the following size: Diameter H. P. piston=35″; stroke=48″; R. P. M.=140; velocity of exhaust from H. P. not greater than 5000 feet per minute; width of piston face (total)=5¼″; width of piston rings=2¼″; piston clearance, top=½″, bottom=¾″; piston ring overruns counterbore ¼″, at top, ⅛″ at bottom; travel of valve =9″; steam lap, top=2″, bottom=1¾″; exhaust lap, top=2¼″; bottom=0″; valve to take steam on inside. Show thickness of liners, both cylinder and valve chest. Least distance possible between cylinder and valve chest axes=3′ 6″.

Design the L. P. piston (cast steel) for an engine having cylinders of 38½″, 63½″ and 2 L. P. of 74″ each. Stroke=48″; boiler pressure =265 p. g. Make a sketch showing all details with dimensions and material used.

Design the I. P. piston (cast steel) with a sketch showing all dimensions and material used. Data same as preceeding question.

Sketch the upper valve-chest cover for an engine, given the following data: Inside diameter of valve-chest flange $=26\frac{1}{2}''$; pressure on cover $=125$ pounds p. g.; travel of valve $=10''$; diameter of balance piston $=6''$; width of balance of piston $=2\frac{1}{2}''$. Find, the thickness of metal (cast iron); number and thickness of ribs; and number and size of studs.

Make a neat sketch of a cylinder cover (double) for the H. P. cylinder of an engine, including the number, size and pitch of the studs. Diameter of H. P. piston $=35''$; steam pressure $=250$ pounds p. g.

Find the sizes of the main steam, receiver and main exhaust pipes for a four-cylinder, triple-expansion engine, from the following data: Distance between centers of cylinders: F. L. P. to H. P $=5'\ 4\frac{1}{2}''$; H. P. to I. P. $=9'\ 10''$; I. P. to A. L. P. $=6'\ 2''$; H. P. valve abaft H. P. cylinder $=3'\ 7''$; I. P. valves forward of I. P. cylinder $=3'\ 7''$; H. P. valve center line to transverse center line of I. P. valves $=2'\ 8''$; L. P. valves forward and abaft their respective cylinders $=4'\ 9''$; cylinder diameters $=32''$, $52''$ and two of $72''$; stroke $=48''$; R. P. M. $=125$. Velocity of steam allowed: Main steam pipe $=8000$ feet per minute; first receiver $=7000$ feet per minute; second receiver $=8000$ feet per minute; main exhaust pipe $=7000$ feet per minute. Make a neat sketch showing connections and lead of pipes (no books). Reduce clearances as much as possible.

Given: the diameter of the I. P. cylinder of an engine $=58''$: R. P. M. $=125$; stroke $=48''$. Two piston valves. Make a sketch of the cylinder showing steam ports, valve liners, etc. Show dimensions of parts.

Find the I. H. P. required for a ship of 20,000 tons displacement and 21 knots speed, from the following data:

	Dimensions of Model Ship.	Dimensions of New Ship.
L.	450 ft.	510 ft.
B.	76 ft. 10 in.	85 ft.
Draft.	24 ft. 6 in.	26 ft. 11 in.
Immersed midship section area	1808 sq. ft.	2245 sq. ft.
Speed.	18 knots	21 knots
Disp.	16,000 tons	20,000 tons
I. H. P.	15,000	?

Find the power (1) by Fronde's "Comparison," (2) by Kirk's "Analysis," using $5\frac{1}{2}$ I. H. P. per 100 square feet of wetted surface.

Find the diameters of the cylinders of an engine from the following data: Triple-expansion, vertical, inverted cylinders, total I. H. P. (both engines) = 26,500; two L. P. cylinders each engine; cut-off in H. P. cylinder at 0.8 of the stroke; total ratio of expansion = 10; stroke = 48″; R. P. M. = 130; boiler pressure = 265 p. g.; vacuum = 26″.

Find the thickness of the cylinder barrels and liners and of the cylinder covers and bottoms for the following engine, using Bureau formulæ. Data: Cylinder diameters = $38\frac{1}{2}$″, $63\frac{1}{2}$″ 2 of 74″, by 48-inch stroke. Boiler pressure = 265 pounds p. g.; first receiver = 140 pounds (absolute); second receiver = 55 pounds (absolute).

From the following data find the size of the piston rods for an engine: cylinder diameters = $38\frac{1}{2}$″, 57″, 2 of 76″; steam pressure at H. P. cylinder = 265 pounds p. g.; in first receiver = 110 pounds (absolute); in second receiver 45 pounds (absolute); vacuum = 26″. All rods to be interchangeable. Make a sketch of the rod, showing threaded part for nut, collar, etc.

Practical Problem.

POWERING, SIZE AND DETAILS OF CYLINDERS, ETC.

Design, the cylinders; valve chests; pistons and rods; steam, receiver and exhaust pipes, for a vessel of the following principal features:

Displacement..................	20,000 tons
Speed..........................	21 knots
Length.........................	510 ft.
Beam...........................	85 ft.
Draft..........................	27 ft.
Area immersed midship section..	2245 sq. ft.
Steam pressure.................	265 lbs. per gauge.
Total ratio of expansion........	10
R. P. M........................	130
Vacuum.........................	26 in.
Stroke.........................	48 in.
Cut-off in H. P. cylinder.........	$\frac{8}{10}$.

The following data is available:

Name.	Displacement.	$\frac{L}{B}$	$\frac{L}{D}$	Block coefficient.	Speed.	I.H.P.
Ohio (class)	12,500	5.37	16.51	.665	18.23	17,000
Connecticut "	16,000	5.88	18.35	.659	18 82	20,442
Michigan "	16,000	5.6	18.35	.636	18.79	16,020
Washington "	14,500	6.9	20.0	.550	22.27	26,800

Use "Michigan" as a model for Froude's; check the cylinder dimensions by a ship whose engines are 32 inches, 52 inches and two of 61 inches, by 48-inch stroke; R. P. M., 120; I. H. P., 16,020; and having the same initial pressure and same total ratio of expansion as the new ship.

Comparison.—All calculations are to be entered on the blank interleaved pages.

Find (1) I. H. P. required (by two methods).

(2) Sizes of cylinders (use card efficiency of 53%).

(3) Details of cylinders, liners, barrels, etc.

(4) Piston and piston rod details.

Make a pencil drawing, showing H. P. cylinder and valve chest in section; a drawing of the L. P. piston and its rod, half in elevation, half in section. Such other details as time permits will be required by the instructor.

Directions and suggestions for the solution of the above problem:

In addition to the data given in the problem the following is needed:

For H. P. steam ports, make width=0.90 of diameter of H. P. cylinder.

To lay down the valve liners, assume travel of valve=10″; steam lap, top=$2\frac{1}{4}$″, bottom=2″; exhaust lap, top=(−)$\frac{1}{2}$″, bottom=0″0. Piston overrides counterbore, at top=$\frac{1}{4}$″, at bottom=$\frac{1}{8}$″; opening of port in valve liners=4″; inside diameter of piston rod=4″; use high-grade steel.

Details as to length, etc., must be determined by laying down the work on the drawing board.

Make steam passages as short and straight as possible. To accomplish this the valve chests may be made longer over-all than the cylinder itself by 9 inches or 10 inches at each end.

To Find the Diameter of the L. P. Cylinder by Comparison.—Let I. H. P., p_e, S and d represent the I. H. P., mean effective pressure, piston speed and diameter of L. P. cylinder of a given engine, and I. H. P.′, P_e', S′ and d′ the same for the required engine. Then

$$\frac{\text{I. H. P}}{\text{I. H. P}'} = \frac{p_e \times S \times d^2}{p_e' \times S' \times (d')^2} \text{ and } (d')^2 = \frac{\text{I. H. P}'}{\text{I. H. P}} \times \frac{p_e}{p_e'} \times \frac{S}{S'} d^2.$$

If the initial pressure and the total ratio of expansion are the same in the two engines, then $\frac{p_e}{p_e'} = 1$ and $(d')^2 = \frac{\text{I. H. P.}'}{\text{I. H. P.}} \times \frac{S}{S'} \times d^2.$

In case there are *two* L. P. cylinders, d and d′ are the diameters corresponding to the *combined* areas of both L. P. cylinders.

Solve and record results as follows:

(1) I. H. P. by Fronde's law.
(2) I. H. P. by Kirk's analysis.
(3) I. H. P., mean of (1) and (2) to be used.
(4) Diameter L. P. by p_e.
(5) Diameter L. P. by comparison.
(6) Diameter L. P., mean of (4) and (5) to be used.
(7) Diameter H. P. cylinder.
(8) Diameter I. P. cylinder.
(9) Thickness H. P. cylinder liner.
(10) Thickness H. P. cylinder barrel.
(11) Thickness H. P. cylinder bottom inner wall.
(12) Thickness H. P. cylinder bottom outer wall.
(13) Number of ribs.
(14) Distance between walls, use about 10 inches.
(15) Thickness H. P. steam port.
(16) Thickness H. P. valve chest.
(17) Thickness H. P. cylinder flange.
(18) Area of steam ports.
(19) Width of steam ports.
(20) Length of steam ports.
(21) Diameter of H. P. valve.
(22) Thickness of valve liners.
(23) Piston rod, diameter solid.
(24) Piston rod, diameter hollow (4-inch hole).
(25) L. P. piston, rise of cone L.
(26) i and a, L. P. piston.
(27) h and d_1, L. P. piston.
(28) T, L. P. piston.
(29) Diameter threaded part of rod.

CHAPTER XVII.

NOTES ON THE DESIGN OF CYLINDRICAL BOILERS.*

133. *General considerations* in connection with the design of the ship determine the type of boilers to be used. The type depends upon the vertical distance between the inner bottoms, or reverse frames, and the beams of the deck above; the total length and breadth allowed for boiler space; but principally upon the vertical space.

The decision depends, somewhat, upon a tentative process. If the ship has small draught and is long, the small, low, or gunboat boilers may be used.

If the ship has great beam compared with the draught, the boilers may be single ended, with a single fore and aft fire room. If the boiler space is sufficient in all respects, the power is divided among three or four furnace, double and single-ended, cylindrical, return-fire, tubular, boilers; or two furnace, cylindrical, single-ended, return-fire, tubular, boilers.

The conditions limiting the diameter are as follows: Sufficient distance must be left between the bottom of the boiler and the inner bottom, or frames, to allow cleaning, painting and caulking. This is generally six inches for small boilers of eleven feet, or less, in diameter, as the rapid curvature allows these points to be reached. For large boilers this distance is increased to ten and sometimes twelve inches. The distance between sides of boilers and bulkheads is about the same as the distance between shell and inner bottom. This is partly taken up by lagging. The distance between the tops of the boilers and the beams immediately above depends upon the spandrel space required for the uptakes. It should, when possible, be not less than 10 inches for boilers up to 11 feet in diameter, and as much as 15 inches for boilers of larger diameters. The uptakes may extend directly up through the protective deck, in which case no extra space is required for them, or they may join below the deck, as in the U. S. S. "Newark."

The distance between adjacent boilers should be the same as, or slightly greater than, that between the boilers and bulkheads.

* From "Notes on Machine Design."

The width of fire room should be at least one foot greater than the length of the grates to allow the easy handling of fire tools.

Selection of the type of boiler, whether high or low, as stated at the beginning of this article, is determined from the class of ship and draught of water, the preference, if there is any choice allowed, being given to the high or return tube type, as less volume of boiler per I. H. P. is required with this type than with the low type.

If the cross section of the ship will allow of boilers as large as 11 feet 6 inches in diameter, the high boiler may be adopted; if a diameter of less than 11 feet 6 inches is all that can be used, the low type must be employed.

This selection of type assumes that the length in the ship for the boilers is not determined, and the type may have to be changed accordingly. If the length of space for one boiler be less than 32 feet and more than 28 feet, the high type must be changed to low, as the double-ended boiler is about 20 feet long and the fire rooms at each end must be at least 7 feet long. Single-ended high boilers, placed back to back, take up more length in a ship than one double-ended, high boiler of the same power, as space must be left between the backs, and also, because one single-ended boiler is longer than one-half of a double-ended boiler, owing to the full width between the back of the combustion chamber and shell, and the thickness of the back heads of the boiler in each case.

The low type must be used for lengths between 32 feet and 28 feet, the low boiler being at least 17 feet long and the fire room 7 feet, while the space behind the boiler proper must be at least 4 feet, to allow for uptakes.

For boiler compartments less than 28 feet long, the single-ended high boiler must be used and this type requires a compartment length of at least 19 feet as the boiler will be nearly 11 feet long, fire room 6 feet and the space behind the boiler, 2 feet.

Grate area is the first definite quantity found in designing boilers for a ship. It was customary at one time in our service to place in the ship an auxiliary boiler for running the auxiliary machinery while in port. These auxiliary boilers, to use only natural draught, were designed primarily to furnish power for a definite number of auxiliary engines running all the time. Auxiliary machinery is extremely wasteful of steam and runs under conditions of very much less efficiency than the main engines. Some auxiliary engines require from 30 to 40 pounds of steam per I. H. P. per hour, while others require from 160 to 175 pounds. If the amount of steam required by the auxiliary machinery were a constant quantity, it would be easy enough to properly proportion the auxiliary boiler for its work; but, as this cannot be done, the auxiliary boiler has to suffer for the wastefulness of the auxiliary machinery and is now seldom or never fitted in our ships. It is customary, at present, to estimate the grate area necessary to furnish power for all machinery in operation on a trial trip, and to use this grate area as that of the main boilers. Where the calculations for a certain ship allow all boilers to be double-ended, one of these is replaced by two single-ended boilers, both together being of the same power as the double-ended boiler. These single-ended boilers are used in port for auxiliary purposes

In general, the machinery in operation during a trial trip is as follows: Main engines; air, circulating, feed, water-service and flushing pumps; fire-room and engine-room blowers and dynamos.

First Method of Calculating Grate Area.—Having determined, from the data obtained on different trial trips, the I. H. P. actually obtained per square foot of grate per hour from boilers similar to those proposed, the grate surface required is determined by dividing the given I. H. P. by the I. H. P. per square foot of grate per hour. The I. H. P. per square foot of grate varies, for one thing, with the air pressure in the fire room, and averages, for an air pressure of $1\frac{1}{2}$ inches of water, 16.3 I. H. P. per square foot of grate per hour.

The air pressure allowed by the naval authorities at present is for our cruisers and battleships about one inch of water.

I. H. P. PER SQUARE FOOT OF GRATE ON TRIAL TRIPS.

FOR AIR PRESSURE ABOUT ONE INCH.

	Air Pressure.	*I. H. P. per sq. ft. of Grate.*
Minneapolis	1.000	14.330
Marblehead	1.080	13.160
Massachusetts	1.000	16.900
Oregon	1.000	18.037
Helena	1.000	15.780
Wilmington	1.000	15.030
Iowa	0.987	16.010
Maine	1.074	16.930
Average for 1″		15.680

FOR AIR PRESSURE ABOUT TWO INCHES.

Texas	1.776	16.20
San Francisco	2.020	17.92
Baltimore	2.085	15.82
Newark (1)	2.230	16.42
Newark (2)	2.250	16.90
New York	2.000	16.55
Brooklyn	2.260	18.47
Average	2.090	16.90

Average of both tables above gives 16.3 I. H. P. per square foot of grate with an air pressure of 1½ inches.

The data above are from official trial trips.

NOTE.—Experience in the English naval service has shown the desirability of reducing the amount to which their boilers are forced, and their latest specifications for cylindrical boilers provide a total heating surface of not less than 2.5 square feet per I. H. P. at natural draught power, and 12 to 12.5 I. H. P. per square foot of grate, while the forced draught power is limited to 20% beyond the natural draught power.

Second Method of Calculating Grate Area.—Having determined from trial trips the number of pounds of coal burned per square foot of grate per hour, and the pounds of coal burned per I. H. P. per hour, the grate surface required is obtained by multiplying the required I. H. P. by the coal required to produce one I. H. P. per hour—this gives the total amount of coal burned—and dividing the product by the number of pounds of coal consumed per hour per square foot of grate.

From 30 to 40 pounds of bituminous coal can be consumed per square foot of grate per hour with an air pressure of 1 inch, and 40 to 50 with an air pressure of 2 inches.

The amount of coal burned depends upon the kind of coal, the amount of pressure of draught, ratio of length of grate to diameter of furnace, ratio of cross-section area of tubes to grate area and height of smoke pipe. The height of smoke pipe has, however, very little effect when forced draught is used.

Reliability of the Two Foregoing Methods.—The first method of determining the grate area is the more accurate. The I. H. P. and the grate area on trials are accurately measured, while the measurement of coal used is not generally made on trial trips, or if it is measured, there is ordinarily not so much accuracy as in the data for the first method.

Another method for grate area, sometimes used as a check upon the first two methods, is as follows:

$$\text{Grate area} = \frac{\text{lbs. water required per I. H. P.} \times \text{I. H. P.}}{\text{lbs. water evaporated per lb. coal} \times \text{lbs. coal burnt per sq. ft. grate.}}$$

Using 21.6 pounds water per I. H. P., and 7 pounds water per pound of coal gives a fair approximation. From this point on the steps in the process of designing can be best illustrated by a problem.

PROBLEM.

Required a set of boilers to supply steam of 160 pounds pressure, to triple-expansion engines, such as are used in modern naval vessels when developing under a maximum forced draught of 2 inches of water, 13,500 I. H. P. for four consecutive hours.

The following points are to be observed in the design of these boilers:

(1) Stays must be placed close enough to prevent tube sheets springing under expansion while steam is being raised.

(2) Stay bolts are to be screwed and the parts between the sheets turned down to the bottom of the thread; the screw stays to be screwed into the sheets and to be further held by nuts at both ends.

(3) The back ends of furnaces to unite to form the lower front plate of the combustion chamber.

(4) All diagonal braces to be fitted with bevelled washers where they intersect the plate and not to be bent.

(5) Tubes to be pitched horizontally and vertically and not zigzag.

(6) Corrugated or some good modern type of strengthened or ribbed furnaces to be used. Least inside diameter 3 feet 3 inches, least outside diameter 3 feet 6 inches, greatest outside diameter 4 feet.

(7) Diameter of tubes outside $2\frac{1}{4}$ inches, ordinary tubes No. 12 B. W. G., stay tubes No. 6 B. W. G.

(8) Ratio of axial length of furnace to rectified length = 1.175.

(9) The front head to be formed of three plates as follows:

Portion above the tube sheet a segment of a circle, then bent back to a cylindrical surface.

The tube sheet straight across at the top and corresponding to the furnace space at the bottom. The furnace sheet of the front head to be flanged externally to receive the furnace ends and to conform to the lower part of the front tube sheet. The tube sheet seams of these pieces to be double riveted zigzag.

(10) Between nests of tubes there must be not less than 6 inches clear passage and between the sides of the combustion chamber and shell at least 4½ inches, this to gradually widen to 8 inches or 9 inches at the top.

(11) Least distance between backs of combustion chambers at bottom about 6 inches, this to widen to about 10 inches at the top.

(12) Manholes to be placed in each spandrel at the bottom, over the middle furnaces, and in upper head sheet. Principal manholes to be not less than 12″ × 15″.

(13) The top row of tubes to be on a line that divides the circular section into $\frac{1}{3}$ and $\frac{2}{3}$, or a little above this if necessary to provide the proper heating surface.

(14) A trial sketch to be made to determine the number and arrangements of tubes.

(15) Two furnaces at each end to be connected to a common combustion chamber.

NOTE.—Some of the foregoing requirements must be met by any marine boiler, others such as (6), (8), (9) and (15) refer to the particular boiler in hand.

CALCULATIONS.

Grate Area—First Method.—Assuming 16.3 I. H. P. per square foot grate.

$$\text{Grate area} = \frac{13{,}500}{16.3} = 828.2 \text{ square feet.} \qquad (1)$$

Grate Area—Second Method.—Assuming 3 pounds of coal per I. H. P. on forced draught trials and 50 pounds per square foot of grate, we have:

$$\text{Grate area} = \frac{13{,}500 \times 3}{50} = 810 \text{ square feet.} \qquad (2)$$

Third Method.—The "Yorktown's" water consumption as accounted for by the indicator was 17.5 pounds, per I. H. P. per hour. The actual water supplied was not ascertained, but supposing a loss from condensation, leakage, etc., of about 20% the water used per I. H. P. $=17.5 \div .80 = 21.8$. Taking 50 pounds coal and assuming 7 pounds of water evaporated per pound coal, we have

$$\text{Grate area} = \frac{13{,}500 \times 21.8}{50 \times 7} = 848 \text{ square feet.}$$

As the first method is the most reliable, and in this case happens to be about a mean between the last two, we will decide upon 828.2 square feet as the necessary amount of grate surface.

Size and Number of Furnaces.—Before it is possible to fix upon the number of furnaces, the length of grate must be decided, for trial at least, and also the diameter of the furnace. Furnaces of cylindrical boilers should not be less than 30 inches in diameter, nor more than 48 inches unless in exceptional cases. Wherever possible, a diameter of not less than 40 inches should be given as, with smaller furnaces (owing to thickness of fire being practically constant for all diameters), the space above the fuel is much contracted and combustion less perfect in consequence. Long grates cannot be properly worked by hand and are therefore not so efficient. The length of grate varies from 5 feet to 7 feet and depends much upon the diameter of the furnace. In our service the average is between 6 feet and 6 feet 6 inches.

The specifications for this ship requires a furnace of 39 inches least inside diameter. Assuming a length of grate of $6\frac{1}{2}$ feet we have: No. furnaces $= \dfrac{828}{3\frac{1}{4} \times 6\frac{1}{2}} = 39+$; taking 40 furnaces we find the actual length of grate $= \dfrac{828}{40 \times 3\frac{1}{4}} = 6.'37 = 6'\ 4''$.

We have, therefore, for the required boilers, 40 furnaces, $6'\ 4'' \times 3'\ 3''$.

NOTE.—The plans of the boiler space for this ship show that a boiler greater than $15\frac{1}{2}$ feet cannot be used; boilers over 15 feet in diameter have four furnaces if single-ended and eight if double-ended. We can then have ten single-ended boilers or five double-ended boilers, provided the boiler space will accommodate this number of double-ended boilers. A reference to the plans of the boiler space shows that four double-ended and two single-ended boilers can be placed. This simply amounts to calculating for five double-ended boilers and cutting one of them in two. In any case the total number of furnaces should be a multiple of twice the number in one head.

Boilers up to	9′	diameter may have	1 furnace.
" "	13′ 6″	" "	2 furnaces.
" "	15′	" "	3 "
" beyond	15′	"	4 "

in each head.

Approximating the Length of One Boiler.—Allow 12 inches in length of furnace for the bridge-wall, and allow for furnace door, flanging, etc., a distance of $4\frac{1}{2}$ inches from the outside of front tube sheet to the commencement of grate. This distance is simply taken in order to get a basis of work from some fixed point of the boiler. In this case the outside of the front tube sheet is taken. The furnace ends will still project beyond this a sufficient amount to allow for riveting the furnace tubes to the head of the boiler.

Length of Combustion Chambers.—When two furnaces empty into the same chamber, this is usually about $\frac{1}{3}$ of the length of the grate.

For this dimension, practice seems to vary within considerable limits. (See Seaton, p. 460.)

Recent practice in the Bureau of Steam Engineering is to make the length of combustion chamber, for furnaces of 40 inches in diameter, two furnaces emptying into the same chamber, about 30 inches, but in any case space enough for a man to work in must be allowed. Make water back between opposite combustion chambers 6 inches at bottom, increasing to 10 inches at top.

Combustion chamber sheets for boilers working with over 70 pounds pressure are made of $\frac{1}{2}$ inch plates.

Then, approximately, the total length of the double-ended boiler is 2[dead plate + furnace + bridge wall + combust. chamber + comb. cham. sheet] + 6″ $= 2[4\frac{1}{2} + 6'\,4'' + 12'' + 2'\,6'' + \frac{1}{2}''] + 6'' = 21$ feet.

From the above, *length of tubes* = dead plate, $4\frac{1}{2}''$ + furnace, $6'\,4''$ + bridge wall, $12'' = 7'\,8\frac{1}{2}''$.

Approximate Diameter of Boiler.—Seaton says good results (for forced draught) can be obtained with an allowance of $1\frac{1}{4}$ to $1\frac{1}{2}$ cubic feet of boiler volume per I. H. P.

$$40 \div 8 = 5 = \text{No. boilers.}$$

$$\frac{13{,}500}{5} \times 1\tfrac{1}{2} = 4050, \text{ volume of one boiler in cubic feet.}$$

$$\frac{13{,}500}{5} \times 1\tfrac{1}{4} = 3375, \text{ volume of one boiler in cubic feet.}$$

$$\frac{4050}{21} = 192.8 = \text{area of head in square feet.}$$

$$\frac{3375}{21} = 160.7 = \text{area of head in square feet.}$$

For an area of 192.8, diameter = 15′ 6″.

For an area of 160.7, diameter = 14′ 3″.

So that a diameter of 15 feet 3 inches just inside the limiting diameter will give good results: ∴ we make the diameter = 15′ 3″.

The *number of furnaces* in one head, by reference to Seaton's tables, or by plotting the circle of the head, and roughly putting in the furnaces, is found to be four.

Thickness of shell:

$$\frac{t}{r} = \frac{p}{f};$$ factor of safety = 4.5. p = 160; f = 58,000 to 62,000 for mild steel; percentage strength of joint = 85.

$$\text{Thickness, } t = \frac{160 \times 183 \times 4.5}{62{,}000 \times 2 \times .85} = 1''.25 = 1\tfrac{1}{4}''.$$

Board of Supervising Inspector's rule:

P = pressure in pounds per square inch.

t = thickness in inches.

f = strength of material.

$$.80P = \frac{ft}{6 \times \text{rad}}.$$

$$t = \frac{.8 \times 160 \times 6 \times 183}{62{,}000 \times 2} + \frac{1}{16} \text{ (for corrosion).}$$

$$t = 1.14 + \tfrac{1}{16}.$$

$$\therefore \text{Thickness of sheets} = 1\tfrac{1}{4}''.$$

Area of Grate: For One Furnace = 6′ 4″ × 3′ 3″ = 20.58 sq. ft. 20.58 × 8 = 164.6 sq. ft. for one boiler.

Heating Surface of Furnace.—The ratio of length of corrugation to length of flue will be about 1.175.

$$\therefore \text{Heating surface} = \frac{8}{2} \times \frac{41 \times 3.1416}{12} \times 7\tfrac{1}{3} \times 1.175 = 370 \text{ sq. ft.*}$$

* There are 8 furnaces in the boiler, but the upper semi-circumference only of the furnace is effective as heating surface, hence $\frac{8}{2}$. 41″ is the mean outside diameter of the furnace.

Heating Surface of Combustion Chamber.—By measuring a rough sketch by means of a planimeter, it is found that the heating surface of the combustion chambers at one end is slightly in excess of the heating surface of the four furnaces emptying into them. This surface for this boiler is estimated at about 400 square feet.

The total heating surface that a cylindrical boiler may contain is = [diameter of shell]^2 × [length of tube + 3] for a single-ended boiler. For a double-ended boiler this will be changed to [2 × length of tube + 3]. From the above rule of Seaton's:

Total heating surface $= (15\frac{1}{4})^2 \times (2 \times 7\frac{17}{24} + 3) = 4981$ square feet.

In modern practice the heating surface varies from 2 to 2.3 square feet per I. H. P. While considerable care and skill are required to keep it as low as 2 feet and still obtain good results, 2.3 is of common occurrence. Taking a mean we have 2.15 square feet per I. H. P., and for one boiler, $2700 \times 2.15 = 5805$ square feet.

Tube heating surface $= 5805 - 370 - 400 = 5135$ square feet.

The area through tubes should be about $\frac{1}{7}$ the grate area, with natural draught (Seaton). In modern boilers designed by the Bureau of Steam Engineering the calorimeter through tubes is nearer $\frac{1}{8}$. For a mean, take ratio of grate surface to area through tubes of 7.5.

Tube area $= 164.6 \div 7.5 = 21.94$.

Diameter tubes, outside $= 2\frac{1}{4}''$ thickness 12 B. W. G. about ".12.

Internal diameter = about 2″.

$$\text{Area through one tube} = \frac{\pi d^2}{4 \times 144} = \frac{3.1416 \times 4}{4 \times 144}.$$

$$\textit{Total number of tubes} = \frac{164.6}{7.5} \times \frac{144}{3.1416} = 1008.$$

Number of tubes in each end = 504.

Pitch of tubes, about 1.5d and greater if possible. $1.5d = 3\frac{3}{5}'' = 3\frac{1}{2}''$.

This distance, $3\frac{1}{2}$ inches, is the standard practice of the Bureau of Steam Engineering for a $2\frac{1}{4}$-inch tube. Pitching these tubes on the tube sheet, about 532 can be placed, as shown, in the four nests. Allowing ample space between nests and over furnaces, number of tubes actually put in each end of a boiler = 532.

In calculating the tube heating surface the total length of the tube was taken and the portion of the tube sheet between the tubes was neglected in calculating the combustion chamber heating surface.

Length of tube $=7'\ 8\frac{1}{2}''$, circumference of $2\frac{1}{4}''=7''.0686$.

$$\text{Tube H. S.}=\frac{2\times 532\times 92.5\times 7.0686}{12\times 12}=4831.$$

Total H. S. $=4831+370+400=5601$ square feet.

$$\text{Ratio of H. S. to grate surface}=\frac{5601}{164.6}=33.82.$$

Laying Down Boiler on Drawing Board.—Draw a transverse view of shell and divide it by a vertical center line. One-half of this view is to be shown in section (the section taken between front head and front combustion chamber sheet), the other half in elevation. From the center, describe the arc of centers of furnaces. Leave 2 inches between the outer surfaces of furnaces. This is practically settled by the fact that the front head sheet must be flanged outward to receive the furnaces and a certain distance is required for this operation. The furnaces are 45″ from center to center, measured on chords of this circumference. The two middle furnaces have their centers $22\frac{1}{2}''$ from the vertical line. Since the corrugations of the furnace alternate, one furnace in the sectional view will be shown through the bottom of the corrugations 39 inches in diameter, and the adjacent furnace will show the maximum diameter 43 inches. Draw the sectional view of the combustion chamber with a water space of 4 inches between it and the shell at the bottom, increasing to 7 inches near the top of the tubes. Between the two combustion chambers allow a 6-inch water leg. Between each nest of tubes over the furnaces allow about 6 inches clear for circulation, examination and cleaning. As this boiler will require three sheets to make up one circumferential belt, we must locate the longitudinal joints. In order to be clear of obstructions, let one joint be about 45° with the vertical, and the remaining joints 120° from this one; the joints of successive belts breaking joints with one another.

The longitudinal joints will be double-strapped butt joints and the details will be found under Rivets and Riveting.

The tubes will be of mild steel $2\frac{1}{4}$ inches external diameter, No. 12 B. W. G. and swelled to $2\ \frac{5}{6}$ inches at front end. The back end of tubes will be expanded into the tube sheet, beaded over into a counterbore, which will be filled with a ring for protecting the ends of the tubes from the action of the flame. Any mode of protection may be used, subject to the following conditions:

1. That it will not interfere with the use of ferrules.

2. Will not cause injury to the tube sheets when the tubes are cut out.

3. Will not reduce the area through tubes.

Stay Tubes.—These will be No. 6 B. W. G. thick and will be reinforced at both ends to an external diameter of $2\frac{3}{8}$ inches, the bore of the tubes being uniform from end to end. They will then be swelled out at the front end to $2\frac{1}{2}$ inches. The stay tubes are threaded parallel at the combustion chamber end and taper at the front end to fit threads in the tube sheets. They will be screwed into the front tube sheet to a tight joint and will be made tight at the back end by expanding and beading.

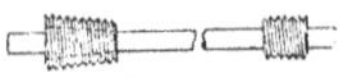

FIG. 122a.

The holes for stay tubes are tapped at one operation by a pair of taps on a rod as shown in Fig. 122a, and a duplicate of this is always carried among the spare machinery of the ship.

Cast-iron ferrules will be used to protect the ends of the stay tubes at the combustion chamber ends. The space between tubes should be sufficient to permit free circulation of water and allow the steam to rise freely to the surface. The tubes should be aranged in horizontal or nearly horizontal rows, *i. e.,* they should not be zigzag. They should not be crowded, as the steaming capacity of the boiler may be seriously impaired thereby.

The top of the top row of tubes should not be placed nearer to the top of the boiler than at a distance equal to one-third of the internal diameter of the boiler.

$$\frac{15'\ 3''-2''5}{3}=5' \text{ to top row of tubes.}$$

From this the center line of the top row of tubes can be obtained and the tubes pitched horizontally and vertically. In laying out the tubes, begin at the lower row, leaving 6 inches clear above the furnace for wing furnaces and 13 inches for the bottom furnaces. This is done for practical considerations and to give free outlet for the bubbles of steam forward on the crown sheet. The top and inboard side of the combustion chamber can then be completed.

Having put in the furnaces, combustion chambers and shell, in the transverse view, lay off the divisions into which the head sheet is divided. The upper sheet is cut with an allowance for lap sufficient for a double riveted lap joint (zigzag) leaving just sufficient metal clear above the tubes to permit of expanding the tubes ($\frac{1}{2}$ inch to 1 inch will be sufficient). Lay out the clear dimensions of the tube sheet for similar riveting at top and bottom.

In cutting the lower part of the front tube sheet let it enclose the manholes to be placed in the wing spandrels and above the middle furnaces, and thus act as a strengthening ring. The lap around these manholes should be sufficient for single riveting. In cutting the tube sheet from the wing to the middle nest of tubes, let the slope be about 30°.

The manholes in the wing spandrels will be three-sided, about 13″×15″, with the corners rounded and the sides curved to conform to the curves of the shell and furnace. The center manholes are to be elliptical, 11″×15″. Between the center furnaces and nests of tubes, support the flat head by braces about 10 inches apart, three over each furnace and one at the corner of the wing manholes. These braces tie the head sheet to the front plates of the combustion chambers. In the lower spandrels cut an 11″×15″ manhole; and at each side of these manholes and at the top strengthen the head by a brace; the upper one to continue through to the opposite end of boiler or to the combustion chamber; the lower ones to be diagonal and connected to the shell.

The back tube plates vary in thickness from $\frac{9}{16}$ inch to $\frac{7}{8}$ inch, Seaton. They are generally made, however, from $\frac{5}{8}$ inch to $\frac{3}{4}$ inch, depending on the size of boiler, as with this thickness the tubes can be made quite tight, and there is less liability of cracking the plates or burning the tube ends than with thicker plates.

It is not usual to make the sides and bottom of the combustion chamber exceed $\frac{1}{2}$ inch in thickness. They are stayed by screwed stays and the bottoms are stiffened by angle irons, about $3\frac{1}{2}''\times 3\frac{1}{2}''\times\frac{1}{2}''$, riveted to them. The screw stays are pitched to suit this thickness of plate by the method given in Unwin, viz:

$$f=\frac{2}{9}\,\frac{a^2}{t^2}\times p;\ f=6000.$$

The increased pressures have led to the adoption of corrugated, ribbed and spiral furnace flues for additional strength against collapse without abnormally increasing the thickness of the plate. There is a great objection among engineers to furnaces over $\frac{1}{2}$ inch or $\frac{5}{8}$ inch thick, on account of the increased resistance to the transmission of heat through the plate and the consequent overheating. While the resistance to the transmission in a laminated material may be abnormally increased by the thickness, yet in a homogeneous metal like steel it would seem that the thickness ordinarily used might with safety be increased, for, from experiment, the *condition of the surfaces* seemed to have a much greater value than the internal conductivity.

From the measured rate of transmission along bars and plates, when fairly clean and free from laminations, it was found that a furnace plate might be increased from $\frac{1}{2}$ inch to $\frac{3}{4}$ inch without increasing the resistance more than 1.25%.

In general, the practical thickness used for 160 pounds pressure is $\frac{1}{2}$ inch and $\frac{5}{8}$ inch.

The following is the formula generally adopted for corrugated flues:

$$p=\frac{C\times t}{d}.$$

t = thickness of plate in inches.

p = working pressure in pounds per square inch.

d = mean diameter of furnaces in inches.

C = constant having values from 12,500 to 15,000.

For Morison furnaces a value of 15,000 is used.

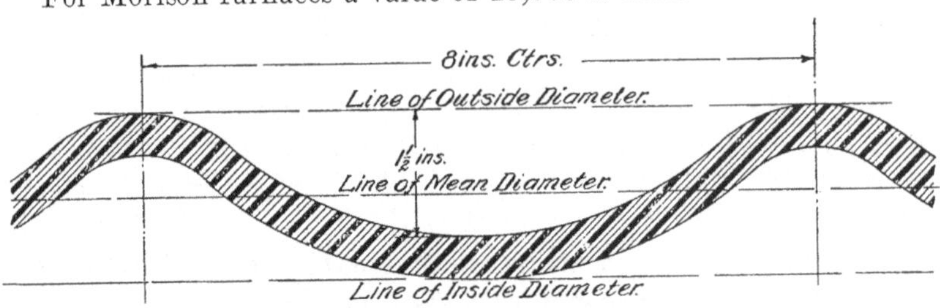

FIG. 123.—Morison Suspension Furnace. Size of Corrugation.

This value of C will depend upon the kind of furnace and of course upon the strength of the metal used. These furnaces are always made of steel of about 27 tons ultimate strength. For an intelligent assignment of a value to C, it is safest to obtain the latest edition of the maker's circular or to consult the value of C as assigned by authorized rules of inspection.

In the case of the boilers in hand,

$$t=\frac{160\times 41}{12{,}500}=\tfrac{9}{16}''.$$

Riveted Joints.—In general, for thin plates, where two plates of unequal thickness come together, it is customary to proportion the joints to the thinner plate. This rule will not always answer for thick plates, and to secure tightness at the joint it is proportioned, if not for the thinner plate, for a mean proportional between the two.

For the method of obtaining the formula used in proportioning joints see notes on riveting, Art. 134. They are mainly from Trail and Foley and are given as representing the best modern practice and are applicable to thick and thin plates.

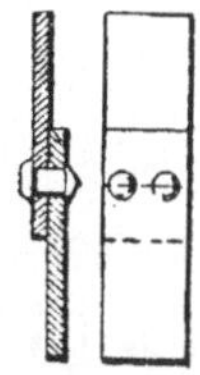

FIG. 124.

For the *connection of back end of furnaces to combustion chamber,* and for the joints of adjacent ring furnaces, use a single riveted lap joint. See Fig. 124.

$$t=\tfrac{1}{2}''; \ \%=54.73.$$

$$d=\frac{1.55\times 54.73\times\frac{1}{2}}{45.27\times 1\times 1}=''937=\tfrac{15}{16}''.$$

$$\frac{100(p-d)}{p}=54.73. \quad \therefore 45.27p=100\times\tfrac{15}{16}''.$$

$$\therefore p=2\tfrac{1}{8}''.$$

$$\text{Lap sheet}=2E=2\times 1.5d=3d=3\times\tfrac{15}{16}=2\tfrac{13}{16}'.$$

For the Upper Head Sheet and Tube Sheet.—Tube sheet $=\frac{7}{8}''$, upper head sheet $=1\frac{1}{4}''$. Making rivet section and plate of equal strength, $\%=\%_1$, and $d=1''25=$ thickness of sheet.

The joint is double riveted, lap (zigzag). See Fig. 125.

$$p=\frac{23\times A\times n\times c}{28\times t}+d=2.308+1.25=3.558=3\tfrac{9}{16}''.$$

In the above, t is taken as $\frac{7}{8}$, that is, the rivet is proportioned to the thinner sheet.

$$\text{Strength of joint}=\frac{3\frac{9}{16}-1\frac{1}{4}}{3\frac{9}{16}}=64.9\%.$$

Rivet section, $\%_1=64.7$.

$$E=\tfrac{3}{2}\times 1\tfrac{1}{4}=1\tfrac{7}{8}''.$$

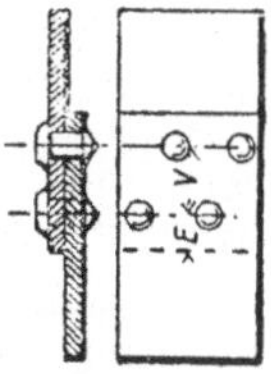

FIG. 125.

$$V=\frac{\sqrt{(11p+4d)(p+4d)}}{10}=\frac{\sqrt{(39\frac{3}{16}+5)(3\frac{9}{16}+5)}}{10}=1''945 \text{ or } 1\tfrac{15}{16}.$$

$V=1\frac{15}{16}''$. Lap of sheets $=2E+V=5\frac{11}{16}''$.

Connection of lower edge of tube sheet to head sheet: double riveted, zigzag, lap, joint.

$$d=t=\tfrac{7}{8}''.$$

$$p=\frac{23\times A\times n\times c}{28\times t}+d\times 1.129+.875=2''.004.$$

The percentage strength of plate and rivet sections are:

$$\%=56.56 \text{ (plate), and } \%_1=56.44 \text{ (rivet)}.$$

If the lower head sheet had been taken at $\frac{3}{4}$ inch and the joint made for equal strength of plate and rivet sections, the joint would more nearly have approached that of the upper one. For example, let $t=\frac{3}{4}''$ (thickness of lower head),

$$\therefore p=\frac{23\times .60132\times c\times n}{28\times .75}+.875=2\tfrac{1}{4}'' \text{ nearly},$$

and the strength of plate $=\left(\frac{2\frac{1}{4}-\frac{7}{8}}{2\frac{1}{4}}\right)100=61.1\%$, while the strength of rivets $=\frac{100\times 23\times 2\times .60132}{28\times 2\frac{1}{4}\times\frac{3}{4}}=60\%$.

This latter proportion for the joint would probably be better, hence the lower plate could be made $\frac{3}{4}$ inch and a very fair joint made. This sheet being thinner would also make the flanging for the furnaces easier.

Longitudinal Shell Joints.—Butt joints, double strapped, treble riveted with alternative rivets omitted in the outer row. See Fig. 126.

Percentage strength of joint $=85$.

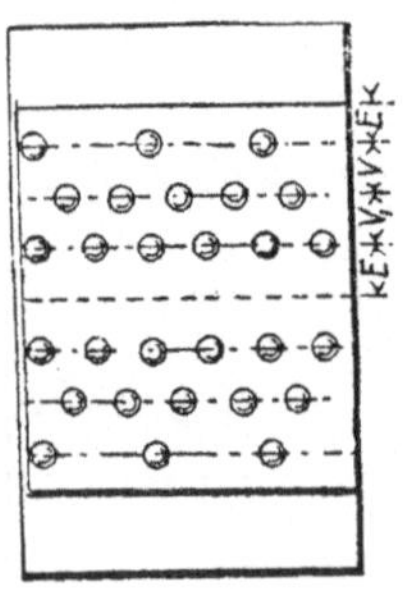

FIG. 126.

$$d=\frac{1.55\times 85\times t}{15\times 5\times 1.75};\ t=1\tfrac{1}{4}''.$$

$$\text{pitch, } p=\frac{1.55\times 85\times 1.25\times 100}{15\times 15\times 5\times 1.75}=8''.36.$$

$$\text{or } \frac{p-d}{p}=.85.$$

$$.15p=1\tfrac{1}{4}.\quad \therefore p=\tfrac{5}{4}\times\tfrac{100}{15}=8''.33.$$

$p=8\frac{3}{8}''$, greatest pitch or pitch of outer row.

Middle Circumferential Joint.—Treble riveted, lap, zigzag.

$t=1\frac{1}{2}''$, $d=1\frac{3}{8}''$. See Fig. 127.

$$p=\frac{23\times 1.484\times 3\times 1}{28\times 1.25}+1\tfrac{3}{8}=2.93+1.375=4.305=4\tfrac{5}{16}''.$$

$$V=\frac{\sqrt{(47\tfrac{7}{16}+5\tfrac{1}{2})\,(4\tfrac{5}{16}+5\tfrac{1}{2})}}{10}=2.28=2\tfrac{5}{16}''.$$

$$E=\tfrac{3}{2}d=2\tfrac{1}{16}''.$$

Lap of sheet $=2\tfrac{5}{16}+2\tfrac{5}{16}+2\tfrac{1}{16}+2\tfrac{1}{16}=8\tfrac{3}{4}''$.

Strength of above joint.

$$\%=\left(\frac{4\tfrac{5}{16}-1\tfrac{3}{8}}{4\tfrac{5}{16}}\right)100=68.1, \text{ plate.}$$

$$\%_1=\frac{100\times 23\times 3\times 1.4849\times 1}{28\times 4\tfrac{5}{16}\times 1\tfrac{1}{4}}=67.9, \text{ rivet.}$$

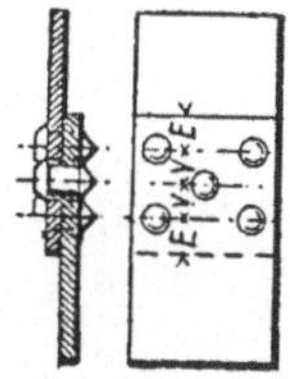

FIG. 127.

Circumferential Seams at Boiler Ends.—Owing to the additional stiffness imparted by the heads of the boiler, these seams are double riveted instead of treble riveted as in the middle seams. The joint is double riveted, zigzag. In this case as the shell is $1\tfrac{1}{4}''$ and the tube sheet $\tfrac{7}{8}$ inch the conditions are the same as for the joint between the tube plate and the head sheet, and the same proportions will be used. $p=3\tfrac{9}{16}''$; $d=1\tfrac{1}{4}''$; lap $=5\tfrac{11}{16}''$; $V=1\tfrac{15}{16}''$.

Width of Butt Straps, Distance between Rows of Rivets, etc.—From center of rivet to edge of sheet $=\tfrac{3}{2}d=1\tfrac{7}{8}''$.

$$V=\frac{\sqrt{(11p+4d)\,(p+4d)}}{10}, \text{ in which p is } \tfrac{1}{2} \text{ greatest pitch}=4\tfrac{3}{16}''.$$

$$V=\frac{\sqrt{(46\tfrac{1}{16}+5)\,(9\tfrac{3}{16})}}{10}=2.168=2\tfrac{3}{16}''.$$

$$V_1=\frac{\sqrt{(11p+20d)\,(p+20d)}}{20}$$ where p is the pitch of the outer row or $8\tfrac{3}{8}''$.

$$V_1=\frac{\sqrt{(92\tfrac{1}{8}+25)\,(8\tfrac{3}{8}+25)}}{20}=3.16=3\tfrac{3}{16}''.$$

Width of butt strap $=4E+2V+2V_1=7\tfrac{1}{2}+4\tfrac{3}{8}+6\tfrac{3}{8}=18\tfrac{1}{4}$.

$$\text{Plate section}=\frac{8.375-1.25}{8.375}=85.37\%.$$

$$\text{Rivet section}=\frac{23\times 1.227\times 5\times 1.75\times 100}{28\times 1.25\times 8.375}=84.26\%.$$

Thickness of Butt Strap.—$T=t\times\tfrac{5}{8}\left(\frac{p-d}{p-2d}\right)=.95$ or $1''$ for each strap.

Longitudinal Section of Boiler.—The length of the boiler is, from the details as fixed in the preceding work, 20 feet 10 inches, and is made up of three rings. The end seams are double riveted and the middle seams treble riveted. The distance from the plane of the front tube sheet to the edge of the shell is 4 inches. The lap of the middle seams, previously calculated, is $8\frac{3}{4}$ inches. Hence, the

$$\text{length of one ring} = \frac{20'\ 10'' = 2 \times 8\frac{3}{4}'' - 2 \times 4''}{3} = \frac{21'\ 7\frac{1}{2}''}{3} = 7'\ 2\frac{1}{2}''.$$

Draw the longitudinal view of the shell, cutting the sheets as shown in Fig. 128.

The upper ends are to be rounded to avoid extensive bracing. Leave a sufficient amount of flat space immediately above the upper joint of the head sheet for one row of braces with their washers, and then round the sheet to an arc of a circle (quadrant) : a radius of about 3 feet 7 inches will effect this.

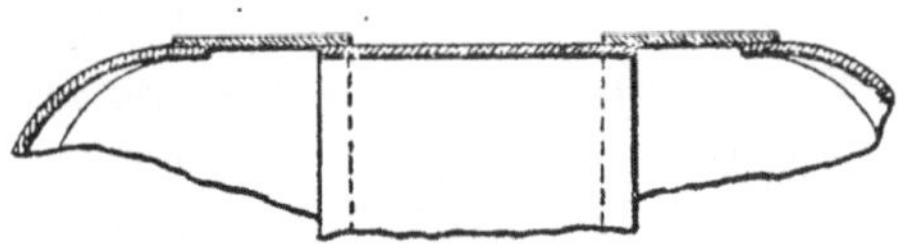

FIG. 128.

NOTE.—Curved ends of cylindrical boilers:

R = Radius of curved end in inches.

T = thickness of tube plate in inches.

d = outside diameter of tubes.

p = pitch of tubes.

r = ratio of solid metal left between tubes to pitch of tubes, *i. e.*,

$$r = \frac{p-d}{p}.$$

P = boiler pressure.

Then $R = \dfrac{C \times r \times T}{P}$, where $C = 13{,}000$.

If the percentage strength of the horizontal seam above the tubes be less than the percentage strength of plate left between the tubes, such lesser percentage should be used in calculating R. If the plate above the tube plate be less in thickness than the tube plate, its thickness should be used in the formula; that is, the smallest of the products $r \times T$ should be used.

Put in the head sheets and make all the joints of the shell. Project the furnace and combustion chambers and lay off the corrugations and laps of the furnace connections. Project the tubes across and put in all screw stays both in the longitudinal and transverse views.

Curved Tops of Combustion Chambers.—The combustion chambers are to have rounded tops, opposite ones to be tied together by gusset stays, and the rounded parts to be further stiffened by angle irons.

On account of the want of continuity, the curved tops of the combustion chambers are not capable of standing the same pressure as a complete cylinder of the same radius, and should be stiffened by angle or T bars. The constants for plain cylindrical furnaces should be considerably reduced when applied to curved tops, as the tops have been known to come down at about one-half the pressure suitable for a round furnace of the same plate thickness and radius. The curved tops should be efficiently stayed to the ends of the boilers, or with double combustion chambers back to back they should be efficiently tied together. The breadth of the combustion chamber in inches multiplied by the radius of curvature in inches is the least surface for which stay power is required. The radius of curvature should not be less than the width of the chamber, so as to avoid flat surfaces on the top and near the tube sheet. The bottoms of the combustion chambers are also stiffened with angle irons.

For determining the size of girders which support the flat top of the combustion chamber, the girder is treated as a beam supported at the ends and loaded at three points, as usually each girder carries three stay bolts. The formula for this is from Church's Mechanic's, page 270, and is:

$$M = fZ = \frac{f \times I}{y} = P_0 l_2 - P_1 (l_2 - l_1), \text{ where,}$$

$f = 8500$ pounds safe stress in outer fiber of girder.

$I =$ moment of inertia of section $= \frac{bd^3}{12}$.

$y = \frac{d}{2}$, distance of outer fiber from neutral axis.

$P_0 =$ load at one end of girder.

$P_1 =$ load on one girder stay bolt.

$l_1 =$ one pitch of girder stay bolts.

$l_2 =$ two pitches of girder stay bolts.

$b =$ breadth of girder $= 1\frac{3}{4}''$ in this case.

$d =$ depth of girder.

Length of girder equals depth of combustion chamber.

The girders are arranged in pairs, so b is the breadth of the pair.

Head Braces.—The area supported by each brace by measurement on the drawings is 15″×14.5″.

$$A=\frac{15\times14.5\times160}{9000}=3.87 \text{ square inches.}$$

d=2″.25 at bottom of thread.

Diameter of washer=6″.

Thickness of washer=$\frac{3}{4}$″.

Braces over Center Furnace.—Area supported by each about 15″×10″.

$$\text{Area brace}=\frac{15\times10\times160}{9000}=2.67 \text{ square inches.}$$

$\therefore$ d=1$\frac{7}{8}$″.

Nuts, threads, etc., from P. I.

Stays below Furnaces.—The positions of these braces are fixed by the relative positions of the shell, furnace flues and manholes.

Having fixed their positions as at XYZ, Fig. 129, the braces may be proportioned by supposing them to support the triangle (XYZ) formed by joining them, together with half the area included between the sides of this triangle and the adjacent shell and flues. To approximate to this area, erect perpendiculars at the middle points of the upper sides of the triangle. Take B, C and E half way between the sides of the triangle and the circumference of flues, etc. Draw ABG, GEF and ACF, so that these lines divide equally the area between the sides of the triangle, XYZ, and the flues or shell. The braces are calculated so that X supports ABDC, Y supports CDEF and Z supports BDEG.

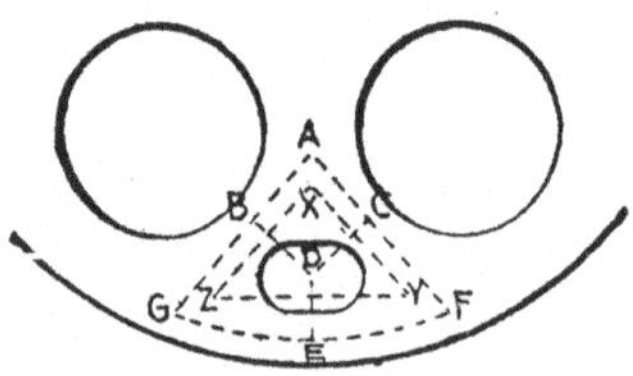

FIG. 129.

For the present case:

Area supported by brace X=128 square inches.

$$\text{Area of brace}=\frac{128\times160}{9000}=2.27 \text{ square inches.}$$

$\therefore$ diameter=1″.7.

Area supported by stays Y and Z=120 square inches.

Area of stay=2.13 square inches. $\therefore$ diameter=1$\frac{5}{8}$″.

Of these three stays the upper one will be a through stay passing between the combustion chambers to the opposite head; the lower stays will be diagonal and secured to the sides of the shell by "palm ends" and rivets.

Diagonal stays are never used if a horizontal stay can be used, as the resultant tension is greater on the diagonal stay.

If a = area of diagonal stay,

$$a = \frac{\text{area supported in square inches} \times \text{pressure}}{f} \times \text{sec. A.}$$

Where A is the angle the diagonal stay makes with the bottom of the boiler, in the same radial plane as the stay.

The braces around other manholes are direct stays fastened to the head and to the front sheets of the combustion chambers.

The washers used for main stays are riveted to the head.

Screw Stays.—The combustion chambers are stayed to each other and to the shell of the boiler by stays screwed into both sheets and fitted with nuts; the nuts to be set up on beveled washers where the stays do not come square with the plates. The holes for screw stays will be tapped in both sheets while they are in place.

All screw stays and braces will have raised threads; all braces will be made without welds.

Pitch of stays $= 7''$.

Area of stay $= \frac{7 \times 7 \times 160}{6000} = 1.306$ square inches.

Diameter $= 1''.25$ or $1\frac{1}{2}$ over threads.

For Pitch of Stays and Braces.—The pitch, l, of the stay bolts and braces depends in a measure on the thickness of the flat plates supported, as will appear by treating the square surface supported as a beam fixed at the ends and uniformly loaded. The formula for this is:

$$\frac{f \times I}{y} = \frac{W \times l}{12}, \text{ where } f = \frac{65{,}000}{4.5}, \; I = \frac{l \times t^3}{12}, \; y = \frac{t}{2}, \; W = p \times l^2.$$

The area supported by each stay or brace is l^2. If p is the pressure per square inch on the flat surface, then the total pressure supported by each stay or bolt will equal $p \times l^2$. The safe load per square inch of section of good steel stays and braces is 8000 pounds, therefore the area of each stay must be such that:

$$8000 \times \frac{\pi d^2}{4} = p \times l^2.$$

The diameter d of the stays is taken at the root of the threads, where the ends are fitted with nuts.

Grate bars are to be of wrought iron, cast iron or steel.

Bridge walls are of cast iron; to be readily removable, finished with fire brick.

Furnace fronts are to be made with double walls of wrought iron, the space between the walls to be in communication with the ash pits. The upper part of inner plate is to be perforated as directed; dead-plate to be of cast iron and easily removable.

Furnace doors must be protected in an approved manner from the heat of the fire, and open inward.

Ash-pit doors to be made of $\frac{1}{8}''$ iron, stiffened with angle irons. False ash pans of $\frac{1}{4}$-inch iron are to be fitted under all grates.

Circulating plates are to be fitted at each side of each nest of tubes and made so as to be readily removed and introduced through manholes.

The uptakes are to be of wrought iron, No. 8 B. W. G., built on channel irons, stiffened with angle irons, and bolted to the boiler heads and shells. Outside of the uptake is to be a jacket enclosing a 3-inch space. This space is to be filled with an approved non-conducting material.

Uptake doors are to be double shells of iron, No. 8 B. W. G.

The space between the shells is to be fitted with the same non-conducting material as used for the uptakes.

Smoke pipes are to be two in number; the top about 85 feet above the grates, the bottom part shaped to join the uptakes. The top is to be finished with angle irons and with a hood covering the casing. The pipes are to be stayed by shackles and guys with turnbuckles for setting up the guys.

A casing enclosing a 3-inch air space extends from the uptake to about 6 inches below the hood at the top. An additional casing is placed for 5 feet or more above smoke pipe hatches.

Boiler attachments are to be supplied as follows:

1 self-closing main steam stop valve.

1 self-closing auxiliary steam stop valve.

1 dry pipe.

1 main feed check valve with internal pipe.

1 auxiliary feed check valve with internal pipe.

1 bottom blow valve with internal pipe.

1 surface blow valve with internal pipe and scum pan.

2 safety valves to be connected with dry pipes, or to have internal pipes.

1 steam gauge, on single-ended; 2 on double-ended boiler.

2 glass gauges with automatic cocks on single-ended, and 3 on double-ended boiler—two of them being at feeding end.

4 gauge cocks on each end of double-ended boiler.

1 sentinel valve of $\frac{1}{2}$-inch area.

1 salinometer pot.

I drain cock.

1 air cock.

1 approved circulating apparatus.

1 cock threaded for the attachment of a syringe.

All external fittings, unless specially stated to the contrary, are to be of composition.

All fittings are to be flanged and through bolted.

All cocks, valves and pipes are to have spigots or nipples passing through the boiler plates.

Internal pipes of brass must not touch the boiler plates except where they join their external fittings.

The stems of all valves on the boilers are to have outside screw threads.

Securing Boilers in Iron and Steel Ships.—The boilers are generally placed on board the ship after launching, as the vessel being waterborne is less liable to be strained from the concentration of

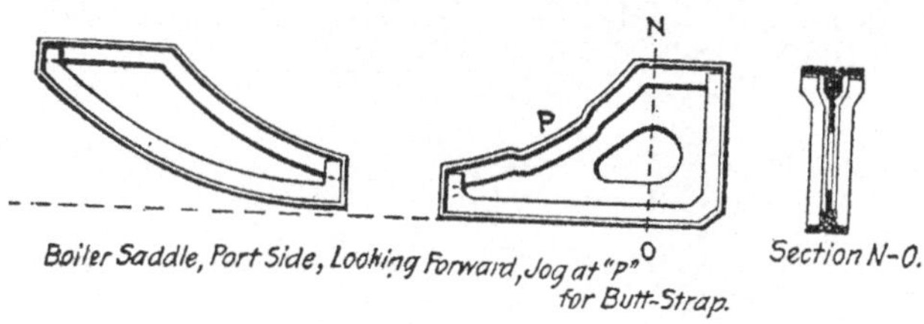

Fig. 130.

weight. The framing of the boiler hatches should be so arranged as to give ample room for lowering the boilers into their compartments. The ship is taken under the shears or derrick, the boilers, slung by straps, lifted on board and lowered on blocking. They are then moved along the floors by jacks until in their proper positions over the saddles and are then lowered into place. The boilers rest on saddles and are secured to them by bolts. The number of saddles depends on the size and kind of boilers used, being generally two for single-ended and four or five for double-ended boilers. They should be spaced so as to allow access to the circumferential seams for caulking. Where double-ended boilers

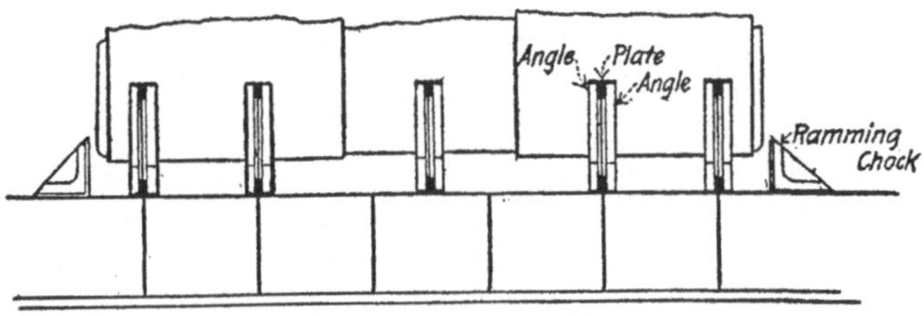

Showing Saddles in Position.

FIG. 131.

are used, the fire rooms are generally athwartships. The saddles are then arranged so as to come directly over the frames. Each saddle consists of a vertical plate secured to the inner bottom and reverse frames by angles. The top of this plate is cut to the curvature of the boiler, to which it is secured by angles on each

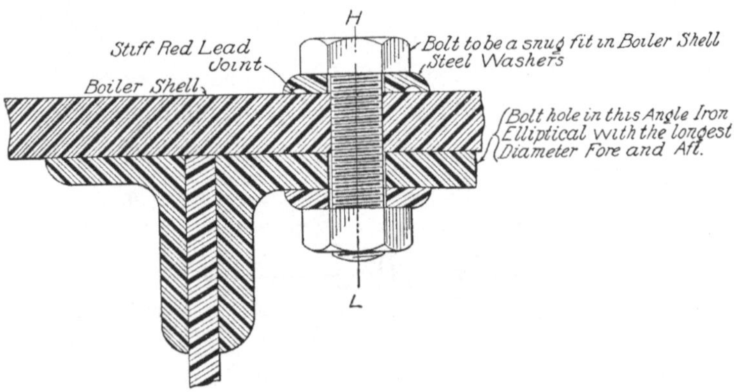

FIG. 132.

side, jogged if necessary to fit over the longitudinal butt straps. The boiler rests on the angles. The boilers are screwed to the saddles by bolts and nuts. The bolts are screwed through the shell from the inside and have a cup washer under the head, the joint being made tight by red lead putty under the washers. The holes through the flanges of the saddles are elongated in a fore and aft direction to allow for expansion. In addition to the above, some means must be adopted to prevent displacement in a fore and aft direction, if the vessel is intended to be used for ramming. The usual method is to fit what are called ramming chocks. These consist of a vertical plate about 1 inch thick fitted to each end of the boiler and secured to the inner bottom by angles. They are placed opposite the lowest point of the boiler and are fitted so as to extend about 6 inches above the lowest point and a clearance of about one-quarter inch is left between them and the boiler. In large boilers two are used at each end. When the boilers are placed with their axes athwartships the saddles should be placed over the longitudinals; when this is not possible, heavier plates should be worked in the inner bottom below the saddles. The saddles and the methods of securing the boilers to them are illustrated in Figs. 130, 131 and 132.

134. *Riveting.—Formulæ for Proportioning Riveted Joints and Calculating the Width and Thickness of Butt Straps for Steel Boilers.*—With the aid of these formulæ the tables for riveting joints in Traill's work are calculated, but for every-day work in the designing room it will undoubtedly be a time-saving operation to obtain the desired information direct from those tables. For convenience in comparing with the above-mentioned tables the same nomenclature is used. The following assumptions are made:

First. That the mean tensile strength of the plate is 28 tons and the mean shearing strength of the rivets is 23 tons per square inch of net section.

Second. That rivets in double shear offer 1.75 times the resistance to shearing exerted by rivets in single shear.

Let p=pitch of rivets in inches (greatest).

d=diameter of rivets in inches.

c=constant whose value is 1 for lap or single butt strap, and 1.75 for double butt-strapped joints.

A=area of one rivet in square inches.

n=number of rivets in the greatest pitch.

$\%$=percentage of plate left between rivets in the greatest pitch.

$\%_1$=percentage of the rivet section compared with that of the solid plate.

$\%_2$=percentage of the combined plate and rivet section when alternate rivets are omitted in the row

p_d = the diagonal pitch.

E = the distance from the center of the rivets to the edge of the plate.

V = the distance between the rows of rivets in ordinary zigzag riveting, and in chain riveting when alternate rivets are omitted in the outer row.

V_1 = the distance between the outer and the next row of rivets in zigzag riveting when alternate rivets are omitted in the outer row.

t = the thickness of the plate in inches.

t_1 = thickness of a single butt strap in inches.

t_2 = the thickness of a double butt strap in inches.

To Find the Percentage Strength of any Given Joint.—The metal left between the rivet holes $= (p-d)t$, and the percentage strength of this as compared with the solid plate is:

$$\frac{100(p-d)}{pt}\times t. \quad \therefore \% = \frac{100(p-d)}{p}. \tag{1}$$

The resistance offered to shearing by the rivets in a single pitch is $23\times A\times n\times c$; comparing this with the strength of the solid plate $=28\times p\times t$, we have:

$$\%_1 = \frac{100\times 23\times A\times n\times c}{28\times p\times t}. \tag{2}$$

Considering now the case where alternate rivets are omitted in the outer row, in the next row there will remain $(p-2d)\times t$ solid metal, and $\frac{100(p-2d)}{p}$ will be the percentage strength of this row as compared with the solid plate; but, supposing that the plate tears along this row of rivets (see Fig. 134), before a total giving away of the joint occurs one rivet for each pitch must also be sheared in the outer row and the percentage strength of this single rivet is $\frac{\%_1}{n}$, which value will be obtained from equation (2). From the above considerations we have

$$\%_2 = \frac{100(p-2d)}{p} + \frac{\%_1}{n}. \tag{3}$$

The lowest of the values obtained from equations (1), (2) and (3) is the percentage strength of the joint. An examination of these equations shows that for double butt-strapped joints, so long as the diameter of the rivets is not less than the thickness of the plates, $\%_2$ is always greater than $\%$ or $\%_1$. This is also the case with lap joints, so long as the diameter is not less than $\frac{t}{\frac{23}{28}\times\frac{3.1416}{4}}=\frac{t}{\cdot 64515}$, and, as both these conditions usually hold, the use of formula (3) will seldom be necessary.

To find d, when p, c, n and t are given, so that $\%$ may equal $\%_1$ (that is, so that plate and rivet section may be equivalent in strength).

From (1) and (2) we have:

$$\frac{100(p-d)}{d}=\frac{100\times 23\times A\times n\times c}{28\times p\times t}. \qquad (4)$$

Substituting for A its value $=\frac{3.1416\times d^2}{4}$ and simplifying, we have:

$$d^2+\frac{1.55td}{n\times c}=\frac{1.55pt}{n\times c}.$$

Solving for d we have:

$$d=\sqrt{\left(\frac{.775t}{n\times c}\right)^2+\frac{1.55pt}{n\times c}}-\frac{.775t}{n\times c},$$

which reduces to:

$$d=\sqrt{\frac{.775t}{n\times c}\left(\frac{.775t}{n\times c}+2p\right)}-\frac{.775t}{n\times c}. \qquad (5)$$

If d, c, n and t are given; from equation (4) we have:

$$p=\frac{23\times A\times n\times c}{28\times t}+d. \qquad (6)$$

To find d and p, when n, c and t and $\%=\%_1$ are known:
From (1) we have

$$p=\frac{100d}{100-\%}. \qquad (7)$$

Substituting this value for p and letting $A=\frac{22d^2}{7\times 4}$ in (2), we have:

$$\%=\%_1=\frac{100\times 23\times 22\times d^2\times n\times c}{28\times 7\times 4\times\frac{100d}{100-\%}\times t}.$$

Solving,

$$d=\frac{1.55\times\%\times t}{(100-\%)\times n\times c}. \tag{8}$$

Substituting this value of d in (7) we have:

$$p=\frac{100\times1.55\times\%\times t}{(100-\%)^2\times n\times c}. \tag{9}$$

It will generally be found simpler to substitute the value for d found from equation (8) in equation (7), thus obtaining a value for p direct, without using equation (9).

Diagonal Pitches and the Width of Butt Straps.—The value of the metal in the diagonal pitch p_d is only about $\frac{5}{6}$ of what it is in the horizontal pitch, or in that part of the horizontal pitch to which it is required to be equivalent in strength. In any case, the diagonal pitch should not be less than that found by the following formulæ:

I.

Ordinary zigzag riveting and chain riveting with alternate rivets omitted in the outer row.

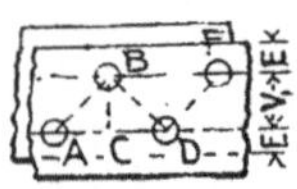

FIG. 133.

Reference to Fig. 133 shows that the same reasoning applies to both cases, as in each case the net section of two diagonal pitches must be made equivalent in strength to the net section contained in the greatest horizontal pitch. The liability of the sheet to tear along AB and BD would be the same as along AD or BF.

Now the metal left along ABD is $2(p_d-d)t$ and along AD or BF is $(p-d)t$.

$$\therefore 2(p_d-d)t=\tfrac{6}{5}(p-d)t, \text{ and } p_d=\frac{6p+4d}{10}. \tag{10}$$

From the right angle ABC we have for the distance between the rows of rivets:

$$V=\sqrt{p_d^2-\left(\frac{p}{2}\right)^2}=\sqrt{\left(\frac{6p+4d}{10}\right)^2-\frac{p^2}{4}}=\sqrt{\frac{11p^2+48pd+16d^2}{100}},$$

$$\text{or } V=\frac{\sqrt{(11p+4d)(p+4d)}}{10}. \tag{11}$$

The authors mentioned before, state that, for chain riveting, this distance should not be less than $\frac{4d+1}{2}$ and, as this result is greater than that obtained from (11), it is the one that is usually found tabulated.

The distance E from center of rivet to edge of plate should not be less than 1.5d; so that the minimum lap of sheets, if lap joint, or half width of butt strap$=2E+V$. (12)

II.

Zigzag riveting with alternate rivets omitted in the outer row.

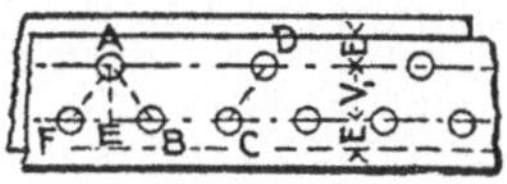

FIG. 134.

An examination of Fig. 134 shows that rupture may occur by the plate giving away along ABCD, or along AD, and to make the joint symmetrical in strength, the net section along ABCD must be equivalent in strength to the net section along AD. Now $BC=\frac{p}{2}$ and the net section along BC is $\left(\frac{p}{2}-d\right)t$, while on AD it is $(p-d)t$; hence the section to which the metal left on the lines AB and CD must be equivalent in strength is: $(p-d)t-\left(\frac{p}{2}-d\right)t=\frac{p}{2}t$, but the net section on AB or CD is $(p_d-d)t$; therefore, $2(p_d-d)t=\frac{6}{5}\left(\frac{p}{2}t\right)$, and

$$p_d=\frac{3p+10d}{10}. \qquad (13)$$

In the triangle AFE we have $FE=\frac{BF}{2}=\frac{p}{4}$, and, for the distance between the rows of rivets, we have:

$$V_1=\sqrt{\left(\frac{3p+10d}{10}\right)^2-\left(\frac{p}{4}\right)^2}=\sqrt{\frac{9p^2+60pd+100d^2}{100}-\frac{p^2}{16}}$$
$$=\sqrt{\frac{11p^2+240pd+400d^2}{400}}=\frac{\sqrt{(11p+20d)(p+20d)}}{20}. \qquad (14)$$

As before, the half breadth of butt strap$=2E+V_1$. (15)

As riveting in all ordinary cases is either of the form shown in Fig. 133 or Fig. 134, or a combination of those forms, equations (10), (11), (12), (13) and (14) have a general application. For instance, consider the case of a double butt-strapped, treble riveted, zigzag joint with alternate rivets omitted in the outer row; the distance V_1 between the outer and second row is obtained from equation (14), the distance V is obtained from equation (11). In this case the half breadth of butt strap$=2E+V_1+V$. (16)

Thickness of Butt Straps.—From considerations of strength and to secure tight joints the aggregate thickness of the butt straps should always be more than that of the plate. For single butt straps $t_1 = \frac{9}{8}t$; for double butt straps $t_2 = \frac{5}{8}t$, are arbitrarily taken as the minimum values allowed.

When alternate rivets are omitted in the outer row, the thickness obtained above must be increased.

The ratio of this increase is expressed by $\frac{p-d}{p-2d}$, which is the ratio of the net section left in the outer row to the net section left in the next row. With this form of riveting, the minimum values of t_1 and t_2 will be found by the following equations:

$$t_1 = \tfrac{9}{8}\left(\frac{p-d}{p-2d}\right)t. \quad t_2 = \tfrac{5}{8}\left(\frac{p-d}{p-2d}\right)t.$$

BOILER PROBLEM.

WORKING SHEET.

Required: A set of boilers to supply steam of......lbs. pressure to........................engines, such as are used in a modern naval vessel of the.....................class, when developing under an air pressure of..........inches of water, a maximum of.........I. H. P. for..........consecutive hours.

Additional specifications ....

BOILER PROBLEM.

....

....

....

....

....

....

....

....

....

....

....

....

....

....

....

....

....

1. Name of ship..

2. Maximum I. H. P...

3. Boiler pressure, pounds per gauge...............................

4. Air pressure, in inches of water................................

5. Coal burned per sq. ft. of grate, assumed from trials.............

6. Coal burned per I. H. P., assumed from trials....................

7. I. H. P. per sq. ft. of grate, assumed..........................

8. Pounds of water per I. H. P. per hour, assumed....................

9. Water evaporated per pound of coal, assumed..................

RESULTS OF PRELIMINARY CALCULATIONS.

10. Grate surface, from (2) and (7)..............................

11. Grate surface from (2), (5) and (6)............................

12. Grate surface from (2), (5), (7) and (8)......................

13. Grate surface used..

14. Trial diameter of inside of furnace...........................

15. Trial length of grate...

16. Number of furnaces, from (13), (14) and (15)...................

17. Number of furnaces used.......................................

18. Type of boiler as determined from tracing of boiler space in ship

....

19. Limiting diameter of boiler as determined from tracing..........

20. Number of furnaces in boiler head.............................

21. Length of bridge wall, assumed................................

22. Length of hanging bridge wall.................................

23. Beginning of grate from outside of front tube-sheet...............

24. Length of combustion chamber..................................

25. Distance between backs of same, to p.......bottom..............

26. Approximate length of boiler..................................

27. Length of tubes...

28. Length of tubes for low boiler, assumed.........................

29. Volume of boiler per I. H. P., assumed..........................

30. Diameter of boiler..

31. Numbers of boilers and class..................................

32. Thickness of shell, Bureau method.............................

33. Thickness of shell, Board of Supervising Inspectors' rule.........

34. Thickness of shell, Lloyd's rules..............................

35. Thickness of shell used..

36. Ratio of corrugated length of flue to axial length................

37. Heating surface of furnace..

38. Heating surface of combustion chamber, from planimeter measurements of rough drawing of combustion chamber..............

39. Total heating surface, Bureau method...........................

40. Total heating surface, Seaton...................................

41. Total heating surface...........method..........................

42. Total heating surface from above methods.......................

43. Tube heating surface..

44. Calorimeter ....

45. Number of tubes in each boiler...................................

46. Number of tubes in each end, by calculation.....................

47. Number of tubes in each end, from rough drawing................

Having completed all of the foregoing preliminary calculations, make a rough drawing, to as large a scale as the paper will admit, determine all the details, and tabulate the results in following columns before commencing the finished drawing on the board.

FINAL DATA OF ONE...........ENDED BOILER, ASSEMBLED TO COMMENCE DRAWING.

48. Diameter of boiler...

49. Radius of arc of furnace centers..................................

50. Distance between furnace centers measured on chords.............

51. Horizontal distance of middle furnace centers from center of head of boiler ....

52. Number of furnaces in head...

53. Inside diameter of furnace..

54. Outside diameter of furnace..

55. Thickness of furnace..

56. Thickness of shell plates..

57. Thickness of front tube sheet..

58. Thickness of back tube sheet..

59. Thickness of top head sheet..

60. Thickness of lower head or furnace sheet........................

61. Thickness of combustion chamber sheets........................

62. Thickness of butt straps..

63. Distance of top of top row of tubes from top of boiler............

64. Water space between shell and combustion chamber, at top........

...................., at bottom ....

65. Water space between shell and nearest furnace..................

66. Water space between two adjacent combustion chambers.........

67. Water space, minimum, between nests of tubes..................

68. Number of combustion chambers in each end.....................

69. Tubes plain, number in one end..............................

70. Tubes stay, No......B. W. G., outside diameter..................

71. Tubes stay, number in one end..............................

72. Tube spacing, all, horizontally............., vertically............

73. Total number of tubes in one end..............................

74. Length between tube sheets..................................

RIVETED JOINTS.

75. Back end of furnaces to combustion chambers, and to adjacent furnaces......................, diameter of rivets............., pitch................, lap of joint.............................

76. Upper head sheet to front tube sheet.............................. diameter of rivets..........., pitch..........., lap.............

77. Lower edge of tube sheet to furnace sheet.......................,

diameter of rivets.............., pitch..........., lap..........

78. Middle circumferential joint...................................,

diameter of rivets..........., pitch..........., lap...........

79. End circumferential seams ....,

diameter of rivets..........., pitch..........., lap...........

80. Butt strap............, width............, thickness............

81. Butt joint, kind of joint.......................................,

diameter of rivets..........., pitch..........., lap.............,

distance between rows of rivets...............................

82. Length of boiler...

83. Length of one ring...

84. Radius of curved head...

85. Braces above tubes, diameter...................................

86. Braces to back tube sheet, diameter...........................

87. Braces above lower manholes, diameter..........................

88. Braces, diagonal, in lower water space........................

89. Screw stays, diameter................., pitch..................

90. Heating surface, plate, square feet............................

91. Heating surface, tube, square feet.............................

92. Heating surface, total, square feet............................

93. Length of grate..

94. Grate area, one boiler...

95. Ratio of heating surface to grate area.........................

96. Volume of steam space, water 6″ above tubes....................

97. Water surface, square feet.....................................

98. Area through tubes, square feet...............................

99. Area over bridge wall, square feet...............................

100. Volume of furnaces and combustion chambers above grates........

101. Combustion chamber head, curved or flat.......................

102. Combustion chamber bracing ....

103. ....

104. ....

105. ....

Questions and Problems.

Find the number and type of cylindrical boilers for a battleship: given dimensions of space available in the ship 72 feet long × 39 feet 6 inches wide × 20 feet high; I. H. P., 11,500; pressure per gauge, 180 pounds; air pressure allowed 1 inch of water. Find the total grate surface necessary and show its distribution among the several boilers. Find diameter and length of boilers.

Find the number and type of cylindrical boilers; the total grate surface necessary, and show its distribution among the several boilers. Given: Space available 128 feet long × 35 feet wide × 20 feet high; I. H. P., 21,000; steam pressure, 180 pounds; air pressure allowed, 1 inch water. Find also the diameter and length of each boiler.

From the preliminary calculations it is found that eight double-ended boilers will be required for the power, about 21,000 I. H. P. for a given ship. Length of boiler 18 feet; diameter (outside), 15 feet 9 inches; each two adjacent furnaces have a common combustion chamber; grate surface, each boiler, 168 square feet; steam pressure, 180 pounds per gauge; outside diameter of tubes, $2\frac{1}{4}$ inches; length, 6 feet 6 inches. Find the number and size of furnaces; thickness of shell, strength of joint being 85% of solid plate; number of tubes; total heating surface.

From the preliminary calculations it is found that three double-ended and two single-ended boilers are necessary for the power, about 11,500 I. H. P., for a given ship. Length of double-ended boilers, 21 feet; diameter (outside), 15 feet 8 inches; each two adjacent furnaces have a common combustion chamber; grate surface, total (all boilers), 685 square feet; steam pressure, 180 pounds per gauge; outside diameter of tubes, $2\frac{1}{4}$ inches; length of tubes, 7 feet $7\frac{3}{8}$ inches. Find, for the double-ended boilers, the number and size of furnaces; thickness of shell, strength of joint being 85% of solid plate; the number of tubes; the total heating surface.

Given the formula, $p=\frac{C\times t}{d}$, C=14,000, find the thickness and make a sketch of a Morison furnace 39 inches mean diameter for a boiler carrying a working pressure of 180 pounds per gauge.

Given a boiler 16 feet outside diameter; thickness of shell plate, $1\frac{1}{4}$ inches; tubes, $2\frac{1}{4}$ inches, outside diameter; pitch of tubes, $2\frac{1}{2}$ inches; percentage strength of joint between front tube sheet and upper head sheet, 65%; thickness of front tube sheet, $\frac{7}{8}$ inch; of upper head sheet $1\frac{1}{4}$ inches. Find the radius of the curve of the upper head sheet. $R=\frac{C\times r\times T}{P}$, C=13,000. Working pressure, 160 pounds per gauge.

Show by neat sketches the method of securing a large double-ended boiler in a ship; boiler axis to be fore and aft. Show ramming chocks and show in detail the method of securing the boiler to the saddles.

Design the girder stays for a flat top combustion chamber of a boiler; working pressure, 180 pounds per gauge; depth of combustion chamber, 30 inches; width of chamber, 5 feet. Each girder to be made of a pair of steel plates 1 inch thick, carrying three stay bolts. The pitch of the bolts to be $7\frac{1}{2}$ inches in each direction. Find the depth of the stays and the diameter of the bolts: f_t for girders =9000; f_t for bolts=8000.

The sketch represents the lower part of the front head of a boiler. The space around the manhole is to be supported by three stays, A, B and C. Of these A is a through stay, perpendicular to the head; B and C are connected to the shell by palms, 5 feet back from the head; f_t for stays=8500 pounds per square inch. Calculate the diameter of each stay; pressure, 180 pounds p. g.

Deduce expressions for the percentage strength of a riveted joint, $\%$, $\%_1$ and $\%_2$. Given p=greatest pitch of rivets; d=diameter of rivets; A=area of one rivet; n=number of rivets in greatest pitch; c=constant=1 for rivets in single shear, =1.75 for rivets in double shear; tensile strength of plate=28 tons; shearing strength of rivets =23 tons; t=thickness of plate.

Given $\% = \frac{100(p-d)}{p}$ and $\%_1 = \frac{100 \times 23 \times A \times n \times c}{28 \times p \times t}$, find values of d and p, when n, c and t are known, and percentage strength of plate = percentage strength of rivet.

Ordinary zigzag riveting, given p and d, find an expression for the distance V between rows of rivets.

Design a double butt-strapped joint, treble riveted, with alternate rivets omitted in outer row. Strength of joint to be 85% of that of

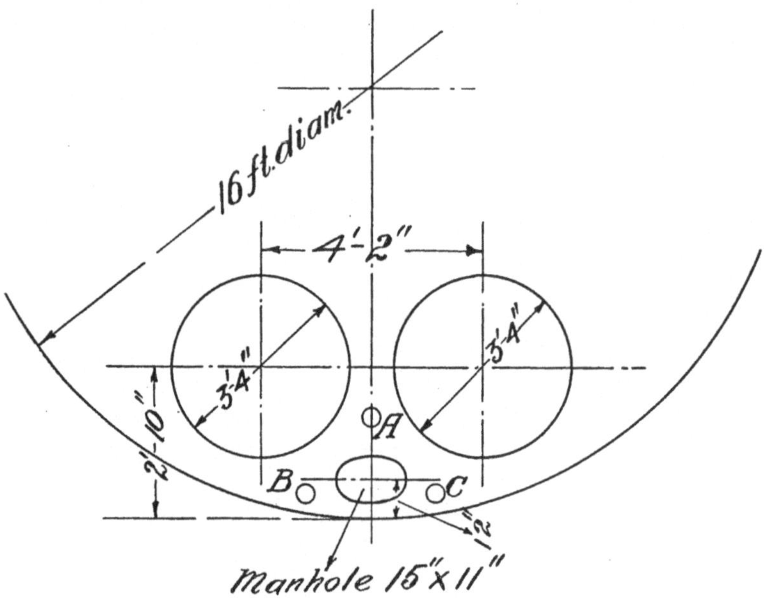

solid plate; thickness of plate, 1 inch. Find diameter and pitch of rivets, distance between rivet rows and total width and thickness of butt straps.

Zigzag riveting with alternate rivets omitted in outer row, deduce an expression for the distance V between rivet rows in terms of d and p.

Design a double-strapped butt joint, treble riveted, with alternate rivets omitted in outer row. Strength of joint to be 80% of that of solid plate; thickness of plate, $\frac{3}{4}$ inch. Find diameter and pitch of rivets, distance between rivet rows and total width and thickness of butt straps. Use nearest $\frac{1}{16}$ inch for detail results. Check percentage of plate and rivet after finding all dimensions.

CHAPTER XVIII.

SCREW PROPELLERS.*

135. *Data Given.*

Name of ship.

Number of screws.

Number of blades.

Kind of screw.

Adjustment of pitch, each side.

Displacement (tons).

Draft, forward.

Draft, aft.

Draft, mean.

Total I. H. P.

Revolutions per minute.

Speed in knots.

Diameter of screw shaft.

Thickness of sleeve on shaft.

Metal of screw.

Working face moved aft on center line.

Blades bent back, or center line bent aft.

Scale of drawing.

(For calculating the diameter by comparison, use the propeller of the U. S. Steamer "Olympia," d=14′ 9″, p=13,500, v=20 knots, r=129.)

NOTES ON PROPELLER DESIGN.

The following notes are compiled from:

Seaton: "A Manual of Marine Engineering."

Seaton and Rounthwaite: "Pocket-book of Marine Engineering."

Chase: "Screw Propellers and Marine Propulsion."

Barnaby: "Marine Propellers."

Foley: "Mechanical Engineer's Reference Book."

Practice of the Bureau of Steam Engineering of the Navy Department.

In all cases, the Department method is given preference

* From "Notes on Machine Design."

DEFINITIONS.

Chase, p. 132: The diameter of a screw propeller is the diameter of a circle swept by the tip of the blades.

The pitch is the axial distance between the same thread at one revolution.

The disc area is the area of the circle described by the tips of the blades including the area of the hub.

The developed or helicoidal area is the actual area of the working face of the blades when flattened out on a plane.

The projected area is the area of the projections of the blade on a plane taken at right angles to the axis of the screw.

The length of the screw is the greatest length of the blade measured parallel to the axis.

The speed of the screw is the pitch multiplied by the number of revolutions.

The pitch ratio is the quotient arising from dividing the pitch by the diameter.

A screw propeller is a true screw when its pitch is uniform; and an expanding pitch screw when the pitch increases from the forward to the after edge, or radially, or both.

The principal feature of the Griffiths screw is the large boss, which, while not impairing the efficiency, enables the blade to be fixed in such a manner that the pitch can be readily altered.

The modified Griffiths screw propeller is generally adopted for vessels of the navies of England and the United States, and is one having blades of any curved shape diminishing in width towards the tips and having the broadest part nearer the boss than the tips.

Chase, p. 139: "The shape of the blade is sometimes oval and sometimes resembles the section exposed by cutting a pear longitudinally. When made pear-shaped, it is for the purpose of getting the greatest possible surface near where the pitch angle is 45°, as that is considered the most efficient angle; and by the blade becoming diminished in width towards the tips, the surface friction is thereby reduced, and room is obtained for the access of the supply water."

CALCULATIONS.

Chase: "In commencing the design of a propeller for a given ship, the diameter and pitch are first decided upon."

Chase, p. 130: "The true secret of success in designing a screw propeller lies in the correct proportion of the three cardinal elements—diameter, pitch and blade surface; and that the shape of the blades (within certain limits), their thickness, and the condition of their surfaces, are of minor importance."

PITCH.

Seaton, p. 329: "Let S=speed of ship in knots; R=number of revolutions per minute; s=slip in knots, generally expressed as a percentage, x. Then S=speed of screw, in knots per hour multiplied by $\left(1-\frac{x}{100}\right)$.

Speed of screw, in knots per hour$=\frac{S}{1-\frac{x}{100}}=\frac{S\times 100}{100-x}$.

But speed of screw, in knots per hour$=\frac{\text{pitch}\times\text{revs.}\times 60}{6080}$.

$$\therefore \frac{\text{pitch}\times\text{revs.}\times 60}{6080}=\frac{S\times 100}{100-x}.$$

$$\text{Then pitch}=\frac{S\times 100\times 6080}{60R(100-x)}=\frac{S}{R}\times\frac{10{,}133}{100-x}.$$

Chase, p. 134: "The percentage of slip now allowed is from sixteen to twenty in large screws for good-sized ships, in which the engines make from 115 to 150 revolutions per minute."

Seaton: "x is generally from 20 to 25% for naval vessels."

From 24 of the latest fast vessels of the United States with engines of Bureau design, the average slip is 20.77%.

Seaton, p. 323: "If the efficiency of the mechanism were the same at any number of revolutions, a fine pitch propeller might be used to advantage, as being more efficient than a coarse one; but since the efficiency of the engine is much higher at a moderate than at a maximum rate of speed, the total efficiency of the screw and engine is often improved by increasing the pitch of the screw."

PITCH RATIO.

Chase, p. 137: "For torpedo-boats and other light-draught boats, which usually run in smooth water and where high-rotative velocities prevail, diameters being restricted in consequence of small draught, the pitch of screws varies from about 1.4 to two times the diameter. For the larger high-speed ocean vessels, which traverse water more or less rough and where ample diameter and immersion can be obtained, the pitch varies from 1.2 to 1.6 times the diameter. For slower cargo vessels, etc., the pitch varies from 1.1 to 1.3 times the diameter."

Foley: "As a rule the pitch ratio should lie between 1 and 1.5 times the diameter; 1.25 times the diameter is usually satisfactory in practice."

From 24 vessels of the U. S. Navy with engines of the latest Bureau design, the pitch averages 1.23 times the diameter.

EXPANDING PITCH.

Chase, p. 32: "The alleged advantages for a propeller of expanding pitch are that, as the screw advances through the water, the forward portion encounters a resistance due to a solid body moving through water at rest; but water, being an exceedingly mobile substance, that with which the screw first comes in contact has motion imparted to it by the pressure exerted by the screw; and this motion of the water is accelerated by the portion of the blade following, and consequently the after-portions of the screw will press upon the water with a motion opposite to that in which the screw is advancing. In order, then, to equalize the pressure over the whole blade, the pitch must be increased proportionately to the motion acquired by the water. The increase in pitch usually employed is from 15 to 25%."

It is questionable whether there is any advantage in expanding pitch.

DIAMETER.

Seaton, p. 323: "The size of a screw depends on so many things that it is very difficult to lay down any rule for guidance and much must always be left to the designer, so as to allow for all the circumstances in each case."

The diameter of the propeller must be less than the draught of the ship, to prevent racing and striking ground.

Barnaby, p. 34: "It is very important that a propeller should have sufficient immersion, since, if it breaks the surface of the water, the efficiency is reduced to a remarkable extent; but, if it is sufficiently far below the surface to prevent its drawing air, any further immersion within the limits that can practically be obtained is of little value."

Chase, p. 140: "The depth of immersion of the upper tips of the blades should increase with the draught of the vessel or the diameter of the propeller. It should not be less than in the following table when the vessel is at the load line:

For a diameter of 4 ft. or less..........	$\frac{1}{11}$ of diameter.
" " " 4 to 8 ft..........	$\frac{1}{10}$ " "
" " " 8 to 12 ft..........	$\frac{1}{9}$ "
" " " 12 to 16 ft..........	$\frac{1}{8}$ "
" " " 16 to 20 ft..........	$\frac{1}{7}$ "
" " " 20 ft. or over........	$\frac{1}{6}$ "

From practice: For large vessels there should be from $1\frac{1}{2}$ to 2 feet of space from the bottom of the keel to the bottom of the blades, and from 4 to $4\frac{1}{2}$ feet from the surface of the water to the blades."

Seaton, p. 323: "The diameter of the screw must bear relation to the turning power of the engine; for, if made of too small dimensions, it cannot absorb the power of the engine, however coarse the pitch may be."

Since the area of surface on which the skin resistance acts increases as the square of the distance from the center, and the arm on which it revolves increases with the distance from the center, the frictional resistance per square foot of surface varies as the cube of the distance from the center, so that any increase in diameter means a large increase in resistance.

If a screw is of sufficient diameter and has ample blade area, any addition beyond the tip will act then as a brake and seriously reduce its efficiency.

Seaton, p. 324: "Diameter of screw $= K\sqrt{\frac{\text{I. H. P.}}{(P \times R)^3}}$.

"I. H. P. = total − 6 to 10% for auxiliaries. P = pitch in feet. R = revolutions per minute. K = constant." Seaton gives for K a value of 20,000.

From trials of similar ships the value of K is determined for the one being designed.

Seaton and Rounthwaite give a value of 22,700 for naval vessels of fine lines.

The following are the average values of K:

For large battleships, K=19,000.
For large armored cruisers, K=22,768.
For small cruisers and gunboats, K=22,000 (about).
For torpedo-boats and destroyers, K=27,000 (about).

PERIPHERAL SPEED.

Foley: "Owing to the great loss of efficiency from friction at high speeds, it is usual not to exceed about 6500 feet per minute at periphery."

This should be applied as a check after the diameter is calculated.

DIAMETER BY COMPARISON.

Barnaby, p. 69: "To find the diameter of a propeller for a given I. H. P. and a given speed from the diameter of another similar propeller at a different I. H. P. and at different speed:

If d=diameter of model, which may be larger or smaller than D;
D=diameter of required propeller;
p=I. H. P. model;
P=I. H. P. of required propeller;
v=speed of vessel with model propeller;
V=speed of vessel with required propeller;
r=revolutions per minute of model propeller;
R=revolutions per minute of required propeller.

Then $D=\sqrt{d^2 \times \frac{v^3}{V^3} \times \frac{P}{p}}$ and $R=r\times \frac{V}{v} + \frac{d}{D}$.

The pitch ratio must be the same as that of the screw, which is treated as a model."

AREA

Seaton, p. 329: "The best area is not easily determined, except by experiment by data derived from the performance of similar ships with similar screws.

Good results are obtained by:

$$\text{Total area of screw blades}=K\sqrt{\frac{\text{I. H. P.}}{R}}.$$

				Single screws.	Twin screws.
The value of K for	four bladed is. .			15	10.5
"	"	three	" ..	13	9.0
"	"	two	" ..	10	"

Chase, p. 158: "For naval vessels with twin screws K=7 to 8.5."

Seaton and Rounthwaite: "For very fine-lined naval vessels with three-bladed screws K=7.75."

From 21 vessels of the U. S. Navy with engines of latest Bureau designs K=8.847.

For large battleships K=11.0.

For large armored cruisers K=9.0.

Seaton, p. 330: "The projected area is often taken as the criterion, and is to some extent a true one."

AREA FROM DISC AREA.

Foley: "The developed surface is usually from 28 to 40% of the disc area. For ordinary cases of moderate or high speeds, where the diameter is well proportioned, 35% is usually satisfactory. It should be remembered that there may be great loss of power from too much surface."

For the disc area the boss is usually spherical, and from $\frac{1}{4}$ the diameter of the screw for small screws to $\frac{1}{5}$ the diameter for large ones.

AREA FROM THRUST.

Seaton, p. 201: "$P = \text{I. H. P.} \times \frac{217}{K}$. P is mean normal thrust; K is speed in knots per hour."

"P varies with the I. H. P. and inversely as the speed in knots, so the thrust may vary considerably. If the speed is from any cause reduced the thrust must increase."

"To allow for the variations of P, the surface should be such that the pressure per square inch from the mean normal thrust does not exceed 70 pounds."

$$\text{Indicated thrust} = \frac{\text{I. H. P.} \times 33{,}000}{\text{revs. per min.} \times \text{pitch in feet}}.$$

"The efficiency of the engine and propeller are each taken at 75%.

Then, I. H. P.×33,000 becomes I. H. P.×33,000×.75×.75 =I. H. P.×18,562."

From six of the vessels of the U. S. Navy with engines of the latest Bureau design, the average thrust per square foot of helicoidal area is 1410.17 pounds, which is much later practice than that of Seaton. This is for propellers of manganese bronze.

The area from thrust is to be a check on the other methods.

BLADES.

Chase, p. 135: "Steadiness in action determines the number of blades. For a single screw, four blades give the greatest satisfaction, especially in rough water."

"Two-bladed propellers cause too great vibration."

"With twin screws where the supply water is not so much interfered with, and where the immersion is more ample, three-bladed screws are preferable."

Seaton and Rounthwaite: "Three blades are generally used when the immersion is sufficient and the revolutions fairly high; but when the calculated diameter of the propeller becomes nearly as great as the draught of water, and the revolutions are only moderate, four blades should be used."

Seaton: "If the necessary disc area can be obtained with a three-bladed propeller, this form is preferable, as less shock is given to the ship; but if the draught is limited, a four-bladed propeller may be necessary."

Practice: Blades are generally separate from the hub in naval vessels.

MATERIAL OF BLADES.

Seaton, p. 335: "Cast iron is generally used because cheap, and when struck breaks clean off and goes out of the way.

"Cast-steel blades are largely used, being thinner and lighter for the same strength than cast iron.

"Wrought-steel blades are strong and efficient. The chief objection to steel is that it corrodes rapidly on the back of the blade, due probably to air detached from the water.

"Phosphor and manganese bronzes are largely employed for making propellers for naval vessels. The blades can be cast thinner than if made of steel, as there is little corrosion. They are objected to on the ground that they cause galvanic action, and injure the iron or steel near by; but this is questioned."

Seaton and Rounthwaite: "For naval vessels propellers are almost invariably of gun-metal or bronze, and except sometimes in torpedo-boats, the blades are always detachable."

Bureau Practice: "For U. S. naval vessels manganese bronze is always used in the latest designs."

FORM OF BLADES.

Seaton, p. 333: "The forms of blades are numerous and varied. Some bend the blades forward, some back; but no general plan has been found to give, always, the best results. The better results are due to a better proportion of blade rather than to the shape."

Barnaby, p. 78: "The standard blade area of the Admiralty standard blade is an ellipse of major axis equal to the radius of the propeller and of minor axis equal to four-tenths of the major axis. It is often found, however, that owing to the diameter being limited, sufficient blade area is not obtained by these proportions. In such cases the elliptical form is adhered to with an increased minor axis of from .5 to .55 of the major."

LENGTH OF BLADES.

Chase, p. 140: "This is important and is determined by practice.

"The length affects the power to move a given columnal mass of water. When made too long, the loss from the divergent and rotary direction given the discharge water and also from the surface friction of the blades more than counterbalances the advantage of the increased mass of water acted upon.

"The length should vary with the rotative speeds. The most efficient length seems to lie between 18 and 20% of the diameter—the former for high and the latter for low speeds. The average of a number of high-speed screws of modified Griffiths shape gives a maximum length of 19%."

BREADTH OF BLADES.

Seaton, p. 334: "The greatest breadth of blade should not, as a rule, be beyond one-third the radius of the disc from the center, and should be approximately as given by the following rule:

$$\text{Max. breadth of blade in inches} = K\sqrt[3]{\frac{\text{I. H. P.}}{\text{Rev.}}}.$$

"For a four-bladed screw, K=14; for three-bladed, K=17; and for two-bladed, K=22.

"The breadth of blade at the tip should be from one-third to two-fifths of the maximum."

THICKNESS OF THE BLADE.

To find the thickness of the blade at the center of the shaft (that is, supposing that the taper of the blade section were continued to the shaft axis):

Let b = maximum breadth of blade, inches.
n = number of blades.
R = number of revolutions per minute.
f = 90.
K = 2.56.

Then, thickness of blade (at center of shaft) $= \sqrt{\dfrac{\text{I. H. P.} \times f}{R \times n \times b}} \times K$,

where f may be taken as 100 for compound engines of the ordinary two and four-cylinder type; 90 for three and six-cylinder compound engines, and 85 for three-crank triple-compound engines.

"The value of K is 4 for cast iron; 2 for ordinary gun-metal; 1.6 for cast steel and 1.5 for forged steel and bronzes of superior make."

Bureau Practice: "From 24 vessels of the U. S. Navy with engines of the latest Bureau design, the average value of K is 2.56."

Chase, p. 162:

"Cast iron	allow	.5 in.	for each foot in dia. of screw.
Cast steel	"	.4	" " " "
Gun metal	"	.45	" " " "
High class bronze...	"	.4	"

THICKNESS AT THE TIP.

Seaton, p. 330: "The thickness of metal at the tip should be .2 of that at the root."

Chase, p. 160: "D is the diameter of the propeller in feet.

Thickness at the tip,	cast iron......................	.04D + .4 in.
" "	cast steel	.03D + .3 in.
" "	gun metal	.03D + .2 in.
" "	highest class bronze...........	.02D + .3 in.

Also, Chase: "The thickness at the tip should be as small as possible consistent with good casting or forging. When gun-metal is used it is generally made about $\frac{3}{64}$ inch for each foot in diameter of the screw."

BOSS OR HUB.

Seaton, p. 332: "When a loose-bladed propeller is designed, the boss is usually spherical in general form with flats or recesses for the blades. The diameter of the sphere is from $\frac{1}{4}$ the diameter of the screw for small propellers to $\frac{1}{5}$ the diameter for large ones."

Chase: "Boss=diameter divided by $3\frac{1}{4}$ or $3\frac{1}{2}$, which will cut away about one-fifth of the standard ellipse."

Seaton, p. 332: "The length of the boss depends on the size of the base of the blades, and is generally about .85 of the diameter of the sphere for two-bladed screws and .75 for four-bladed.

"In most navies gun-metal is used for the hubs of propellers."

BOSS AND FLANGES ON BLADES.

Chase:

"Diameter of flange=2.25×diameter of shaft.

"Thickness of flange=.85×diameter of stud for bronze or steel.

"Length of boss=2.7×diameter of shaft.

"Thickness of boss=.65×thickness of blade at shaft axis.

"Diameter of boss=3.3×diameter of shaft.

"Taper of boss=1″ to 1′.

"The fore and aft section of the boss should be oval—the principal radius being .8×diameter of boss."

Seaton: "A well-designed and carefully made screw should have the base of the blade conforming to the general outline of the boss, and the nuts or bolt heads recessed into the blade base and covered with a metal case or cemented flush with the surface."

BOLTS.

Chase, p. 161: "The bolts are usually of naval brass (muntz-metal with the addition of a small percentage of tin), or are of one of the stronger bronzes for gun-metal or bronze propellers, and of mild steel for cast-iron or cast-steel propellers.

"The size is from the formula: $a \times N \times r = \frac{T \times L}{K}$.

a=area in square inches of one stud bolt at bottom of thread.

N=No. of studs or bolts for one blade (usually 7 to 11).

r=radius of pitch circle of studs or bolts in inches.

T=indicated thrust per blade

$$= \frac{\text{I. H. P.} \times 33{,}000}{\text{mean pitch} \times \text{revolutions per min.} \times \text{No. of blades}}.$$

L=.6×total length of blade in inches (flange joint to tip).
K=1700 for steel bolts or studs.
1400 for naval brass or bronze studs or bolts.

TAPER OF SHAFT, FEATHER, THREAD, NUT.

Chase, p. 161: "The taper of the part of the shaft within the boss should be about one inch in diameter for each foot of length, but never less than three-fourths of an inch.

"The thread of the large nut that holds the propeller on the shaft is $2\frac{1}{2}$ threads per inch, regardless of the diameter. It should be left handed when the propeller is right handed, and vice versa.

"The nut should be securely locked, preferably by a plate fixed to the after end of the boss by set screws.

"The propeller should be secured by one feather or key extending the whole length of the boss, the proportions of which may be:

Breadth of key $=.22\times$ largest dia. of shaft $+''.25$.
Thickness of key $=.55\times$ breadth.

"The diameter of the screwed end of the shaft should be sufficiently reduced to allow the key to be fitted in from the after end clear of the thread."

The key should have clearance at the after end, so that the nut will not come tight against it, and should be partly imbedded in the shaft, the forward end being rounded.

In Bureau drawings, sometimes the key is arranged to be secured in place in the shaft and the propeller slid on from aft, in which case the top of the thread on the shaft should be less in diameter than the smallest bore of the propeller.

ALTERING THE PITCH.

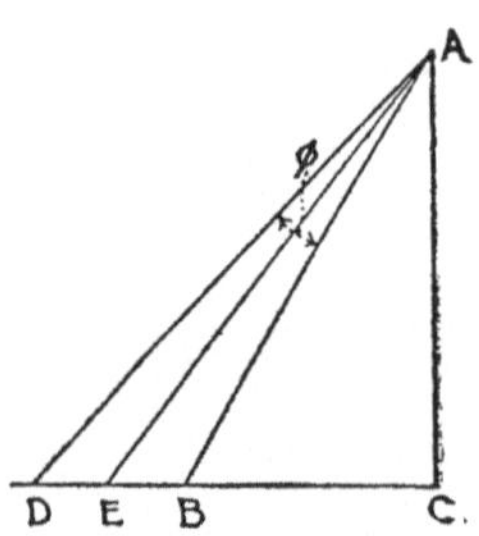

Barnaby, p. 91: "For the sake of possible adjustments that may be desired on the trials of machinery, and also for finding the pitch when it becomes necessary to reduce the steam pressure in the boilers, it is customary to make the bolt holes in the blade flanges oval, in order that the inclination of the blades to the axis may be altered. The amount of the oval may be determined thus: 'Let the true pitch of the blade be 2πCE and let the desired range of pitch be $2\pi EB$ and $2\pi ED$ respectively on each side of this pitch. Take a radius CA, so that A is about half way up the blade. Join AD, AE, AB. Then, if ϕ=the angle DAB, the holes in the flange must be so elongated as to admit of the blade being turned through $\frac{\phi}{2}$ degrees on each side. The amount of this elongation on each side is, then, $\frac{\pi C}{360} \times \frac{\phi}{2} = \pi C \frac{\phi}{720}$, ϕ being measured in degrees.

$2 \times$this length+the diameter of the bolt or stud=length of slot. C=diameter of bolt circle.' "

136. *Calculations Required.*

Percentage of slip.
Pitch, mean.
Pitch, greatest.
Pitch, least.
Pitch, ratio.
Diameter from pitch ratio.
Depth of immersion (Chase).
Depth of immersion (Practice).
Distance below blades.
I. H. P. of auxiliaries.
I. H. P. of each engine used in calculations.
Value of K USED.
Diameter, Seaton.
Diameter, Foley.
Diameter, Foley, 2d method.
Diameter by comparison.
Average of methods.
Peripheral speed.
DIAMETER USED.
Value of K used.
Helicoidal area, Seaton.
Diameter of boss, Foley.
Diameter of boss, Chase.
Diameter of boss, Chase, 2d method.
DIAMETER OF BOSS, USED.
Disc area.
Helicoidal area from disc area.
Thrust, Seaton.
Thrust, Foley.

THRUST USED.
Helicoidal area from thrust, Seaton.
Helicoidal area from thrust, Bureau.
Helicoidal area, average.
HELICOIDAL AREA USED.
Blades, number.
Blades, material.
Bend back of blades.
Helicoidal area of each blade.
Major axis of trial ellipse.
Minor axis of trial ellipse.
Major axis of ellipse USED.
Minor axis of ellipse USED.
Length of blade, Chase.
Length of blade, Seaton.
LENGTH, USED.
Distance of maximum breadth of blade from center.
Breadth of blade at the tip.
Thickness of blade at $1\frac{1}{2}d$, Seaton, 1st method.
Same, Seaton, 2d method.
Same, Chase.
THICKNESS OF BLADES, USED.
Thickness at tip, Seaton.
Same, Chase, 1st method.
Same, Chase, 2d method.
THICKNESS AT TIP, USED.
Boss, diameter.
Boss, length, Seaton.
Boss, length, Chase.
BOSS, LENGTH USED.
Boss, thickness, Chase.
Boss, taper.
Boss, radius.
Flange, diameter, Chase.
Flange, thickness, Chase.
Bolts, material, Chase.
Bolts, number, Chase.
Bolts, diameter, Chase.
Bolts, diameter, Chase, 2d method.
Taper of shaft.

Threads per inch on shaft.
Right or left handed.
Breadth of key.
Thickness of key.
Length of key.
Diameter of screwed end of shaft (outside).
Depth of thread.
No. of threads to 1″.
Diameter of plug in the end of shaft.
Length of same.
No. of threads to 1″ of same.
Length of slots in flanges of blade.

137. *Geometry of the Screw.*—It is first noted that the face only of the blade is drawn of the true geometrical form and is the driving face of the blade; the form of the back depends on the thickness given at the different radii.

An exception is sometimes made in the case of torpedo-boats, where rapid manœuvring is required and rapid speed astern, where the front and back of the blades are alike.

Barnaby, p. 175: "If a point moves on the surface of a cylinder in such a way that, while moving uniformly around the cylinder, it advances uniformly in the direction of its axis, it will trace a curve known as the helix. Imagine the cylinder cut on one side by a straight line, BC, parallel to the axis meeting the helix in consecutive points A and C, and then unrolled and laid flat, the circumference through A will become a straight line AB; BC at right angles to it will represent the direction of the axis, while that part of the helix formed during the complete revolution of the tracing point will be represented by the line AC. Since the distances moved by the point in the directions AB, BC are proportional, AC is a straight line; BC, being the distance moved through in the direction of the axis while the point goes once entirely round the cylinder, is the pitch; while the angle BAC, which the unrolled helix makes with the plane at right angles to the axis, is the angle of the helix or screw.

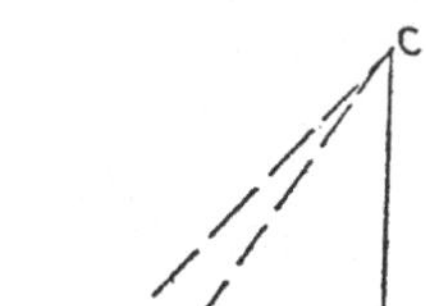

"If a straight line moves uniformly around an axis which it intersects, and to which it is always at right angles, advancing at the same time uniformly in the direction of the axis, it will sweep out a surface known as a helicoid, and every point in the generating line will describe a helix as shown above, necessarily lying on this helicoid. Since, during a complete revolution of the generating line, every point moves through the same distance in the direction of the axis, the helicoid is the surface of uniform pitch; that is, BC is constant for the helices of all points in the generating line. A helicoid can, therefore, be, and often is, used for the acting face of a screw blade of uniform pitch. It is not necessary, however, that the generating line should be at right angles to the axis; such a surface may be generated by any line, straight or curved, moving uniformly along and revolving uniformly around an axis, intersecting and always making the same angle with it.

"The helix traced by any point in the generating line will also be the curve of intersection with the screw surface of a coaxial cylinder of radius equal to the perpendicular distance of the point from the axis. The larger the radius of the cylinder, the larger of course the length of the circumference, as A′B; and, as the pitch is constant, it follows that the angle of the helix must decrease as the radius of the intersecting cylinder increases. We thus arrive at the fundamental geometrical property of a surface of uniform pitch, viz: Co-axial cylinders intersect it in helices, all of which have the same pitch, but whose angles vary, decreasing as the radius of the cylinder increases. Near the axis, therefore, the helices will approximate in direction to that of the axis, and as the distance from the axis increases, they will lie more and more at right angles to it. If θ be the angle of the helix, p the pitch and r the radius of the intersecting cylinder, $\tan \theta = \frac{p}{2\pi r}$."

Barnaby, p. 78: "The expanded blade area, which may be described as a flat surface of approximately equivalent area to that of the blades, both as to amount and disposition, is derived as follows:

"A co-axial cylinder will intersect the screw surface in a helical curve making a certain angle with the axis, and it will intersect a plane passing through that diameter of the cylinder which passes through the middle point of the helical curve and making the same angle with the axis, in an elliptical arc. The length of the screw being small compared with the pitch, these two arcs will nearly coincide, and no great error will be involved by assuming that they do coincide. Imagine these elliptical arcs at all radii to be swung around a common center line till they all lie in the same plane with their major and minor axes respectively coincident (though necessarily of different lengths), then the curve passing through their extremities will form the expanded blade area. This area is very nearly equal to the actual working face of the blade, being in fact somewhat less than it."

Drawing.—The following is a general form for laying down the screw to be observed in the designing-room of the Department of Marine Engineering and Naval Construction at the United States Naval Academy:

The drawing is to consist of vertical elevation of the propeller from aft; one-third of the hub and one flange of the blade in section; a side elevation; a side elevation of one blade showing the section of the hub and the thickness of the blade; a drawing of the developed area and sections of the blade either on the angular lines or at right angles to the central line of the blade on one of the other figures; guide-irons and sections of flanges and bolts as required.

The side elevation of the blade showing the sections of the hub may be combined with the side elevation of the whole propeller by making part section and part elevation, though it is better to separate them for clearness.

Having determined most of the dimensions from the previous notes, we first begin Fig. 1, Sheet 1, a vertical elevation of the blade looking forwards. Draw the center line OPO′ representing a line at right angles to the center line of the shaft, O. Lay off the circle showing the greatest diameter of the hub, and draw a vertical center line through O, the center of the shaft, for the center line of the blade. Lay off OA equal to the radius of the propeller. Lay off equally on each side of this line the developed area of one blade required. Since about one-fifth of this area will be inclosed in the hub, it is well to lay off at first the area on OA as an axis equal to six-fifths of the area required. This area may be laid off entirely by trial suiting the curves to the ideas of the designer, but it is better to use as a basis for a trial area an ellipse with OA as a major axis and a minor axis found as follows:

Area of an ellipse $= \pi ab$, when a and b are the *semi* major and minor axes. In this case, the area must equal $\frac{6}{5} \times$ developed area calculated for one blade. The major axis is equal to OA. To find, then, the minor axis, we have $\pi ab = \frac{6}{5} \times$ helicoidal area of one blade.

Then,

$$\frac{6 \times \text{helicoidal area of one blade}}{5\pi a} = b.$$

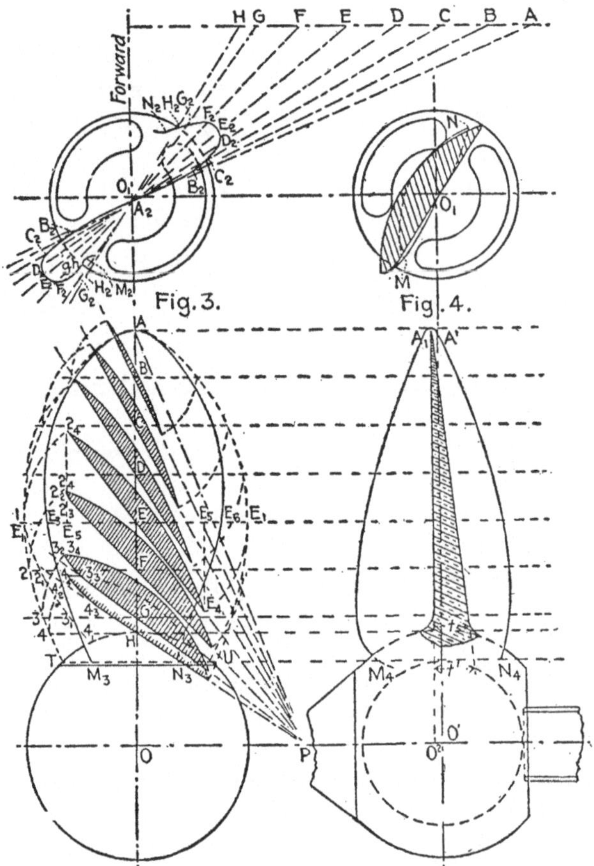

SHEET 1.

The ellipse is then constructed according to any approximate method, and the area TE_1 AE_1 UHT is measured by planimeter. If the area does not agree with that required, it may be enlarged or diminished by changing the length of the minor axis until the required area is obtained. If the area be too small, one way of adding to the surface is to enlarge the lower part, thus making the blade more pear-shaped, according to the ideas of the designer. If the area be too large, sometimes a portion is removed from the top of the blade, thus again producing a pear-shaped blade.

In Fig. 1, Sheet 1, the lower part of the blade has been made pear-shaped, the lower portion of the original ellipse not being drawn.

Some designers use for the semi-major axis of the ellipse about 85% of the radius of the propeller, and take four-tenths or more, as required, for the semi-minor axis. The ellipse then is not so much within the circle of the hub and a less proportion is cut off.

Divide the center line HA into any number of parts—five or six, and preferably equal, or, as generally constructed, make the outer divisions, AB, BC, etc., to FG equal and a definite number of feet and half feet long and the divisions GH nearest to the hub longer or shorter than the others, if necessary. Draw horizontal lines through these points.

Lay off $OP = \frac{\text{pitch}}{2\pi}$ and draw lines through PA, PB, etc., to PH. $\text{Tan}\,\theta = \frac{\text{pitch}}{\text{circumference}} = \frac{\text{pitch}}{2\pi r} = \frac{OP}{OA}$.

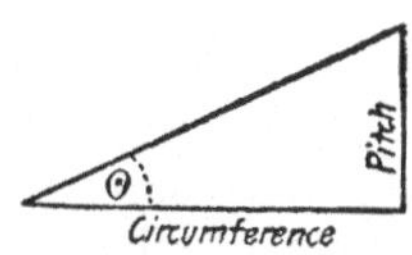

r is a varying quantity, OE, OD, etc., to OA. Since the tangent of the angle which any one of these lines PA, PB, etc., passing through a point in OA distant r from O makes with AO is $\frac{p}{2\pi r}$, these lines represent the angles of the blade of the various sections at distances OB, OC, etc., from the axis.

From the points B, C, D, etc., set off on the lines PB, PC, etc., the lengths of the sections made by the horizontal lines through B, C, D, etc. on the expanded blades, as shown by the arcs.

Next, in Fig. 2, Sheet 1, the side elevation, or fore and aft projection, draw the vertical center line O′A′ for the center line of the hub. The blade is seldom laid off on this line, as it is frequently found that, if the thickness of the blade be laid off from this line, the back of the blade will interfere with the bolts used for securing the blade to the hub. Consequently, the working face is moved back, as the thickness is placed on the forward side. This distance of moving back the center line of the blade varies with the diameter of the propeller, and in this case is 1½ inches. The lines O^2A_1, $1\frac{1}{2}$ inches back, is then the center line of the blade.

From previous calculations we have the thickness of the blade at the tip and at the center of the shaft. Lay off from O^2A_1 these distances (the one at the hub, t, and the one at $1\frac{1}{2}$ diameter of shaft, t′, will be checks on each other), and connect the points of thickness at tip and hub by a straight line, the extension of this line (dotted) to the thickness at $1\frac{1}{2}$d, being drawn.

Where the line showing the center line of the blade and the one showing the thickness meet the hub easy curves are drawn as shown.

We have then a section of the blade through the center line of the developed blade.

This thin sectional strip showing the thickness at the different distances from the center is supposed to represent a section through the center line of the flattened or developed blade, as it is apparent that only on this assumption could a plane section be obtained giving the thickness of metal normal to the blade.

Extend the horizontal lines through A, B, etc., of Fig. 1, Sheet 1, to Fig. 2, which gives the thickness of the blade for the various sections represented by the horizontal lines. Transfer these thicknesses to Fig. 1 by laying them off at B, C, etc., perpendicular to the lines representing the angles of the blades, as shown at E and F, and draw through these points arcs of circles touching the edges of the sections already obtained. These arcs should really be portions of ellipses, but are so near portions of circles that they are practically correct. This will give the shapes of the sections of the blade at different radii.

In some drawings these sections are laid out horizontally on the lines B, C, etc. Either way may be followed.

Those sections near the base are sometimes modified in order to allow the water to escape freely from the following edge of the blade, in which case the outlines of these sections are not finished until the amount of this modification is determined. The following edge is frequently curved upwards or back, and the amount of this flare is governed by the designer. The portion to be rounded back is shown by the assumed dotted line $12_1 3_1$ on the developed blade. By measuring the distances 22_1, 33_1, 44_1 and setting them off as $2_2 2_3$, $3_2 3_3$, $4_2 4_3$, on the lines showing the angles at the different distances from the center, we find the points where the blades begin to flare. Draw easy curves $2_3 2_4$, $3_3 3_4$, $4_3 4_4$, thus dropping back the working edge near the hub and still preserving about the same helicoidal area. The perpendicular distances from 2_4, 3_4 and 4_4 to the corresponding lines for the angles of the blades give the actual amounts of the setting back of the edges to be used in the plan view.

These points 2_4, 3_4 and 4_4 are connected by arcs of circles with the points showing the thickness of the blade at the center and the other edge as found.

The plan, Fig. 3, Sheet 1, is next begun. A circle of the diameter calculated for the flange of the blade is drawn. If the working face of the blade were not moved back from the center line of the hub, Fig. 2, the projection of the central line of the blade would be at O_1, Fig. 3. In this case, the center having been moved back, the center of the blade in the plan view will be at A_2, a distance aft equal to that in Fig. 2. Through the point A_2 draw lines representing the angles of the blade for the sections A, B, C, etc., as in Fig. 1, and lay off on these lines the developed lengths at the different sections. For instance, $E_1EE_1 = E_2A_2E_2$. Note that the lengths of the sections of the lower portions of the blade where cut away are measured to the actual length on the after edge of the blade, as $A_2G_2 = G3_2$. Then lay off perpendicular to A_2G_2 the distance from the section, Fig. 1, that the blade is bent back from the section, as $G_2g = 3_2 3_4$.

Before completing the plan view, the intersection of the blade with the spherical surface of the flange is obtained. In Fig. 1 it is seen that the lower edges of the developed blade, T and U are just within the circumference of the flange, so that the edges of the intersection of the flange and hub should be on this line. To get the intersection it is convenient to draw Fig. 4, though the same operation may be carried on in Fig. 3. Draw the circle of the flange, Fig. 4, and place in its proper position the section of the blade at H of Fig. 1. This section is, of course, tangent to the round up of the flange. It is seen that this section overlaps the circle of the flange, so that from this point downwards to the blade the width must be gradually shortened, so that the edges may fall within the circumference of the flange; and these edges must be distant from the center of the flange a distance equal to the radius at TU, Fig. 1. Another consideration that affects the positions of the points of intersection of the edges of the flange and the blade is the position of the bolts, covering plates, etc. These are drawn in, according to experience. The number of bolts has already been decided upon and also their diameters, so that the distance apart and the clearance must be determined to the best advantage. The outlines of the covering plates being determined, the section at H is worked down in Fig. 4, so as to shorten and possibly twist a little one way or the other as required. Drawings from the Bureau of Steam Engineering are to be consulted for this. The points M and N will be determined from the required distance from the center, and the best arrangement of the line of intersection of the flange and blade as shown by the unhatched lines, taking care that the greatest thickness at the middle point of the intersection is not less than the corresponding thickness of the section H.

This theoretical intersection is not shown on the final drawing, as the fillets at the bottom of the blade cause easy curves of intersection.

To find where the ends of the blades intersect the flange in the different projections:

The points M and N are transferred to Fig. 3, thus giving M_2 and N_2, and are projected to Fig. 1 to the line TU, thus giving M_3 and N_3. Project M and N, Fig. 4, to a vertical line and measure the distances from the center of the blade and transfer these distances on either side of A_1C_2, Fig. 2, on the line TU extended for the points M_4 and N_4.

To obtain the projections of the working edge of the blade:

Having determined M_2 and N_2, Fig. 3, we have now all the points of the plan, so we may draw in the curve $N_2E_2A_2E_2M_2$. In Fig. 1, project from Fig. 3 the ends of the lines $E_2A_2E_2$, etc., to their corresponding lines on Fig. 1 and draw a curve through these points and extend to M_3 and N_3, taking care to note which edge falls behind the round up of the flange.

For Fig. 2 project the ends of the angular lines of Fig. 3 to a vertical plane and measure on Fig. 2 from the center line A_1O^2 and on the corresponding lines the distances on this vertical plane from the projection of the center A_2. Connect the points so found with a curve, extending it to M_4 and N_4, noting which curve falls behind the round up of the flange.

Another method of finding the projections is as follows:

In Fig. 1 project E_4 vertically to E_5. Measure E_5E_4 from E to E_6. This will be then one of the points of the intersection. In Fig. 1 measure EE_5 and lay it off on Fig. 2 on the corresponding line and edge of the blade from the center line O^2A_1 of the blade.

The same methods are followed for all points of the projections.

OUTLINE OF THE BACK OF THE BLADE.

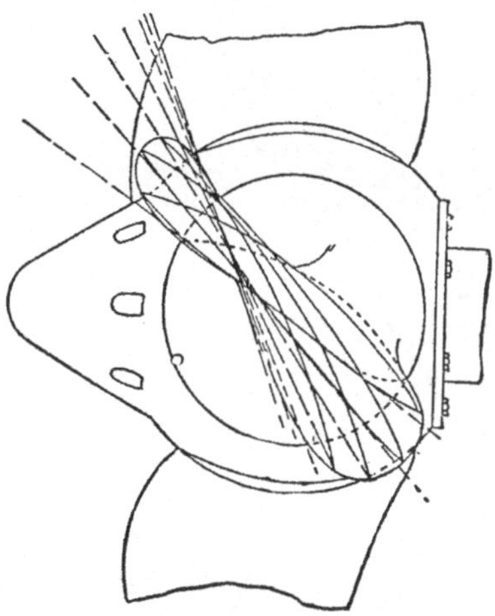

SHEET 2.

In the plan, Fig. 3, Sheet 1, it is evident that a portion of the back of the blade is seen. The outline is obtained as follows: On the elements B_2B_2, C_2C_2, etc., lay off the thickness strips as determined in Fig. 1. Then the highest portions of the different thicknesses will be points in the outline of the view of the back of the blade, so that a curve drawn tangent to these curves will give the outline of the back of the blade.

Sheet 2 shows this method of finding the outline of the back of the blade. At some point this outline gradually joins the curve of the flange. This is represented approximately by a curve.

Having determined the general outline of the blades for the different views, the hub may be drawn and finished according to the dimensions and details as determined upon. Some of the methods as used in the designs of machinery for vessels of the United States Navy are shown in Sheets 3 and 4.

In the case of a screw with expanding pitch the above method of drawing the propeller is somewhat changed. See Sheet 5.

Divide the center line as before and set off $OP = \dfrac{\text{entering pitch}}{2\pi}$ and $OP' = \dfrac{\text{leaving pitch}}{2\pi}$. From P draw lines through B, C,

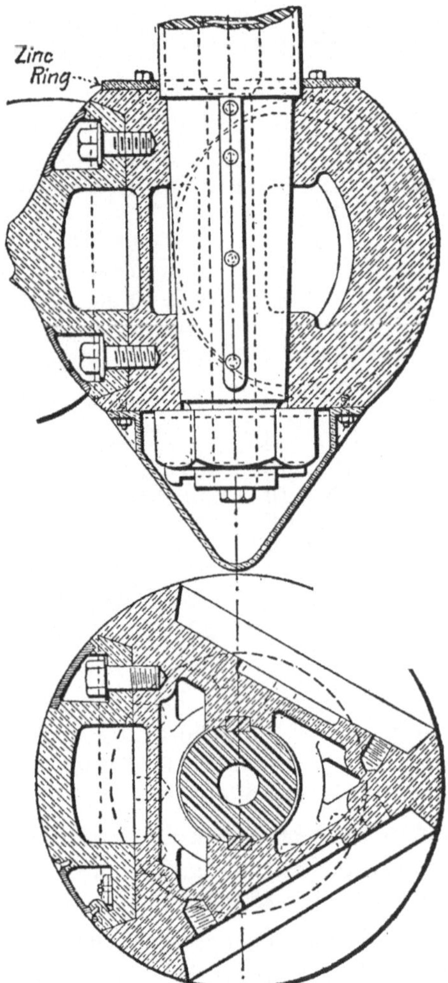

SHEET 4.

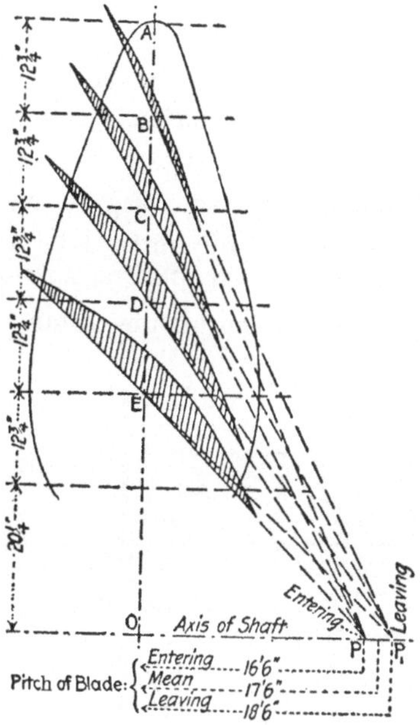

SHEET 5.

etc., these lines giving the angles of the blades for the entering edges. From P′ draw lines through B, C, etc., for the angles of the blade for the leaving edge. These lines intersect at the center, and for the entering half of the blade the lines from P form the bases of the sections cut out, while lines from P′ form the bases of the sections cut from the after half of the blade as shown.

The rest of the drawing conforms to this method.

GENERATRIX INCLINED.

Barnaby, p. 85: "Blades are sometimes made with the generating line inclined to the axis; or, in technical terms, they are made with a skew. Let the two straight lines, AB, AC, of the figure, the former at right angles to the axis, the latter inclined to AB at an angle α, moving together, generate screw surfaces of uniform and equal pitch. Then the helices of intersection of these two surfaces will be exactly similar, and one will be always a constant distance from the other; this distance being at a radius r, equal to r tan α. Imagine these two surfaces so far similar that when AC at any radius leaves the surface, AB at the same radius leaves its surface, then the expanded area of the surface so formed by AB will represent what may be termed the effective expanded area of that formed by AC, and this should be of the elliptical or other form which would have been used if the generating line had been at right angles to the axis. A blade generated by AC would therefore be formed from a blade generated by AB, simply by setting the helices of intersection definite distances aft; the distance at M, for example, being MN. It follows, therefore, that the athwartships projection of the blade

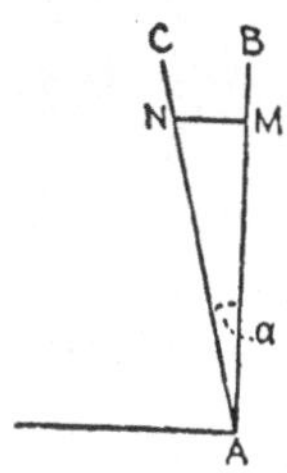

for the same "effective" expanded area is independent of the skew, and consequently for a skew blade with effective expanded area as for the blades shown in Figs. 1, 2 and 3, Sheet 6, the athwartships projection will be as in Fig. 1, no matter what the skew may be:

"The fore and aft projection of the upper blade, Fig. 4, will be formed by using a center line inclined to the vertical at an angle equal to the inclination to the vertical of the generating line, and proceeding as in Fig. 2, Sheet 1, setting off the distances

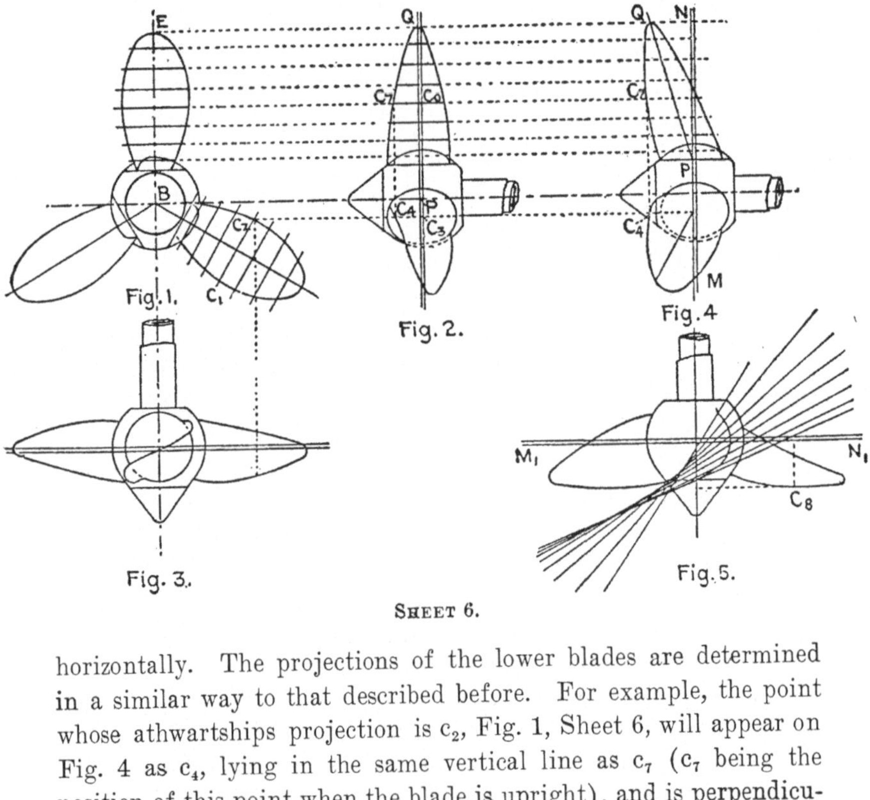

SHEET 6.

horizontally. The projections of the lower blades are determined in a similar way to that described before. For example, the point whose athwartships projection is c_2, Fig. 1, Sheet 6, will appear on Fig. 4 as c_4, lying in the same vertical line as c_7 (c_7 being the position of this point when the blade is upright), and is perpendicularly away from the axis, a distance equal to that of c_2 from the horizontal plane through the axis.

"For the horizontal projections, Fig. 5.—For the top blade the pitch lines are not all drawn through the same point as in Fig. 3, but each line is drawn at a corresponding angle to the axis through a point on the axis at a distance from M_1N_1, equal to the distance of the corresponding point on the generating line PQ from MN, Fig. 4, the process then being as previously shown. For the lower blade we proceed as follows: The point (c_2, Fig. 1) corresponding to c_7, Fig. 4, when the blade is upright, will appear in Fig. 5 as c_8 at a distance from the axis equal to the distance of c_2, Fig. 1, from the vertical plane through the axis and from M_1N_1 equal to the distance of c_7 from MN, Fig. 4.

"The projections of a three-bladed propeller with skew blades are shown in Figs. 1, 4 and 5, except that the left-hand lower blade of Fig. 1 has not been shown in Fig. 4 to avoid confusion.

"Where the generating line is curved, the method is now obvious."

138. The above described method of designing from an assumed elliptical form of developed area is subject to the following criticism:

If the pitch ratio varies the form of the projection of the blade on the disk (that is, on the transverse plane of the ship) for a given developed area changes. If the pitch ratio is constant and the width of the developed blade is changed from the standard on which the experimental formula are based, in order to obtain the required area, the distance of the center of pressure from the shaft axis is changed. When the blade is wider this distance increases, when it is narrower the distance decreases, and the resistance of the screw to turning is thus increased with greater width, not only on account of the increased area, but also on account of the greater arm at which the resistance acts. When the blade is narrowed the resistance is decreased by the decrease in area and also by the shortening of the resistance arm. Thus the rough assumption that the thrust will vary inversely as the developed area for the absorption of equal powers is very considerably in error.

To overcome this objection the Bureau of Steam Engineering has very recently adopted a standard form of *projected* area, which has been obtained by a combination of the best of a very great number of forms that have been carefully investigated. In connection with this standard form of projected blade the Bureau uses a chart of curves which has been developed by Commander C. W. Dyson, U. S. N. For a full description of the method see an article by the above-mentioned officer in the Journal of the American Society of Naval Engineers, Vol. XXII, No. 3, August, 1910.

Questions and Problems.

Define the following terms as applied to the screw propeller:

1. Diameter.
2. Pitch.
3. Disk area.
4. Developed area.
5. Projected area.
6. Length of screw.
7. Speed of screw.
8. Pitch ratio.
9. Slip; and how expressed.
10. Expanding pitch.

In designing a screw propeller what are the three elements of greatest importance?

Given the total I. H. P. (both main engines) = 16,500; speed of ship = 18 knots; R. P. M. = 120; find the pitch and diameter of the screws (using Bureau constants). Check by comparison with the following screw: I. H. P. (both main engines) = 10,890; R. P. M. = 128.25; speed of ship = 16.79 knots; diameter of screw = 15 feet; pitch ratio = 1.03. Assume a slip of 14%. After finding diameter check by the following: speed at periphery $<$ 6500 feet per minute.

Given total I. H. P. (both main engines) = 16,000; speed of ship = $18\frac{1}{2}$ knots; R. P. M. = 120; diameter of screw = 17 feet 9 inches pitch = 17 feet 9 inches; find necessary developed area of each screw. Check by indicated thrust. Efficiency of engine = 90% efficiency of screw = 62%.

Given the total I. H. P. (both main engines) = 11,000; R. P. M. = 128; each screw has three blades. Find the maximum breadth of the screw and the thickness of the blade at shaft axis and at tip. Diameter of screw = 15 feet.

Given a screw propeller 15 feet diameter and 16 feet pitch; three blades; diameter of pitch circle of blade studs = 25 inches; distance between blade flange and shaft axis = 14 inches; number of studs in each blade = 7; I. H. P. (one engine) = 4500; R. P. M. = 128; Tobin bronze studs. Find the nominal diameter of each stud.

A ship whose total I. H. P. (both main engines) is 9000; speed 15 knots at 128 R. P. M., has twin screws each 15 feet diameter and 16 feet pitch. What value of the constant K was used in determining the diameter of the screws? What is the % slip?

A ship whose total I. H. P. (both engines) is 9000 at 128 R. P. M., has a total helicoidal area of *both* screws of 132 square feet. What value of the constant K was used in determining the developed area?

Explain how the surface of a screw propeller is generated, showing how the angle changes with the radius. Show the value of the angle of the helix. Explain the approximate method of finding the expanded blade surface.

Make a sketch (plan, front and side elevations) of one blade of the following propeller: Helicoidal area (each blade) $=28$ square feet; diameter of screw $=16.75$ feet; thickness of blade at shaft axis $=9\frac{1}{4}''$; thickness of blade at tip $=1''$; diameter of hub $=\frac{1}{5}$ diameter of screw; pitch $=17.25$ feet. Assume that $\frac{1}{5}$ of the area of the ellipse falls within the hub.

Make a sketch (plan, front and side elevations) of one blade of the following propeller: helicoidal area (each blade) $=32$ square feet; diameter of hub $=\frac{2}{9}$ of diameter of screw; diameter of screw $=$ 18 feet; mean pitch $=19$ feet; thickness of blade (considered as produced) at shaft axis $=12''$; thickness of blade at tip $=1''$. Assume that $\frac{1}{5}$ of the area of the ellipse falls within the hub.

Make a sketch (two views) of the hub for a three-bladed screw propeller, showing clearly the method of securing the hub to the shaft and of securing the blades to the hub.

Make a sketch (plan, front and side elevations) of one blade of the following propeller: helicoidal area (each blade) $=22$ square feet; diameter of screw $=15$ feet; mean pitch $=16$ feet; diameter of hub $=\frac{2}{9}$ of diameter of screw; thickness of blade (considered as produced) at shaft axis $=7\frac{3}{4}''$; thickness of blade at tip $=\frac{3}{4}''$. Assume that $\frac{1}{5}$ of the area of the ellipse falls within the hub.

USEFUL TABLES.

STANDARD DIMENSIONS OF BOLTS AND NUTS FOR THE UNITED STATES NAVY.

Diameter.		Area.		Thr'ds.	Long Diameter.		Short D.	Depth.	
Nom.	Effect.	Nom.	Effect.	No.	Hex.	Square.	H. & Sq.	Head.	Nut.
1/4	.185	.049	.026	20	9/16	23/32	1/2	1/4	1/4
5/16	.240	.077	.045	18	11/16	27/32	19/32	19/64	5/16
3/8	.294	.110	.067	16	25/32	31/32	11/16	11/32	3/8
7/16	.345	.150	.093	14	29/32	1 3/32	25/32	25/64	7/16
1/2	.400	.196	.125	13	1	1 1/4	7/8	7/16	1/2
9/16	.454	.249	.162	12	1 1/8	1 3/8	31/32	31/64	9/16
5/8	.507	.307	.202	11	1 7/32	1 1/2	1 1/16	17/32	5/8
3/4	.620	.442	.302	10	1 7/16	1 3/4	1 1/4	5/8	3/4
7/8	.731	.601	.419	9	1 21/32	2 1/32	1 7/16	23/32	7/8
1	.837	.785	.550	8	1 7/8	2 5/16	1 5/8	13/16	1
1 1/8	.940	.994	.694	7	2 3/32	2 9/16	1 13/16	29/32	1 1/8
1 1/4	1.065	1.227	.891	7	2 5/16	2 27/32	2	1	1 1/4
1 3/8	1.160	1.485	1.057	6	2 17/32	3 3/32	2 3/16	1 3/32	1 3/8
1 1/2	1.284	1.767	1.294	6	2 3/4	3 11/32	2 3/8	1 3/16	1 1/2
1 5/8	1.389	2.074	1.515	5 1/2	2 31/32	3 5/8	2 9/16	1 9/32	1 5/8
1 3/4	1.491	2.405	1.746	5	3 3/16	3 7/8	2 3/4	1 3/8	1 3/4
1 7/8	1.616	2.761	2.051	5	3 13/32	4 5/32	2 15/16	1 15/32	1 7/8
2	1.712	3.142	2.302	4 1/2	3 19/32	4 13/32	3 1/8	1 9/16	2
2 1/4	1.962	3.976	3.023	4 1/2	4 1/32	4 15/16	3 1/2	1 3/4	2 1/4
2 1/2	2.176	4.909	3.719	4	4 15/32	5 15/32	3 7/8	1 15/16	2 1/2
2 3/4	2.426	5.940	4.622	4	4 29/32	6	4 1/4	2 1/8	2 3/4
3	2.676	7.069	5.624	4	5 11/32	6 17/32	4 5/8	2 5/16	3
3 1/4	2.926	8.296	6.724	4	5 25/32	7 1/16	5	2 1/2	3 1/4
3 1/2	3.176	9.621	7.922	4	6 7/32	7 19/32	5 3/8	2 11/16	3 1/2
3 3/4	3.426	11.04	9.219	4	6 5/8	8 1/8	5 3/4	2 7/8	3 3/4
4	3.676	12.57	10.61	4	7 1/16	8 21/32	6 1/8	3 1/16	4
4 1/4	3.926	14.19	12.11	4	7 1/2	9 3/16	6 1/2	3 1/4	4 1/4
4 1/2	4.176	15.90	13.70	4	7 15/16	9 23/32	6 7/8	3 7/16	4 1/2
4 3/4	4.426	17.72	15.39	4	8 3/8	10 1/4	7 1/4	3 5/8	4 3/4
5	4.676	19.64	17.17	4	8 13/16	10 25/32	7 5/8	3 13/16	5
5 1/4	4.926	21.65	19.06	4	9 1/4	11 5/16	8	4	5 1/4
5 1/2	5.176	23.76	21.04	4	9 11/16	11 27/32	8 3/8	4 3/16	5 1/2
5 3/4	5.426	25.97	23.12	4	10 3/32	12 3/8	8 3/4	4 3/8	5 3/4
6	5.676	28.27	25.30	4	10 17/32	12 29/32	9 1/8	4 9/16	6

Decimals and Fractions of an Inch.

1/32	.031	9/32	.281	17/32	.531	25/32	.781
1/16	.062	5/16	.312	9/16	.562	13/16	.812
3/32	.094	11/32	.344	19/32	.594	27/32	.844
1/8	.125	3/8	.375	5/8	.625	7/8	.875
5/32	.156	13/32	.406	21/32	.656	29/32	.906
3/16	.188	7/16	.438	11/16	.688	15/16	.938
7/32	.219	15/32	.469	23/32	.719	31/32	.969
1/4	.250	1/2	.500	3/4	.750	1	1.000

Table of Areas of Circles, advancing by Eighths.

Diam.		+⅛″	+¼″	+⅜″	+½″	+⅝″	+¾″	+⅞″
0″	.000	.012	.049	.110	.196	.307	.441	.601
1″	.785	.994	1.227	1.485	1.767	2.074	2.405	2.761
2″	3.142	3.547	3.976	4.430	4.909	5.412	5.940	6.492
3″	7.069	7.670	8.296	8.946	9.621	10.32	11.05	11.79
4″	12.57	13.36	14.19	15.03	15.90	16.80	17.73	18.67
5″	19.64	20.63	21.65	22.69	23.76	24.85	25.97	27.11
6″	28.27	29.46	30.68	31.92	33.18	34.47	35.73	37.12
7″	38.48	39.87	41.28	42.72	44.18	45.66	47.17	48.71
8″	50.27	51.85	53.46	55.09	56.74	58.43	60.13	61.86
9″	63.62	65.40	67.20	69.03	70.88	72.76	74.66	76.59
10″	78.54	80.52	82.52	84.54	86.59	88.66	90.76	92.89

Values of $\frac{d^2}{16}$, when d advances by Eighths.

d		+⅛″	+¼″	+⅜″	+½″	+⅝″	+¾″	+⅞″
0″	.0000	.0010	.0039	.0088	.0156	.0244	.0352	.0479
1″	.0625	.0791	.0977	.1182	.1406	.1650	.1914	.2197
2″	.2500	.2822	.3164	.3526	.3906	.4307	.4727	.5166
3″	.5625	.6104	.6596	.7119	.7656	.8213	.8789	.9385
4″	1.000	1.064	1.129	1.196	1.266	1.337	1.410	1.485
5″	1.563	1.642	1.723	1.806	1.891	1.978	2.066	2.157
6″	2.250	2.345	2.441	2.540	2.641	2.743	2.848	2.954
7″	3.063	3.173	3.285	3.399	3.516	3.634	3.754	3.876
8″	4.000	4.136	4.254	4.384	4.516	4.649	4.785	4.923
9″	5.063	5.204	5.348	5.493	5.641	5.777	5.941	6.095
10″	6.250	6.407	6.566	6.727	6.891	7.056	7.222	7.391

TRIGONOMETRIC FORMULÆ.

$\sin(x+y)=\sin x \cos y+\cos x \sin y.$

$\cos(x+y)=\cos x \cos y-\sin x \sin y.$

$\sin(x-y)=\sin x \cos y-\cos x \sin y.$

$\cos(x-y)=\cos x \cos y+\sin x \sin y.$

$2 \sin x \cos y=\sin(x+y)+\sin(x-y).$

$2 \cos x \sin y=\sin(x+y)-\sin(x-y).$

$2 \cos x \cos y=\cos(x+y)+\cos(x-y).$

$2 \sin x \sin y=\text{eos}(x-y)-\cos(x+y).$

$\sin x+\sin y \doteq 2 \sin \frac{1}{2}(x+y) \cos \frac{1}{2}(x-y).$

$\sin x-\sin y=2 \cos \frac{1}{2}(x+y) \sin \frac{1}{2}(x-y).$

$\cos y-\cos x=2 \sin \frac{1}{2}(x+y) \sin \frac{1}{2}(x-y).$

$\cos x+\cos y=2 \cos \frac{1}{2}(x+y) \cos \frac{1}{2}(x-y).$

$$\frac{\sin x-\sin y}{\sin x-\sin y}=\frac{\tan \frac{1}{2}(x+y)}{\tan \frac{1}{2}(x-y)}.$$

$$\tan(x+y)=\frac{\tan x+\tan y}{1-\tan x \tan y}.$$

$$\tan(x-y)=\frac{\tan x-\tan y}{1+\tan x \tan y}.$$

$$\cot(x+y)=\frac{\cot x \cot y-1}{\cot x+\cot y}.$$

$$\cot(x-y)=\frac{\cot x \cot y+1}{\cot x-\cot y}.$$

$$\tan(x \pm 45^\circ)=\frac{\tan x \mp 1}{\tan x \pm 1}.$$

$\sin(x+y) \sin(x-y)=\sin^2 x-\sin^2 y=\cos^2 y-\cos^2 x.$

$\cos(x+y) \cos(x-y)=\cos^2 x-\sin^2 y=\cos^2 y-\sin^2 x.$

$$\tan x \pm \tan y=\frac{\sin(x \pm y)}{\cos x \cos y}.$$

$$\cot x \pm \cot y=\frac{\sin(y \pm x)}{\sin x \sin y}.$$

$$\sin 2x=2 \sin x \cos x=\frac{2 \tan x}{1+\tan^2 x}.$$

$$\cos 2x=\cos^2 x-\sin^2 x=1-2 \sin^2 x=2 \cos^2 x-1=\frac{1-\tan^2 x}{1+\tan^2 x}.$$

$$\frac{1-\cos 2x}{1+\cos 2x}=\frac{2\sin^2 x}{2\cos^2 x}=\tan^2 x.$$

$$\tan 2x=\frac{2\tan x}{1-\tan^2 x}.$$

$$\cot 2x=\frac{\cot^2 x-1}{2\cot x}.$$

$$\sin 3x=3\sin x-4\sin^3 x.$$

$$\cos 3x=4\cos^3 x-3\cos x.$$

$$\tan 3x=\frac{3\tan x-\tan^3 x}{1-3\tan^2 x}.$$

$$\sin^2\left(\frac{x}{2}\right)=\frac{1-\cos x}{2}.$$

$$\cos^2\left(\frac{x}{2}\right)=\frac{1+\cos x}{2}.$$

$$\tan^{-1} x+\tan^{-1} y=\tan^{-1}\left(\frac{x+y}{1-2y}\right).$$

Integrals.

$$\int x^n dx=\frac{x^{n+1}}{n+1}.$$

$$\int\frac{dx}{x}=\log x.$$

$$\int a^x dx=\frac{a^x}{\log a}.$$

$$\int\cos\theta d\theta=\sin\theta.$$

$$\int\sin\theta d\theta=-\cos\theta.$$

$$\int\sec^2\theta d\theta=\tan\theta.$$

$$\int\operatorname{cosec}^2\theta d\theta=-\cot\theta.$$

$$\int\sec\theta\tan\theta d\theta=\sec\theta.$$

$$\int\operatorname{cosec}\theta\cot\theta d\theta=-\operatorname{cosec}\theta.$$

$$\int\frac{dx}{\sqrt{1-x^2}}=\sin^{-1} x+c=-\cos^{-1} x+c'.$$

$$\int\frac{dx}{\sqrt{a^2-x^2}}=\sin^{-1}\frac{x}{a}+c=-\cos^{-1}\frac{x}{a}+c'.$$

$$\int\frac{dx}{1+x^2}=\tan^{-1}x+c=-\cot^{-1}x+c'.$$

$$\int\frac{dx}{a^2+x^2}=\frac{1}{a}\tan^{-1}\frac{x}{a}+c=-\frac{1}{a}\cot^{-1}\frac{x}{a}+c'.$$

$$\int\frac{dx}{x\sqrt{x^2-1}}=\sec^{-1}x+c=-\mathrm{cosec}^{-1}x+c'.$$

$$\int\frac{dx}{x\sqrt{x^2-a^2}}=\frac{1}{a}\sec^{-1}\frac{x}{a}+c=-\frac{1}{a}\mathrm{cosec}^{-1}\frac{x}{a}+c'.$$

$$\int\frac{dx}{\sqrt{2x-x^2}}=\mathrm{vers}^{-1}x.$$

$$\int\frac{dx}{\sqrt{2ax-x^2}}=\mathrm{vers}^{-1}\frac{x}{a}.$$

$$\int\tan x\,dx=-\log\cos x.$$

$$\int\tan^2\theta\,d\theta=\tan\theta-\theta.$$

$$\int u\,dv=uv-\int v\,du.$$

$$\int\frac{dx}{(x-a)(x-b)}=\frac{1}{a-b}\log\left(\frac{x-a}{x-b}\right).$$

$$\int\frac{dx}{x^2-a^2}=\frac{1}{2a}\log\left(\frac{x-a}{x-b}\right).$$

$$\int\sin^2\theta\,d\theta=\tfrac{1}{2}(\theta-\sin\theta\cos\theta).$$

$$\int\cos^2\theta\,d\theta=\tfrac{1}{2}(\theta+\sin\theta\cos\theta).$$

$$\int\frac{d\theta}{\sin\theta\cos\theta}=\log\tan\theta.$$

$$\int\frac{d\theta}{\sin\theta}=\log\tan\frac{\theta}{2}=\log\left(\frac{1-\cos\theta}{\sin\theta}\right).$$

$$\int\frac{d\theta}{\cos\theta}=\log\tan\left(\frac{\pi}{4}+\frac{\theta}{2}\right)=\log\left(\frac{1+\sin\theta}{\cos\theta}\right)=\log(\sec\theta+\tan\theta).$$

$$\int\frac{d\theta}{a+b\cos\theta}=\frac{2}{\sqrt{a^2-b^2}}\tan^{-1}\left[\sqrt{\frac{a-b}{a+b}}\left(\tan\frac{\theta}{2}\right)\right]$$

$$=\frac{1}{\sqrt{b^2-a^2}}\log\left[\frac{\sqrt{b+a}+\sqrt{b-a}(\tan\frac{1}{2}\theta)}{\sqrt{b+a}-\sqrt{b-a}(\tan\frac{1}{2}\theta)}\right].$$

$$\frac{dx}{x\sqrt{a^2+x^2}} = \frac{1}{a}\log\left[\frac{\sqrt{a^2+x^2}-a}{x}\right].$$

$$\int\frac{dx}{x\sqrt{a^2-x^2}} = \frac{1}{a}\log\left[\frac{a-\sqrt{a^2-x^2}}{x}\right].$$

$$\int\frac{dx}{\sqrt{x^2\pm a^2}} = \log\left[x+\sqrt{x^2\pm a^2}\right].$$

$$\int\sin^m\theta d\theta=\int\cos^m\theta d\theta \text{ (when m is positive and even)}$$

$$=\frac{(m-1)(m-3)\ \ldots\ 1}{m(m-2)\ \ldots\ 2}\theta\Big]_{limit}^{limit}$$

(when m is positive and odd)

$$=\frac{(m-1)(m-3)\ \ldots\ 2}{m(m-2)\ \ldots\ 1}\cos\theta\Big]_{limit}^{limit}.$$

$$\int_0^{\frac{\pi}{2}}\sin^m\theta\cos^n\theta d\theta=\frac{(m-1)(m-3)\ \ldots\ (n-1)(n-3)\ \ldots}{(m+n)(m+n-2)\ \ldots}\times a,$$

where $a=1$, unless m and n are both *even*, in which case $a=\frac{\pi}{2}$.